Matthew J. Hatami, P.E.

# SHALE

## OIL and GAS

# OPERATIONS

# MAXIMIZE

# CASH FLOW

## WITH

# COST REDUCTION

**CAPEX & OPEX ECONOMIC ACTIONS**

# SHALE OIL & GAS OPERATIONS:

# MAXIMIZE CASH FLOW

*with*

# COST REDUCTION

Oilfield Books
Practical Oil & Gas Publications

*Published 2024 by Oilfield Books*

Shale Oil and Gas Operations: Maximize Cash Flow with Cost Reduction (also known as Shale Cash Flow)

A catalog record for this book is available from the Library of Congress. Library of Congress Control Number: 2024915947

*Concept, format, and design by Matthew J. Hatami*
*Cover design by Matthew J. Hatami*
*Tables, charts, and graphs by Matthew J. Hatami*
*Models and diagrams by Matthew J. Hatami*

The publisher and author do not accept responsibility for anything that occurs because of the use or misuse of any actions, techniques, tactics, procedures, checklists, or anything written or quoted in this book.

Never take an "Action," procedure, tactic, recommendation, suggestion, or anything else and force it into use in your area of operations. This is dangerous, and you will have problems.

ISBN 979-8-9907640-0-2 (paperback)
ISBN 979-8-9907640-1-9 (hardback)
ISBN 979-8-9907640-2-6 (ebook)

First printing
Paperback Edition: August 2024

*For my Family*

## WARNING

Never apply any actions, techniques, tactics, procedures, checklists, or anything written or quoted in this book without performing thorough calculations, hazard/risk assessments, statistical analysis, and due diligence with your company, team, and everyone involved with your operation.

Accordingly, the publisher and author cannot accept any responsibility for anything that occurs due to the use or misuse of any action, techniques, tactics, procedures, checklists, or anything written or quoted in this book. As stated in the book, never take an "Action," procedure, tactic, recommendation, suggestion, or anything else and force it into use in your area of operations. This is dangerous, and you will have problems.

# CONTENTS

# SHALE

# OIL and GAS

# OPERATIONS

—

# MAXIMIZE

# CASH FLOW

# with

# COST

# REDUCTION

# 700 ACTIONS

—

This book is focused on generating exponential stakeholder returns by maximizing free cash flow with cost reduction. By reviewing over 700 cost reduction actions across all oil and natural gas operations—geology, land, resource development, supply chain, drilling, completion, and production—we will attack CAPEX and OPEX from every possible angle.

Our goal is not only to reduce costs but also to reduce risks. The foundation of an aggressive cost reduction program is risk management—a topic we will address in detail.

To ensure success, we will identify opportunities to systemize and automate large segments of our business. By incorporating current and cutting-edge technologies with statistics, probability, and artificial intelligence, we will combine time-tested methods with new innovative solutions to propose practical actions to maximize free cash flow.

Many cost reduction actions are presented here for the first time. This information cannot be found anywhere else.

Since we do not control the price of our products, we must control the cost. Let's start with the most critical aspect of maximizing free cash flow with cost reduction: our people.

# 1

## PEOPLE FIRST

—

Energy and natural resources are commodities. The people who work in these industries are not. This is especially true for the highly complex and technically challenging shale oil and natural gas business, in which economic success primarily depends on the decisions and actions of individuals working together towards a common goal. The question is, do we have the right people in every position in the field and office to maximize value creation? Probably not.

Finding, hiring, and retaining top talent is incredibly difficult but necessary, particularly when maximizing cash flow with cost reduction. If someone on the team is not interested in the business strategy of maximizing cash flow, regardless of what the issue is, they can and will distract, derail, and destroy our game plan, our operation, and our company. It is easy for one person to cause significant damage. Therefore, we must identify and remove undesirable, high-risk individuals as quickly as possible. Do not hesitate to eliminate them fast. Our success depends on it.

Competitors in the global oil and natural gas industry produce relatively indistinguishable products. Oil and gas molecules vary by reservoir and depth. However, the percentage of primary elements, carbon and hydrogen, are relatively equivalent. What is not equivalent is the risk and cost per unit to bring these elements up to the surface.

The lowest-cost producer holds a valuable competitive advantage, with major players including not only public and private companies but also governments, countries, and competing sources of energy. The capacity for shale oil and gas to deliver sustainable positive free cash flow is one of the primary factors determining long-term economic success.

## Building a 100-to-1 Shale Machine

If we agree that the value of a business is primarily based on the present value of future cash flows, then it is a noble pursuit to maximize cash flow with cost reduction. Growth is a primary driver, but so is cost, in the form of CAPEX (drilling and completion costs) and OPEX (production costs).

Humans are critical to success because people drive all aspects of cash flow generation, including capital allocation, risk management, growth, and cost. The market places a premium on companies that can sustain free cash flow (FCF) growth and high returns on invested capital (ROIC), which compound over long periods. A company that can do this extremely well over many years is sometimes referred to as a potential 100-bagger or 100-to-1 investment opportunity.[1] Meaning, if we invest $10,000 it grows to $1,000,000. A 100X return is difficult but not impossible.

## Shale Compounding Machine Model

Free Cash Flow = **C**ash **F**low from **O**perations (CFO) – **CAPEX**

| Components of the Shale Machine | • Top Talent & Management |
|---|---|
| | • Low-Cost Owner Operator |
| | • Prudent Capital Allocation |
| | • High Returns on Capital |
| | • Strong Risk Controls |
| | • Scalable Assets |
| | • Time, Patience |

**CFO** = (Commodity Price – **OPEX**) x (Production + **Growth**)

For public companies, free cash flow and earnings growth over many years should lead to the market providing larger multiples on valuation metrics, resulting in an expansion in the price-to-cash flow (P/CF) and price-to-earnings (P/E) ratios. From a stock price perspective, we want to open the choke on our 100-to-1 machine by growing cash flow with a larger multiple on that cash flow. We stimulate and enhance our shale machine with CAPEX and OPEX reductions per unit. For maximum impact, we scale our efforts across drilling, completions, and production while reinvesting the cash at high rates of return. Doing this consistently over many years will lead us to exponential value creation for stakeholders.

For example, let's say we work for an operator that has 250 producing wells, growing with a 5-rig program, drilling one well per rig per month. On the production side, our team identifies an opportunity to reduce costs $50 a month on one well. That small savings might not sound like much, but it is significant if we scale it and place the cash in our shale compounding machine.

## OPEX Reduction of $50 Creates Significant Value

*Small OPEX reduction during oil and gas production of $50/month can create massive value over time if it can be scaled across the wells with savings reinvested in drilling, acquisitions, or other projects at high rates of return over long periods.*

### How Saving $50 Grows to +$100,000,000

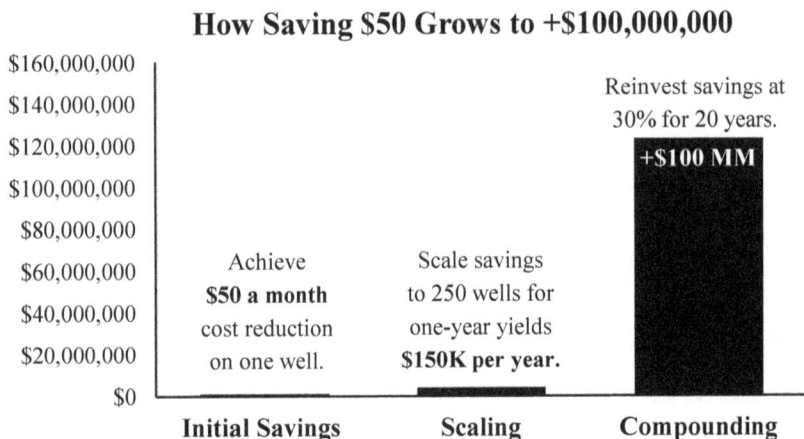

When initiating cost reduction efforts, I prefer to start with small, low-risk actions. Hitting consistent singles and doubles is essential when building confidence and stability while managing risk during cost reduction campaigns. We will get highly technical and detailed regarding specific engineering and operational cost reduction actions in which the strategy will make more sense.

In general, my preference is to collect many small wins that build over time while simultaneously hunting for big-game reductions. The core of this strategy is focused on capturing opportunities that collectively add up to massive cost reduction results. However, the largest opportunities, in terms of total cost reduction dollar amounts, are typically found on the CAPEX side of the business during drilling and completion operations. While the more intricate opportunities are on the production side of the business.

Continuing with our 5-rig development program example, if our team identifies opportunities to collectively save $1,000,000 per well, which is then reinvested in drilling, acquisitions, or other projects at high rates of return over long periods, the impact can be billions of dollars.

## CAPEX Savings of $1.0 Millon Grows to $20.0 Billion

*Large CAPEX savings of $1.0MM/well can create massive value if it can be reinvested at high rates of return over long periods. This example assumes $60MM/year total savings for only the first two years.*

### How Saving $1.0MM Grows to +$20.0 Billion

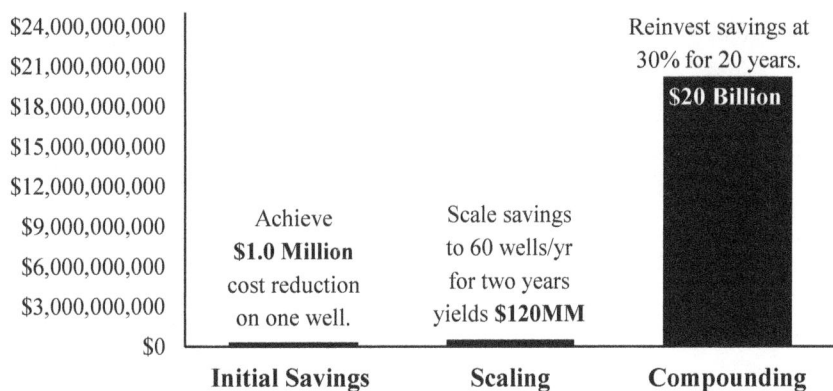

| | |
|---|---|
| $24,000,000,000 | Reinvest savings at |
| $21,000,000,000 | 30% for 20 years. |
| $18,000,000,000 | $20 Billion |
| $15,000,000,000 | |
| $12,000,000,000 | |
| $9,000,000,000 | Achieve / Scale savings |
| $6,000,000,000 | $1.0 Million / to 60 wells/yr |
| $3,000,000,000 | cost reduction / for two years |
| $0 | on one well. / yields $120MM |

Initial Savings  Scaling  Compounding

Saving money on large projects can create significant value and change our company's future opportunities for both employees and shareholders. Entire industries have been founded, expanded, and evolved by reducing costs and making things economic. The groundwork for business revolutions is often based on cost reduction. Many of the world's best operators, engineers, executives, and entrepreneurs spend considerable time and energy on cost reduction and deciding the optimal strategy to reinvest the proceeds.

Finding opportunities that return 30% for 20 years is difficult, especially with fluctuating commodity prices, increasing costs, variable geology, and changing regulations. Therefore, let's look at what happens at various rates of return with our previous example, achieving a cost reduction savings of $60 million per year for only the first two years.

## CAPEX Savings Reinvested at Different Rates

*Large CAPEX savings of $1.0MM/well can create massive value if it can be reinvested over long periods. This example assumes $60MM/year total savings for only the first two years.*

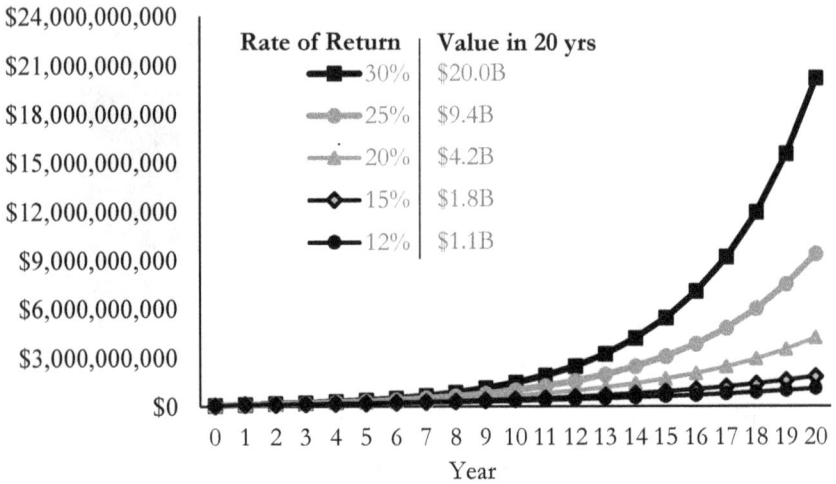

| Rate of Return | Value in 20 yrs |
|---|---|
| 30% | $20.0B |
| 25% | $9.4B |
| 20% | $4.2B |
| 15% | $1.8B |
| 12% | $1.1B |

Even if we only find opportunities that generate a 12% return, our $120 million in savings will grow to over $1.0 billion in 20 years. Unfortunately, the reverse is also true. If we make bad decisions and do not adequately manage risk, significant value can be destroyed, and it can happen a lot faster than 20 years. Bad decisions can destroy a company or industry in seconds. Yes, in seconds. The downside is a lot more violent and unforgiving. Therefore, enhancing our decision-making intelligence and risk management skills must be incorporated into our strategy. I have no interest in achieving massive cost reductions if we must push the risk into the upper right-hand corner of the risk assessment matrix.[2]

## Risk Assessment Matrix

| Level of Risk | | Impact | | | | |
|---|---|---|---|---|---|---|
| | | Tiny | Minor | Moderate | Major | Catastrophic |
| **Probability** | **Very Likely** | Medium | High | High | Extreme | Extreme |
| | **Likely** | Medium | Medium | High | High | Extreme |
| | **Moderate** | Low | Medium | Medium | High | High |
| | **Unlikely** | Low | Low | Medium | Medium | High |
| | **Rare** | Low | Low | Low | Medium | Medium |

Risk management is a crucial aspect of our plan. Everyone who works in shale operations is a professional risk manager, as far as I am concerned. We are paid to manage risk regardless of position or rank. Therefore, if you take responsibility and ownership and do whatever it takes to address safety and environmental impact to protect yourself and your company, you have my permission to add "Risk Manager" to your title. For example, if you are a pumper/lease operator, you can upgrade your title to Risk Manager and Pumper. That's how important it is to manage risk in this business.

Every one of us can shape the success or failure of our operation. It only takes one bad decision or lack of attention to detail to cause considerable shareholder damage. It is important to understand that even with our robust processes, procedures, and advanced technology, the key to risk management lies in the hands of our people. Good decisions and prudent actions are what make the difference.

# Pumper Destroys Market Capitalization

In Texas, one Friday afternoon, a pumper runs his route for Big Shale Corporation. Due to cold weather, it is taking longer than usual, and he is running late for a family dinner. A heater treater on one of the new multi-well pads needs attention, which he plans to address on his way home. The burner needs to be turned off, and a regulator needs to be flushed. However, due to his dinner plans, the pumper heads home instead and does not mention the situation to his supervisor.

Later that evening, the SCADA alarms send text alerts that are missed or ignored. Soon after, a landowner calls the emergency line to report a fire. It takes several days to put the fire out. The accident investigation reveals one treater failed, causing a domino effect, leading to 25 tanks burning down.

Meanwhile, in New York, a fund manager working for one of the top institutional investors in Big Shale receives alerts on the company. Several were from social media sites with video of the fire, and it made the local news. The fund manager starts to feel uncomfortable. He begins to doubt his position and decides to slowly sell over the course of a month.

Big Shale's investor relations VP observes that one of the top institutional holders has exited the stock. A meeting is requested. During the meeting, the fund manager does not mention the fire, just that he changed his mind on the investment. However, the fund manager does mention his sentiment to other investors at lunch meetings and investment events. During the next market pullback, the relative performance of Big Shale's stock begins to deteriorate.

Without the right people, the ability to create value is almost impossible. There are many things we can incorporate into our operation to prevent the situation that occurred with Big Shale Corporation—we will address all of them. I share the story to show how easy it is to destroy value and investor confidence. Do you think the pumper had any idea of the impact of his decision on the value of the company?

We will need the focus, energy, and commitment of the entire team, from the office to the field, including every single person at all service companies, vendors, suppliers, consultants, and company team members, to accomplish our goals. Management alone cannot deliver new wells and daily production with total costs dramatically lower than current levels without dedicated front line personnel. Maximizing cash flow by reducing cost, the primary goal of this book, is no easy task. You might even think it is impossible to achieve on the horizontal wells that you are currently working on. It may be impossible for a few people working alone, but not for a highly talented, determined team working together.

Field personnel have the detailed knowledge necessary to drive improvements because they spend most of their time embedded within operational processes. Therefore, a critical aspect of cost reduction is selecting the right people and the right combination of people, particularly on location. Building a low-cost operations team is like building a winning sports team. Each player has strengths and weaknesses. Knowing what they are is essential to determine if they are a good fit. To create a horizontal well in the United States, it takes the teamwork of 300 to 400 people.

# Workers on One Horizontal Well in the United States

*\*30-days spud to RR, 50-stage PNP, CTDO, gas lift install, 30-day flowback, 3-phase facility, gas pipeline*

| Operation: Work Performed* | # of Workers |
|---|---|
| **Pre-Spud:** Geology, Mapping, Land, Legal, Title, Buyers, Engineering, Field Ops, Surveying & Staking, Regulatory, Permits, Surface & Water Use Agreements, Pad & Road Construction, Rock, Fence, Cattleguard, Gate, Signs. | 30 |
| **Drilling:** Cellar, Conductor, Water Transfer, Pre-set Rig Mob and Operation, Primary Rig Mob, Primary Rig Crews, Mechanics, Technicians, Trainees, Supervisors, Wellsite Consultants, Fuel Delivery, Bits, Reamers, Stabilizers, Mud, Additives, Directional Drillers, MWD Services, Downhole & Surface Equipment, Trailer Houses, Water, Septic, Tanks, Pits, Lights, Trash, Well Control, RCD, Chokes, Flare, Mud Logging, Geo-Steering, Casing Crews, Torque-Turns, Thread Reps, Casing Inspection, Cementers, Centralizers, Floats, Sleeves, Closed Loop, Centrifuge, Welders, NU/ND, Testing, Trucking, Wellhead Install, Repairs, Sales Reps, Safety Hands, Regulatory Inspectors. | 130 |
| **Completion:** Pad Reconditioning, Cellar Prep, CBL, Water Transfer, Frac Crew, Trainees, Trucking and Storage of Proppant, Materials, Fuel, Wireline, Plugs, Guns, Wellsite Consultants, Surface Equip., Frac and Acid Tanks, Containment, Stack Install and Greasing, Flowbacks, Coil Tubing Drill Out, Coil Tubing BHA, Drill Out Fluids, Completion Packer, Work Over Rig Tubing Install, Gas Lift Valves Install, Production Tree, Sales Reps, Safety Hands. | 125 |
| **Facility:** Screenings Base, Secondary Containment, Liner, Construction, Roustabouts, Tanks, Coating, Stairs, Vessels, Pipe, Fittings, Valves, Compressors, Gauges, Telemetry, Automation, SCADA, Heat Trace, Insulation, Lightning Protection, Static Protection, Grid Power, Chemical Pumps, Transfer Pumps, Combustors, Sales Reps, Safety System, Welders, Walls, Signs, Trucking, Equipment Setting. | 35 |
| **Production:** Pipeline, Meter Run, Calibration, Testing, Water & Oil Haulers, Flowbacks, Combo Units, Foreman, Pumper, Roustabouts, Compression, Regulatory Inspectors. | 25 |
| **Back Office:** Invoicing, Processing, Taxes, Mineral / Landowner Relations, Division Orders, Accounting. | 5 |
| **Total Number of Workers**** | 350 |

*\*\*Based on 24-hour ops, day shift, night shift workers, traveling crews, plus all personnel working a rotational nature across multiple positions including working back-to-back and relief on 20/10, 14/14, 14/7, 7/7 schedules and other various work arrangements common in the American oil and natural gas industry.*

# The Power of Cost Reduction

Cost reduction fuels our ability to deliver attractive returns not only by increasing organic free cash flow but also by enabling us to make acquisitions that immediately add value. For example, if we are the lowest-cost operator in our area of interest, every competitor becomes more valuable under our control. Of course, this does not mean we acquire all our competitors, but we should at least consider it. Our lower cost structure makes a compelling consolidation case if the deals are accretive and our company grows stronger.

We will review over 700 specific cost reduction actions during our journey together. However, we only need to get a small group of them to work on a scale large enough to drive our costs lower than our competitors and put us in a position of power. Having the right people focused day after day for years allows us to achieve this and leverages the full power of free cash flow in our shale compounding machine.

Although generating massive returns can take decades, achieving significant cost reductions should not. The most effective campaigns happen over the course of months, not years. If we set a period to achieve our goals, we should not allow too much time to pass to the point that there is no sense of urgency. The most impactful cost reductions, beyond the typical continuous cost pruning, occur as the result of a focused effort over a short period. Having the right people enables this to occur; having the wrong people makes it impossible. The right people, combined with short-term targets and long-term value creation goals, are a powerful combination.

## Human Machine Interface

The interface between our shale machine and our workforce is the structure of our company, which is derived from our leadership, goals, and strategy. Since most of our team is comprised of non-employee personnel, a strong structure is critical to elevating our performance.

Analysis from almost every industry indicates that 60-90% of undesirable events are due to human error.[3] Every time we make a mistake, whether drilling out of zone, incurring a casing failure, getting stuck, or simply forgetting something, we must undo our mistake operationally and financially to get back to even. Only then can we start making progress.

There is no substitute for great people. However, our people are constantly changing, and everyone cannot be great all the time. To address this dynamic and stop costly mistakes, we must leverage the full power of Artificial Intelligence (AI).

AI is a broad term that is defined differently depending on who you talk to. In this book, AI is any system that enables a computer to replicate a human action.[4]

What's driving the expansion of AI is the convergence of multiple technologies, including improved sensors, wireless technology, mobile phone applications, advances in machine learning, and faster data transmission; combined with a decline in costs across all technologies.[5]

AI helps prevent costly mistakes by providing our team with increased information on the task at hand, institutional knowledge from our company's experiences, and industry intelligence for every step of our operation.

## Human + AI is More Powerful than Human Alone

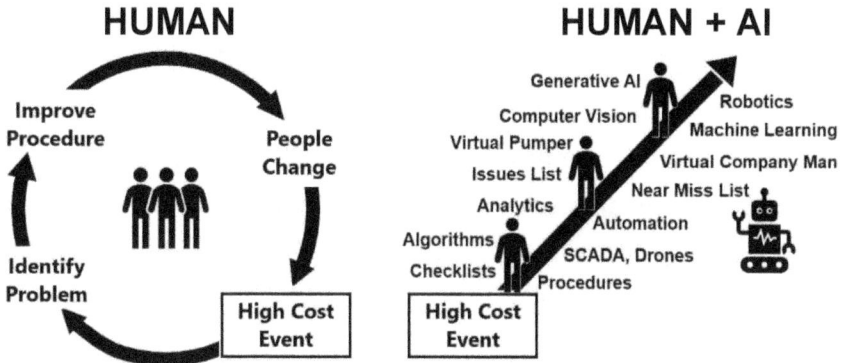

**HUMAN**

Improve Procedure

Identify Problem

People Change

High Cost Event

**HUMAN + AI**

Generative AI
Computer Vision
Virtual Pumper
Issues List
Analytics
Algorithms
Checklists

Robotics
Machine Learning
Virtual Company Man
Near Miss List
Automation
SCADA, Drones
Procedures

High Cost Event

As the above diagram illustrates, without AI, we repeat expensive mistakes. With AI, we systemize and automate every aspect of our operation. The right people combined with the right AI is the key.

The most cost-effective way to set this up is to manually enter enough oilfield knowledge so that our AI can operate at the edge of what is known and learn on its own at some point in the future, once it has proven itself. This is similar to bringing on a new person. We must get them up to speed by educating them up to the point of what is known. The problem is that it takes years to do human-to-human training, and it isn't easy to find and retain great people.

Even if we have excellent, highly experienced people, it is impossible to have someone who has seen everything. For example, on Deepwater Horizon, the crew and company men were highly experienced and had decades of knowledge. However, they did not have experience dealing with a failed negative pressure test, and the mistake cost $67.4 billion and 11 deaths.[6] It also damaged the value of our entire industry.

A simple classical AI system would have prevented it, avoiding 11 deaths and saving billions. We will develop such a system together in this book. In conjunction with our cost reduction actions, we will build an artificial intelligence system called "Virtual Company Man" and "Virtual Pumper."

People who are the best at doing something are much better than the average. Our AI will help raise the bar and get everyone to the highest performance levels. There is no substitute for the best boots-on-the-ground intelligence. However, there is often too much uncertainty and variables for a human to process in real-time, all the time.

AI amplifies our intelligence, helping us make and execute good decisions. It is particularly helpful when we have a small amount of information, a high amount of uncertainty, or multiple issues occurring simultaneously. Our Virtual Company Man will centralize disparate or siloed information across our organization to address these issues.

It is common within organizations to experience a costly mistake only to find that someone else, either another employee, contractor, or manager, has experienced the same thing and could have prevented it if they were involved.

This is the cost of siloed knowledge. AI eliminates silos by centralizing expertise and communicating it to the team, either on demand or automatically. The choice is yours.

AI is our secret weapon to avoiding costly mistakes, maximizing cash flow with cost reduction, and enabling us to reinvest in high-return projects over long periods that help us reach our aspirational goal of 100X returns.

# The Principle of 100X

When we find a system or design that we are comfortable with, it is human nature to not want to change. This is understandable, especially in the oil industry, because it is expensive to get to that point. However, what can happen is that we stop making progress. Setting a long-term goal that seems impossible forces us to avoid getting comfortable with our current achievements and continue to push for more.

The principle behind an aspirational 100X returns goal is to have our team operate from a mindset that is working to create exponential value while protecting the company over the many years it takes to get there. Thinking both short-term and long-term is a risk management mindset I hope to instill with this book. We never want to take an action that could jeopardize the company's value, reputation, or long-term vision. Some people might push the risk for a short-term reduction that is not aligned with our long-term goal. Thinking from a 100X mental model is designed to prevent that.

It might be tempting to implement every FCF action we review. However, I strongly advise against it. That would be a mistake, like going on a crash diet and starving yourself out of existence. Many actions in this book are aggressive, requiring extreme engineering and operational tactics to be implemented correctly under the right circumstances.

Ideas are important, but execution is more important, so having the right people to deliver a safe, low-cost operation, that's continuously innovating is one of the most critical aspects of our journey to 100X shareholder returns.

To achieve significant cost reductions, our team must be safety-focused, highly motivated, willing to work hard, and value attention to detail. The best results happen when each team member is interested in achieving the targets, not for money, recognition, or promotion, but for themselves, for their own satisfaction, pride, reputation, and legacy.

We will embrace a People-First strategy, upon which we will build a lower-cost operation to deliver maximum economic benefit. Ideally, if every person involved in the operation acted as an owner, with their net worth on the line, achieving our cost reduction goals would be uncomplicated.

Unfortunately, this is not possible because most of the 300+ people we depend on are on the other side of the table, making money off us regardless of the project's economics. To address this, selecting the right service providers is crucial, and we explore this dynamic throughout the book. But first, let's review the different personalities we may encounter on our journey to maximize cash flow with cost reduction.

## The 10 Cost Reduction Personality Types

Identifying and selecting the right people to achieve success is the most critical aspect of maximizing cash flow with cost reduction. The people we select will be our greatest asset or greatest liability.

Working on money issues can be highly emotional. Spending money is fun and easy while reducing spending can result in the opposite reaction: stress and pain. To address this, we will implement a plan that makes cost reduction enjoyable. Regardless, different people react differently toward cost reduction. Some are highly interested and jump on board, while others fight back. We can waste a lot of time and energy trying to engage someone who is not interested.

Identifying and understanding the different personality types within our teams can significantly aid us in making informed personnel decisions. For example, if we identify a team member as a "Cost Reduction Hater," one of our ten cost reduction personality types, it may be best for that individual not to be directly involved.

As someone who has worked on cost reduction efforts for over two decades, on the services side and operator side, working within teams and leading efforts, my experience is that people either want to or do not want to maximize cash flow with cost reduction. These two categories can be further divided into 10 personality types. Most people are a mix.

Similar to the well-known 16 Meyers-Briggs Personality Types,[7] the 10 Cost Reduction Personality Types slowly reveal themselves, especially when efforts become difficult or stressful. Increasing our personality type awareness and self-awareness will improve our workplace effectiveness.

## Cost Reduction Owners

When initiating cost reduction efforts, having owners on the team is critical to success. People who have a significant amount of their net worth on the line, are direct working interest owners, or act as if they have their own capital at stake, are deeply invested in achieving success. If you have, or act as if you have, your financial future on the line, you will be focused on results. There isn't room or time for anything else.

Owners are focused on delivering quantifiable results as fast as possible. Failure is not an option. The owner cannot afford to fail or quit and will do whatever is necessary to succeed. If it requires working seven days a week, skipping meals, not sleeping, canceling vacations, living in the office or on location, the owner will do that. The owner's mindset is consumed with success and will do whatever it takes.

At the highest levels within a corporation, the chief executives, i.e., CEO, COO, and CFO, must be Cost Reduction Owners. If leadership is not interested in the operational aspects of reducing costs, it is highly unlikely that the company will succeed. A company adapts the personality traits of its most senior executives as their behavior flows through the ranks and onto the field. If leadership has more passion for the latest exploration play, high 24-hr IPs, or number of locations, that is what the primary focus of the company will be, all the way down to the field level.

Senior executives would be amazed to learn how much the field team observes their every move and dissects their every word. When I say field team, I do not only mean the operations team in the office. I am talking about the guys in the field, whom, most likely, the executives have never met and, whose names they probably do not know. These front line workers are more responsible for an energy company's economic success or failure than most people realize.

Wellsite consultants, vendors, truck drivers, tool pushers, drillers, frac hands, pumpers, cementers, mud men, equipment operators, sales representatives, and all the people we reviewed earlier—these hard-working team members closely watch the actions of senior leadership. They follow the investor presentations, download the slide decks, and discuss what model car or plane the executives enjoy. Gossip about executive leadership is a common conversation on location.

Every action senior leadership takes influences the perception, respect, and motivation the field team has toward the company and our cash flow strategy. Suppose senior leadership buys a multi-million-dollar house or fancy car or takes a lavish vacation during a cost reduction initiative. In those cases, I guarantee everyone on location will lose respect for the company and be demotivated to fight hard to lower operational costs. I know this because I have seen each one of these situations play out on the operator side and service company side. A lavish purchase by a manager or executive can derail an entire cost reduction strategy because it sets the tone and culture of a company more than any corporate initiative, speech, or email. Actions speak louder than words. The field understands this more than anything. Talk is cheap, but actions carry the real weight of a corporate strategy. A company that desires to establish a lean operation and culture of efficiency will have a difficult time if leadership does not walk the walk.

The factory floor in the oilfield is at the wellhead on location. Due to the nature of oil industry work, leadership is not present on the factory floor unlike other manufacturing industries, where leadership often has an office or constant presence at the factory where operations are taking place, and the majority of the company's capital is being spent. Therefore, in the oilfield, perception and respect for leadership are more important compared to other industries due to the vast spread-out assets, remoteness, high turnover, and dependency on vendors and contractors, who are all on the other side of the table. Vendors and contractors do not make more money if an oil company is more efficient. They make less, and they are firmly aware of this. Increased volume and stability of work are of course the carrot for vendors in becoming more efficient and working with a more organized and respected operator.

People do what they are incentivized to do. If executives are more incentivized to grow production, with minor incentives on cost control, what do you think will take precedence? Vendors are only directly incentivized to cut costs if the operator demands it, aggressively pushes for it, and makes decisions based on it.

As an owner, no one cares more about your business than you do. Cost Reduction Owners benefit from being directly or indirectly incentivized to reduce cost. It does not have to be a monetary incentive. Acknowledgement goes a long way.

The best cost reduction efforts work when Cost Reduction Owners have vendors who are onboard with the game plan. Although vendors, including company men / wellsite consultants, are on the other side of the table, if they do not take some form of ownership, it will be hard to get the cost significantly down. A sustainable reduction in current costs will be tough if our service providers are pushing against our efforts. Consider vendors and contractors who are determined to help the operator with its strategy and objectives, Cost Reduction Partners.

## Cost Reduction Partners

Selecting the right vendors and contractors will make or break the best of our intentions. Accepting that we may have to make vendor changes to achieve our goals is part of the process if we desire or need deep reductions. In a way, selecting our vendors is similar to a sports team selecting its players. For each entity, strengths and weaknesses are evaluated, and a decision to put them on the team or in the game is made. After each well, individual performance is assessed, and management must decide whether to keep or release them.

Often, a team has good players, but they are not committed to certain aspects of the team's big picture strategy. They are not put on the field because they are not aligned with the game plan. In our case, the game plan is to maximize cash flow with cost reduction. Only some vendors are low-cost or want to be low-cost. Several well-known vendors have a reputation for good work, but they are high-cost and have no desire to be more efficient or reduce their costs and pass those savings to the customer. Sometimes, we must let good vendors go because they are not economical, similar to a professional sports team letting good players go.

As sports fans, we often see excellent players getting cut or traded. It seems irrational, but it is not uncommon, because good players don't always fit with the strategy, objectives, or budget. The ideal vendors and contractors reduce costs and risks without being asked, either through tightening their margins or through ideas, innovation, execution, and actions.

When commodity prices collapse and operators are not incentivized to drill or complete wells, our Cost Reduction Partners should call us up and offer significant relief. These folks do this without being asked or having to be beaten down because they are true partners.

Always remember when a high-value vendor volunteers to reduce their costs to help. It goes a long way with me. Smart service providers do this because they know the operator is in pain and want to help. Additionally, the most innovative service providers are working on their costs to be more efficient and pass along some of the savings. Lazy vendors don't do anything. They don't work on their costs and rarely volunteer to reduce costs until after beating it out of them. Who enjoys that? Well, some people do, but I don't.

Cost Reduction Partners are aligned with low-cost operations because it is part of how they differentiate themselves from competitors. When bidding services, vendors often say, "If you want the lowest cost, that is not me." Okay, then, what value do you provide that demands a premium price? If they cannot answer with something that makes sense, then we cannot select them. If they cannot deliver on what was promised, then it's time to make a change.

Where I find the most value is with individuals who provide excellent brain power. When selecting Partners, I don't always choose the lowest-cost option. Often, the low bid is fake, which we will review in more detail in Chapter 5: Supply Chain.

Seek brain power regarding personal experience, ideas, personality, innovation, troubleshooting, cost, cash flow, and risk management. Ultimately, is this someone who is a genuinely engaged person continuously helping to deliver the job safely who lives and breathes this business day and night?

Below is a mental checklist to consider when meeting with vendors and after each operation as an evaluation. People change over time, and performance is not robotic. Shale oil and gas manufacturing is not traditional manufacturing. It is much more dependent on people being able to react to a constantly changing "factory" due to the Earth's inherent rock heterogeneity, our limited ability to see what is happening in the subsurface, and our industry's complexity, including personnel rotation, personnel turnover, topography, and weather.

## Cost Reduction Partner Checklist

**1) Are they low-risk?**
We do not want high-risk, high-reward people. We want low-risk, high-reward. We are looking for value.

**2) Are they experienced in what we are hiring them to do?**
We want people to have experience in the task at hand. It is not uncommon for vendors to use customer wells as a training ground. Try to avoid this. Always ask about experience and who they are working for in our area.

**3) Do they have a low-cost approach and mentality?**
Getting to know people is critical, especially with the onsite drilling and completions consultants, wellsite managers, onsite leaders, company men, or whatever name you use for this critical onsite management position. During the drilling

and completions operational phase, most of our capital is being deployed, and they are managing the process on location. If the consultants do not have a low-cost mentality, it will be challenging to deal with them. Due to the high day rates the position provides, some of our consultants live high on the hog in their personal lives, with multiple homes, $100,000 trucks, planes, sports cars, extravagant vacations, and every toy imaginable. Don't get me wrong, I think it's great that high paying jobs in the oil industry provide people with the ability to do what they want. The problem arises when people carry their extravagant spending habits into our operation. This is an issue that needs to be addressed.

Many company men are savvy at managing relationships with office leadership over the phone but act and do something drastically different on location. They often think they are running their own private empire out there, as most oilfield locations are remote, far removed from direct oversight. Of course, it's not with their money; it's with our money. If someone spends their money carelessly, how do you think they are making decisions about spending our money? Many spend so voraciously that their money quickly runs out when times slow down. To fill the gap, it is not uncommon for folks to join or file Fair Labor Standards Act (FLSA) lawsuits, claiming they were employees when they know they are independent contractors. Think about that for a moment. We have a highly paid onsite consultant independent contractor whom we trusted with multi-million-dollar operations, and as soon as things slow down, they decide to sue claiming they were an employee to extract even more money. They are suing based on lies. What is the probability they have been lying about operations-related issues? Let me tell you, it's high, very high. There are legitimate FLSA lawsuits; however, if someone is willing to file or join a lawsuit based on lies, they most likely have been lying about anything and everything while working on location. This is why it is so critical to have good, honest people.

**4) How do estimated costs compare to actual costs?**

Once the final invoice comes in, we must analyze it and discuss how it compares to our AFE with each vendor.

**5) Are they cost-competitive?**

Benchmarking relative to competitors is a must.

**6) Are they a problem-solving resource?**

An excellent way to test this is to propose hypothetical situations and see what they suggest. Please do not wait until we are dealing with actual problems to find out they do not have any experience or cannot think on their feet.

**7) Are they interested in continuous improvement?**

If they are content with their current performance and not striving for better results, our costs will gradually increase. We need to work with vendors who are always seeking to improve.

**8) Do they provide ideas on how to reduce costs?**

It makes our job easier if our vendors present ideas on cost reduction without having to pull it out of them, force them to do it, or come up with all of the ideas internally.

**9) Are they happy, friendly, accessible team players?**

Working with arrogant, aggressive, know-it-alls is not fun. Although confidence is comforting, they usually make expensive mistakes that they do not take responsibly for.

**10) Will they help achieve our cost reduction objectives and contribute to profitable ideas?**

A lot of people in this business are not interested in helping an operator reduce costs. People have all kinds of motives that we do not know about and that are directly opposed to achieving significant reductions. Get to know people on a personal basis, including service provider owners, account managers, field managers, and field technicians. Spend the time to find out who has the brainpower and interest to help us with efficiency and optimization.

---

Consider reviewing this checklist or something close to it regularly. The stakes are too high not to be constantly evaluating people and service companies to determine if a change needs to be made.

Spend time with our service providers as much as possible because they play a significant role in our success. If a service company has internal problems, they will spill out onto the field, driving up our risk and costs.

Things happen quickly in the world of oil. A good service company today can go bad tomorrow. People move around a lot, and we want to know exactly what is happening. If a key person leaves a service company, we may have to cut the company. That's how important people are in this business.

## Cost Reduction Leaders

When cost reduction efforts take center stage, those who step up to lead the effort are Cost Reduction Leaders. They are good at motivating others, especially service providers. Leaders are often managers, engineers, foremen, and superintendents. Full-time non-rotational employees that live with the wells, start to finish. A good leader leads by example. People know whether someone is genuinely passionate about efficiency and low cost. When working on costs and speaking about it, the team can see if the leader believes what they say or is simply leading a cost reduction effort because their boss told them to.

## Cost Reduction Idea Generators

Cost reduction efforts often start with an idea. It is ideas on a variety of topics that could lead to cash flow results. In this highly complex business, which draws on multiple disciplines, implementing ideas successfully requires a team of individuals to make modifications, turning additional ideas into reality to get the initial bigger-picture idea to work. Sometimes ideas are revolutionary and change an entire industry. However, a single idea, by itself, is often not as valuable as one might think.

Some people have a new cost reduction idea every week, which is good. That is what we want. We want to encourage Idea Generators to provide proposals that reduce costs. The key is getting them to develop the concept all the way through to fruition, so we can realize the cash flow contribution.

The problem is that certain people cannot focus on one thing long enough to realize value. This dynamic exists across all industries, especially with medical professionals. Some doctors are excellent at developing new ideas on how to do things. However, if that same doctor had to focus for 10 hours on surgery to implement the idea, they would fail, and the patient would die.

The point is that ideas are a good starting point but putting them into practice is where things often become challenging and fail. Every cost reduction team needs Idea Generators, but there are levels on which we are working. For example, our big picture idea might be to reduce hydraulic fracturing costs by 40% to address an uneconomic project. That would be a high-level initiative based on the idea that fracturing cost needs to be reduced since it has been identified as the highest cost discipline with inefficiency.

After careful analysis, we decide to reduce fuel and proppant costs, so the team decouples them from the frac provider and sources the items directly. Of course, this has been done successfully by many competitors, but to implement it, several logistical issues need to be addressed to get it to work in a challenging mountainous area in the middle of winter, where it has proven to provide mixed results due to downtime issues. Solving those problems is where the real value is created.

In this scenario, the engineer who proposed decoupling services and selected the vendors might get all the credit, but that person was successful because field operators applied additional ideas to make the big picture idea work. The real innovation occurs three or four levels down from the initial big-picture idea.

When moving up in an organization, changing jobs, or changing companies, it is common for consultants, engineers, managers, and executives to get humbled when trying to implement a value-added method that previously worked for them. This is because they do not fully know how to implement the method. They were managing the process but did not work it on location. Often, what is missing are the seemingly small nuances that occurred in the field to get the issues addressed successfully and get the method to deliver consistent results. Those ideas are where the real value is.

Personnel in the field figuring out how to make things work don't always understand the importance of their discoveries, and their discoveries are not always documented and shared up the chain because no one is taking the time to engage with them. Or the ideas and methods are hoarded because they do not want to share them with management since they are not recognized for their efforts. To get a big-picture idea to work, multiple levels of idea generation need to occur and be documented, with the right personnel recognized for their efforts.

## Cost Reduction Operators

People who put ideas into action and see them through to fruition are Cost Reduction Operators. These highly valuable people often do not get the credit they deserve. Most of these folks are field personnel.

It will be hard to reduce costs without management spending time on location. If for no other reason than to show the field team that management considers cost a top priority.

If we are focused on generating value by producing oil and gas, then our factory floor is at the wellsite. In this industry, we often refer to the field as the real world as opposed to office work, which is essential but not as imminent. If we do not have top-tier Cost Reduction Operators, then operational mistakes will financially overwhelm our cost reduction efforts.

## Good systems prevent problems.

We must put good systems in place to prevent expensive mistakes. Once we have systems, we can automate them. Our field operators are central to our cash flow strategy which leverages systemization and automation.

The key with Cost Reduction Operators is that they must be interested in delivering safe, low-cost wells. They must want it more than anything else.

Field personnel often feel overworked and underpaid, while office personnel are making the big money doing nothing. Of course, this is not true, but that is the mentality.

This thinking is not just in the oil industry but in all sectors where large groups perform manual labor in complex environments. Since operational execution can be one of the most difficult and dangerous aspects for any enterprise, we must do everything we can to make our field teams' job easier. This one action will materially increase FCF.

## Cost Reduction Zombies

No matter what we do, there will be people who do not care about running a low-cost operation. In fact, they do not really care about anything work-related except their paycheck. The mindset is, "How can I get the most out of this company while doing the least amount of work." Efficiency in reverse.

Believe it or not, I have had employees and executives admit this to me. Yes, I said executives, including those at the highest levels. If disengaged zombie executives, employees, contractors, and vendors do not change their ways, they must be let go. It's just that simple. It is hard to run a cost reduction program with people that don't care because eventually they will go negative.

## Cost Reduction Doubters

When initiating cost reduction efforts, if we are pushing for costs that are industry-leading, in other words, costs that no one else has achieved, we will have Cost Reduction Doubters. The Doubter is someone who does not think it is possible to achieve the reduction goal. For example, if we currently have $10 million wells and we are targeting $5.0 million wells, the Doubter would be someone who does not think it is possible but who is not unwilling to be involved a little in trying to achieve our objectives.

We will have Doubters if we are pushing hard. It's good to have a Doubter on the team, as long as they are constructive. For example, if an idea is proposed and the Doubter explains why they think it will not work, then they are engaged and adding value. If the person proposes an alternative or adjustment to make it work, then they are not a Doubter. The Doubter is someone who only pushes back or portrays the facade of being interested in making changes to reduce cost.

Often, Doubters want to be left alone and do not want anyone in their business. The key with Doubters is making sure they do not transition into Haters. If the Doubter can transition into an Operator, that is what we want. However, in an aggressive push, when we get into the Doubters' business sphere of influence or question their work, it is common for the Doubter to turn into the Hater.

## Cost Reduction Haters

The Hater wants cost reduction efforts to fail. They get joy from this. One might think, why would someone not want a lower cost well which could help the company? The reason has to do with human elements and money. Anything that involves money is highly emotional.

When working to reduce costs, service providers will be writing tickets that are lower than before. They might grow to hate this if the volume of work does not make up for it or if they cannot lower their costs.

Ultimately, in many circumstances, if well costs do not decline, everyone will lose because the operator will stop drilling, go out of business, or management will be replaced. As a result, the vendor will lose their source of income.

However, the vendor may have difficulty seeing the big picture in the middle of a cost reduction program. The service company owner might see the big picture, but the workers on location, who might be asked to take a pay cut or change how they operate, may have animosity towards the process and, ultimately, the operator.

It is not only non-employee Haters that we need to worry about, but also employee Haters, who pose more risk and, believe it or not, are more common.

During cost reduction efforts, when money is addressed, and it is suggested a certain area has an opportunity to reduce costs, the people managing that area might take offense. It is a common reaction to take a directive to reduce costs as a personal insult, implying the team is inefficient or wasteful. Knowledgeable people understand that it's not personal, only business, and part of the continuous improvement process of a company and industry that is constantly evolving.

Since working on the subsurface is very challenging, once technical folks find a method that works, getting them to change or at least consider changing is hard. The problem is that although the method may work, it's uneconomic.

One dynamic I discuss with operations teams is that anyone with deep pockets can drill a well. We select vendors and go from there. It's not magic that no one else can do. It is common for the Hater to think they are God's gift to the oilfield and they are the only one with a particular skill.

It's true that people develop area-specific knowledge that no one else has. Often, these same people ensure no one else has this knowledge by hoarding all the information or keeping the critical methods to themselves. Regardless, the real value is with a team that can drill and complete a project safely and economically.

Haters get personally offended when someone tells them to reduce costs. In their opinion, the cost is already as low as it can go. The most common response I hear is:

*"I already reduced it, and that's it. It's not going any lower."*

The problem with this statement is that the Hater is positioning themselves against progress in opposition to the company's goals. Most of the time, they hope the cost reduction efforts focus on another area and leave them alone.

When Haters take a strong position that costs can no longer be reduced, they transition into an obstacle. They block progress because they are incentivized to do so by their own words. Remember, they said publicly that we cannot reduce costs any lower, so if we get costs lower, it makes them look bad. Haters will often do anything to protect their ego and reputation, and I mean anything. They will commit sabotage.

Haters are not interested in FCF for the company; they are interested in self-gratification or increasing cash flow to themselves. This can exist at all levels, from the field hands to the senior executives. Haters can be good workers, too, but they are not interested in prioritizing the company's success. The primary objective for them is to be proven right. They may even know the company is failing, but instead of helping, they bleed the company and drive it into the ground.

If we do not do things "their way," the Hater may want our efforts to fail and the company to fail to prove that they are right. They get joy from this. The Germans have a word for it, *"Schadenfreude." Schaden* (damage/harm) and *Freude* (joy).[8]

## Cost Reduction Blockers

The Blocker takes action to prevent cost reduction. They may try to undermine efforts by gathering supporters from consultants, vendors, and employees. We will review many actions to reduce costs. The Blocker will try to stop all of them.

The Blocker does not want a lower cost well. They have other motives. It may be that they do not like people questioning what they are doing or getting into their business. Blockers do not accept that every aspect of the business is reviewed for improvement, and improvement ideas do not always come from the people who are managing the process or working in the daily grind of ongoing operations. It is not uncommon for someone from outside the group to propose an idea to reduce costs, which is a good idea. But the idea needs buy-in from the operations team. Since this idea did not come from them, they automatically hate the idea and will do whatever it takes to kill it as fast as possible.

This discourages everyone else from helping because they see what happened. The Blockers don't realize that senior management understands that the initial idea is only the beginning, and although the ops team did not propose the idea, they must figure out how to make it work consistently at scale, which is often where the real value is. Getting an idea to work consistently in the field with all the daily challenges is very hard. Any new idea will be filled with opportunities to enhance the process and make it work. The Blocker does not see any of this. They just block everything. Once a Blocker is identified, they must be cut immediately.

## Cost Reduction Saboteurs

Every company deals with employees or vendor personnel that they wish would quit or go away. These folks are often referred to as "nightmare" employees, your basic toxic employees. They do their job but do it in a way that damages the culture, the company, and the enjoyment of work. These folks make others dread coming to work. Nightmare personnel should, of course, be let go. However, we might hesitate to release them until the timing is right.

As bad as "nightmare employees" are, they are nowhere near as bad as the Cost Reduction Saboteur. The reason is that the nightmare employee has the personality to do everything out in the open and directly to our face, so we know where they stand and who they are. However, the Saboteur is working behind our back, and we may not even know that they are there, secretly operating against the company and its cost reduction goals. The Saboteur is savvy, highly sophisticated, and hard to spot. One cost reduction Saboteur can derail and destroy an entire cost reduction campaign.

In Gallup's State of the Global Workplace: 2023 Report, 18% of employees are taking actions that "directly harm the organization, undercutting its goals and opposing its leaders."[9] If we have a cost reduction effort going on and the company is downsizing, certain employees and vendors know they will probably get cut. At this point, they may enact sabotage. Before starting a cost reduction effort, consider cutting all the deadweight personnel, contractors, consultants, managers, and executives.

# Worker Statistics

## The Bad News

- 77% of employees are not engaged at work.
- Low worker engagement costs the global economy $8.8 trillion. This is 9% of global GDP.
- 18% of employees are actively sabotaging their employers.
- Nearly 6 in 10 employees are quiet quitting.
- 51% of currently employed workers are seeking a new job.
- The average worker is only productive for 60% of the day across all professions. However, the average office worker is only productive for 2 hours and 53 minutes per day.
- 75% of employees have stolen from their employer. 90% of significant theft losses come from employees.
- 91% of workers daydream during meetings.
- 80% of workers are stressed from poor company communication.
- 64% of oil & gas workers do not feel "very comfortable" in raising safety issues with management.

## The Good News

- +70% of employees say that a strongly engaged culture makes them do their best work.
- In highly complex occupations, high performers are 800% more productive than average performers.
- 69% of employees would work harder if they were appreciated.
- Companies with high engagement are 21% more profitable.
- Strong company culture increases revenue by 4X.
- Companies with happy and engaged employees have 70% fewer accidents and 40% fewer quality defects.
- High-potential employees (HIPOs) are 91% more valuable to business operations compared to their peers. However, companies are bad at identifying who the HIPOs are.

*This bullet list was constructed referencing multiple sources. See Sources, Chapter 1, note 10, infra*

# Human Capital

People are the most valuable asset of a company. If we do not have the right people, we will not achieve success. Working across the U.S. and the world, the one thing I have found that every human has in common is complexity.

Humans do not fit neatly into one personality type. Life is not that simple. We might have a person who doubts certain aspects of our strategy but is contributing to profitable ideas. Overall, this person is adding value.

Additionally, assembling the right people on an individual basis misses a critical aspect of the group dynamic in which the right combination of people makes things work. Removing or adding the wrong person can impact the entire team. For example, some individuals are aggressive on cost reduction while others are hesitant. It is the combination of people that makes it work. Let's say we want to eliminate the intermediate casing to save $750,000 per well. Super-aggressive people are ready to do it immediately. However, a few people explain that changes to the mud system and BHA probably need to occur to pull it off without an issue. This delays implementation for a couple of months but ultimately leads to a first attempt success with the new design rather than a possible failure which could lead to abandoning the effort.

In this situation, who added more value? The aggressive people who proposed eliminating the intermediate casing or the hesitant people who provided ideas on how to make it work? If we are not careful, we could add or remove the wrong person from the team, impacting a successful dynamic.

# Economic Actions

During our time together, we will review economic actions to achieve massive cost reductions. These include big-picture strategic actions and intricate technical actions. The ideas are highlighted as ACTIONS and titled with a brief description. The details for select cost reduction actions are summarized under each ACTION item throughout the book.

Within each titled cost reduction action, many additional actions and ideas are often presented to reduce costs. Therefore, even if you are already implementing a titled action, you may find value in the associated sub-actions presented.

You should not agree with every action, nor is every action for every operational situation. Please perform your own due diligence before attempting any suggested actions.

As stated in the warning at the beginning of the book, never apply any actions, techniques, tactics, procedures, checklists, or anything written or quoted in this book without performing thorough calculations, hazard/risk assessments, statistical analysis, and due diligence with your company, team, and everyone involved with your operation. Never take an action, procedure, tactic, recommendation, suggestion, or anything else and force it into use in your area of operations.

A safe operation is the foundation for all actions. High risk is high cost. Safety incidents are expensive. When initiating a cost reduction campaign, some people automatically think that implies we will cut corners and do something unsafe to save money. Actually, it is the opposite. A low-cost operation is a safer operation compared to a high-cost operation. A dynamic we will explore from multiple angles during our time together.

When making changes to an operation to reduce costs, we must be extra careful to ensure everyone is aware of what we are doing to prevent any unintended consequences. Stop Work Authority is part of the foundation of a safe operation, and it must be discussed at each safety meeting to remind the team to stop unsafe or questionable actions. The safe success of a cost reduction effort is based on the dedication of our people to deliver risk-adjusted economic value.

The individual knowledge and collective wisdom of our team can be referred to as human capital. Encyclopedia Britannica defines human capital as follows:

*"Intangible collective resources possessed by individuals and groups within a given population. These resources include all the knowledge, talents, skills, abilities, experience, intelligence, training, judgment, and wisdom possessed individually and collectively, the cumulative total of which represents a form of wealth available to nations and organizations to accomplish their goals."* [11]

We must quantify human capital using multiple metrics and figures. As we have established, people are the most important element. Therefore, we must measure what matters most. What gets measured gets improved, and although measuring the value of individuals and groups of people is difficult, it will add significant value to our efforts and goal of maximizing cash flow with cost reduction. Let's dig into our initial **ACTIONS**.

———————— **ACTION** ————————

## Human Capital Due Diligence

In theory, it would be ideal and minimally disruptive if our current team of employees and service providers were the optimal group of people to deliver maximum cash flow with cost reduction. Unfortunately, this is usually not the case. Many good people in the oil and gas business are not interested in efficiency.

Historically, the shale industry has not focused on and rewarded cost reduction and cash flow. Valuations and value creation have focused on acreage size, number of locations, initial production, total potential resource, and production growth. However, recently, there has been a change. The industry is now focused on free cash flow.

Old habits are hard to break. Many individuals who are excellent at growing acreage positions, growing number of locations, and growing production, regardless of cost, are not necessarily strong at cost reduction. Spending money to grow a company is more fun than the drudgery of reducing costs.

Many do not enjoy cost reduction. In fact, they hate it. I can't blame them. For years, we have been incentivized based on growth, regardless of cost. Many team members are not excited about cost reduction and the diligence it takes to run a lean operation day in and day out. It's a grind.

Knowing how each employee and vendor feels about cost reduction is essential. Depending on what the job alternatives are at the time the cost reduction initiatives are taking place will determine how many employees and vendors self-select out of being involved by resigning, taking another job, transferring within the company, or not bidding or working with our company, as in the case of vendors that are not interested in cost reduction. When employees and vendors leave early on their own accord, it is best for both parties.

Unfortunately, this does not typically happen. High-cost people stick around for as long as possible, to continue extracting as much as possible from the company until someone forces them to leave. They can't help themselves. They have a supervisor to please, as in the case of service providers, and have operated with the philosophy of viewing the operator as a personal ATM, extracting money on demand.

*"If they're paying, I'm staying!"*

This is a common phrase among day rate and hourly oilfield workers. Meaning that even though they are not needed or not doing anything of value and should head out voluntarily, they will stay and bill lavishly until forced to leave.

In the case of high-cost employees (when I say high cost, I don't mean their salary, I mean how they think, work, and make decisions); they have been doing things in a certain way for years and believe it has been working, so why change? A suborn engineer or manager can consume a lot of time, capital, and energy in trying to get them on board for a cost reduction program.

Taking the time to think about each team member (employee and vendor) to determine if they are a good fit for our strategy will set the stage for success or failure.

> **To deliver game-changing cost reduction,**
> **we need impact players.**

We will review specific metrics in the following action items to help make human capital decisions to construct a cost reduction team that delivers results. However, most managers probably already know who is a liability and who is an asset. Everyone has an opportunity to show they can contribute to profitable ideas, are focused on delivering safe, low-cost solutions, and participate in the company's cash flow goals.

However, certain employees, vendors, and managers must be removed from the team if things are not working out. Any team, from sports team to management team, that is not delivering winning solutions must make personnel changes.

To generate industry-leading free cash flow with cost reduction, we must have impact players. These are individuals who have a positive cost reduction personality type and are a force on the team, helping others and contributing to the success of the company. Their contributions are powerful.

Running a cost reduction campaign to deliver specific CAPEX and OPEX targets by incorporating cost reduction into ongoing operations is not a single event. It is a core part of the company's culture, and it is not for everyone.

--------------- ACTION ---------------
# Track Human Capital Metrics

Structuring our team with the right people is essential to an efficient operation. Therefore, we need to look at our goals and operational performance in relation to individuals and groups of individuals.

Consider analyzing costs segmented by personnel. This includes all vendors, employees, and managers. For example, let's say we are running five drilling rigs or frac crews. We need to aggregate actual costs incurred and segment them by individuals associated with each crew. When asked the question:

**Specifically, who should we be tracking?**

*The answer is:*

**Everyone, including all vendors and subcontractors.**

This means we need to know all our service providers' personnel by name. Of course, the CEO will not know the individual names of the rig manager, directional drillers, frac leaders, wireline engineers, or wellhead technicians. However, field consultants and operating company engineers should know who these good folks are by their names.

Although there are multiple layers of oversight and crossover, in general, the service leaders are tracking the field operators, the consultants are tracking the service providers, the engineers are tracking the consultants, managers are tracking engineers, executives are tracking managers, and well performance in terms of safety, cost, and production is keeping track of us all.

Since we agree that people will determine our success or failure, controlling who is on location is essential. From a safety, environment, cost, and production perspective, we need to track precisely who is working on location and what value or lack of value they deliver.

Due to the rotational nature of our business and high turnover, it is common to feel that luck determines who is going to show up on location. I used to hope for good, experienced people to show up—they didn't.

It was only when we started showing an extreme interest in who was coming to our locations and requesting certain people based on performance that we were able to get control of this crucial aspect of operational success. Service providers were aware that we were carefully watching this critical aspect of the operation, and they would not send folks on the lower end of the performance or experience spectrum, especially trainees who could destroy our economics and our company.

The simple action of showing interest in each person involved in our operation will reduce costs. Show that we value individual people, know who the best people are, and request the best by name. Tell them that we asked for them specifically when they show up on our locations, and we will not only get the best people working for us, but we will also get the best they have to offer.

The foundational human capital metric is our ability to control who works on our wells and when they are working on our wells. We need specific names and dates. We will then correlate those names and dates with a variety of metrics. We are looking to tie performance to individuals.

> **Since people are our most valuable asset, we must control who is on our locations.**

Are we in control, or are we not in control? Do we have the best people, or do we not have the best people? We must know who is operating efficiently and who is not. Once we know how we are incurring costs relative to individuals and groups of individuals, we can identify potential improvement opportunities.

For example, let's say each drilling engineer drilled twelve wells last year. We look at drilling metrics segmented by engineer, including well cost, cost trends throughout the year, drilling performance curves, and other high-level metrics.

Then, we dig deeper into individual line items. Which engineers are spending the most money on specific items, and are those items delivering value? However, this is more complex than it may appear. If one engineer is focused on a deeper, more geologically difficult part of the basin, his or her costs will probably be higher than another engineer working on a shallower part of the basin.

Furthermore, we want to look at cost by geologist. Is there a correlation between each geologist and cost? Some geologists are more operationally focused than others. A poor geologic prognosis that ends up being incorrect can cause drilling costs to skyrocket. A strong engineer with a poor geologist will reveal itself in well cost and/or well production performance. Therefore, we must look at cost and production performance segmented by geologists.

Next, we must look at each well segmented by drilling consultant and all other service providers. Here is where it gets complicated due to the rotational nature of the positions. Evaluating metrics relative to our workforce will help identify the most efficient tactics, individuals, and teams executing on those tactics to achieve consistent success.

Segmenting metrics by personnel will help us identify those operating at a high level, those needing help, areas requiring personnel changes, and crews that are dysfunctional. The following pages provide examples of personnel metrics to consider segmenting by drilling, completion, and production.

Remember, 77% of the workforce is disengaged.[12] Nearly 6 out of 10 are quiet quitting[13] (doing the bare minimum). 51% are seeking a new job[14] and probably not focused on the task at hand. A dangerous situation in our business.

18% are actively sabotaging.[15] Shocking but true. The average office worker is only productive for three hours per day.[16] What are they doing the other 5 to 9 hours? Let me tell you, it's nothing good.

We will not achieve industry-leading cash flow metrics if we have an operating team full of these people. We must find out who they are and help them find a new job, at a company far, far away.

# Drilling Metrics to Segment by Workforce

| Drilling Metric | Details |
| --- | --- |
| **Safety and Environmental Incidents** | Segment incidents, spills, well control events by personnel on duty to determine if certain individuals are unsafe, high-risk, high-cost liabilities. |
| **Undesirable Drilling Situations** | Lost-in-hole, failures, sticking / stuck events, lost circulation, instability, downtime, high-cost events segmented by associated personnel. |
| **Rig Performance** | Drilling rig operations and issues (trip speed, connections, equipment failures, etc.) segmented by crew. |
| **Geologic Accuracy** | GeoProg precision including depths, structure, thickness, hazards, faulting relative to actual segmented by geologist. |
| **Geosteering Performance** | In-zone statistics, prediction accuracy, target change recommendation correctness, segmented by geologist. |
| **Directional Performance** | Plan vs. Actual, control and target change execution, slide and rotary statistics, daily cost segmented by personnel. |
| **Footage Performance** | Rate of Penetration (ROP), total drilled depth, Depth vs. Days segmented by personnel. |

# Drilling Metrics continued...

| Drilling Metric | Details |
|---|---|
| **Operations Time** | Total time analysis, productive and non-production time (NPT) segmented by operation and personnel. |
| **Bit Performance** | Parameters employed relative to recommendations (WOB, dP, GPM, RPM), pull decision compared to bit breakdown, segmented by personnel. |
| **BHA Performance** | Drilling performance and condition segmented by personnel during install & ops. |
| **Problem Prevention and Problem Resolution** | Who is proactive, able to catch issues before they become problems. When problems occur, who can identify and solve them successfully? |
| **Drilling Mud Performance** | Daily costs, losses, additions, properties, additives, problems segmented by personnel. |
| **Casing Performance** | Running issues, wear or other related problems during drilling, failures during completion. Segment by personnel involved during inspection & installation. |
| **Cementing and Equipment Performance** | Cementing results relative to plan, testing, issues, quality, performance by personnel. |

# Drilling Metrics continued...

| Drilling Metric | Details |
|---|---|
| **Pad Stability** | Issues segmented by builder. |
| **Mob / Demob Performance** | Statistics segmented by crew. |
| **Fuel Logistics and Usage** | Delivery performance and daily fuel usage to identify logistics, efficiency, theft, and contamination situations. |
| **Accommodations Management** | Call outs, cleaning, deliveries, and other expenses incurred, segmented by personnel. |
| **Transportation & Trucking Management** | Cost segmented by personnel to determine who is accurate, efficient, and optimally managing transportation costs. |
| **Invoice Approval Performance** | Who is coding correctly and performing invoice due diligence thoroughly to prevent inaccurate invoices from being approved for payment. |
| **Field Estimated Costs** | Identify who is accurately capturing field estimated cost each day compared to actuals. |
| **Daily Report Quality** | Ability to keep operations activities documented in daily reports with detailed accuracy, value added content and knowledge segmented by consultant. |

# Completion Metrics to Segment by Personnel

| Completion Metric | Details |
|---|---|
| **Safety and Environmental Incidents** | Segment incidents, spills, well control events by personnel on duty to determine if certain individuals are unsafe, high-risk, high-cost liabilities. |
| **Undesirable Completion Situations** | Max pressure violations, annular pressure, downtime, lost-in-hole, failures, high-cost events by associated personnel. |
| **Pre-Frac Geologic Analysis Validity and Accuracy** | Geologic evaluation to determine stage strategy, pumpability, design optimization, fault recognition relative to actual treatment response segmented by geologist. |
| **Pre-Frac Reservoir Analysis Validity and Accuracy** | Reservoir assessment based on drilling data to determine frac design optimization, real-time reactions and adjustments, chemical program, volumetrics, production projections segmented by engineer. |
| **Offset Well Action Plan Validity and Accuracy** | Offset well monitoring plan versus execution, offset well incidents, real-time response to avoid unsafe, high-cost situations segmented by personnel. |

# Completion Metrics continued...

| Completion Metric | Details |
|---|---|
| **Wellhead & Frac Stack Performance** | Operational execution, events segmented by personnel. |
| **Fracturing Performance** | Stages/day, stages/shift, pump time, rate, prop ppg, chemical rates, plan versus actual segmented by personnel. |
| **Diversion, ISIP, Water Hammer Implementation** | Ability to administer and react to real-time events segmented by consultant. |
| **Placement Performance** | Percentage of frac job placed segmented by consultant. |
| **Screenout Avoidance** | Screenouts by consultant. |
| **Wireline Accuracy, Efficiency, Problems** | Stats segmented by personnel. |
| **Operations Time** | Total time analysis, productive and non-production time (NPT) segmented by minute level operations and personnel. |
| **Problem Prevention and Problem Resolution** | Who is proactive, able to catch issues before they become problems? When problems occur, who can identify and solve them successfully? |
| **Fuel Management** | Usage by consultant. |
| **Surface Rentals** | Costs segmented by personnel. |
| **Water Transfer** | Issues by personnel. |

# Completion Metrics continued...

| Completion Metric | Details |
|---|---|
| **Millout Performance** | Performance by personnel. |
| **Flowback Performance** | Leaks, safety, monitoring, issue avoidance segmented by personnel. |
| **Accommodations Management** | Call outs, cleaning, deliveries, and other expenses incurred segmented by personnel. |
| **Transportation & Trucking Management** | Cost segmented by personnel to determine who is accurate, efficient, and optimally managing transportation costs. |
| **Facility Construction Performance** | Construction statistics (safety, time, costs, leaks, quality, failures) segmented by builders. |
| **Invoice Approval** | Who is coding correctly and performing invoice diligence thoroughly to prevent inaccurate invoices from being approved for payment. |
| **Field Estimated Costs** | Identify who is accurately capturing field estimated cost each day compared to actuals. |
| **Daily Report Quality** | Ability to keep operations activities documented in daily reports with detailed accuracy, value added content and knowledge segmented by consultant. |

# Production Metrics to Segment by Personnel

| Production Metric | Details |
|---|---|
| **Safety and Environmental Incidents** | Segment incidents, spills, well control events by lease operator to determine if certain individuals are unsafe, high-risk, high-cost liabilities. |
| **Production Downtime** | Segment downtime by issues and lease operator to determine if lack of preventative maintenance or attention to detail is resulting in downtime. |
| **Artificial Lift Management** | Ability to manage and optimize lift parameters for production and cost segmented by lease operator and engineer. |
| **Oil and Water Hauling Performance** | Hauling efficiency and cost segmented by lease operator & logistics supervisor. |
| **Oil and Water Spills** | Tank overfill incidents and spill events segmented by hauler personnel and lease operator. |
| **Oil Load Rejects** | Rejected loads segmented by driver and lease operator. |
| **Secondary Containment Condition and Maintenance** | Containment maintenance segmented by lease operator. |
| **Facility Equipment Operation** | Vessels, units, valves, chokes, scrubbers, flares, combustors, etc. safe, proper operation segmented by personnel. |

# Production Metrics continued...

| Production Metric | Details |
|---|---|
| **Repairs & Maintenance** | Performance and cost segmented by personnel. |
| **Chemical Injection and Usage** | Pump operation, chemical performance and total cost segmented by personnel. |
| **Wellhead and Production Tree Operations** | Cleanliness, maintenance, leaks, washouts, valve function, safety valve operation, general performance segmented by personnel. |
| **Telemetry and SCADA Management** | Data accuracy (transducers, tank gauges, rate, condition), maintenance, alarm & ESD setpoints / function segmented by personnel. |
| **Gas and Liquid Measurement Accuracy** | Regular inspection, calibration of orifice plates, meter tubes, flow computers, tank gauges, liquid meters to minimize measurement inaccuracy, lost and unaccounted for oil & gas volumes, and theft, segmented by personnel. |
| **Location Condition** | Cleanliness, holes, signs, weeds, fence, lights, gate, locks, and overall condition segmented by lease operator. |
| **Overtime Hours and Callouts** | Number of overtime hours and afterhours callouts by personnel. |

# Production Metrics continued...

| Production Metric | Details |
|---|---|
| **Pipeline Shut-in Events and Associated Equipment Performance** | Effective communication, operator presence on location, equipment safe functioning (alarms, valves, flares) segmented by personnel. |
| **Theft Incidents** | Documenting oil and equipment theft by potential personnel involved can help identify the criminals. It is not uncommon for a pumper, superintendent, vendor, former vendor, or a former employee to be involved. |
| **Weather Event Protection** | Weather damage / situations segmented by lease operator to see who is proactive about preventing costly weather-related events. |
| **Solar Power Management** | Solar panel, battery, wiring, controller condition, maintenance, monitoring segmented by lease operator. |
| **Electricity Usage** | Ability to cost effectively manage electricity usage segmented by lease operator. |
| **Daily Production Report Quality and Accuracy** | Ability to measure, confirm, and enter daily production into software. Include descriptive notes and pictures to add value to report segmented by lease operator. |

# Cash Flow Focused Workforce Performance

People are not robots that consistently make sound, well-thought-out, logical decisions. Health, focus, emotion, pride, politics, vengeance, greed, anger, and a thousand other variables impact human performance. Using human capital metrics customized for our operation will help identify desirable team members. The goal is to determine which individuals and teams are adding value and which are destroying value. With objective data on human performance, analysis can be performed, and decisions can be made. Removing high-risk, high-cost vendors, engineers, managers, and executives will beneficially impact cost reduction and free cash flow, allowing us to reach our goals.

Our approach to maximizing cash flow with cost reduction is comprehensive, covering systems, procedures, designs, technology, implementation, and execution. We measure success using finance and accounting principles, ensuring a robust economic strategy. However...

> **The best designs, engineering methods, and systems will fail to deliver success, with the wrong people in place.**

For this reason, optimizing the workforce provides the foundation for a safe, low-cost operation, enabling us to meet our objectives. Implementing a system that measures human performance at an extremely detailed, unbiased level is a critical component to quantitatively and qualitatively analyze our workforce, ensuring we have the right people in place.

Everyone in the field knows of individuals who are continuously responsible for high-cost operations, but never seem to change their methods or be released by the operator. Part of the reason for this is due to the remoteness of oilfield operations and the ability to manage relationships.

For example, a drilling or completions consultant who inefficiently manages operations but is friendly with the right people or very good at dealing with engineers and management over the phone, can hang around for a long time, contributing to the company's inability to achieve positive free cash flow.

These people need to be identified and removed, regardless of the strong relationship they may have. Unlike other industries, due to the capital intensity of oil and gas operations, a single weak link can cause an entire corporation to fail. Using data driven metrics to categorize personnel performance is the key to cut through the dynamics of real-world operations.

## Build a Quant-based Workforce Analysis Tool

A simple spreadsheet could be utilized to analyze our workforce. However, consider using AI to gather data from multiple sources, including daily reports, telemetry, EDRs, post-job reports, engineering reports, post-well reviews, vendor reviews, end-of-well reports, accounting reports, production, and economic analysis. The data is mined for pertinent information and organized to quantify individual and team performance. Digital transformation vendors exist to help build these systems, or we can build them ourselves.

## ——————— ACTION ———————

# Begin Each Meeting
# Discussing Human Capital Dynamics

When conducting leadership meetings at each level across our operation, consider beginning the discussion with a focus on human capital. Our philosophy to maximize cash flow with cost reduction and create value is a people-first approach. Therefore, during meetings, let's focus on people first. This is entirely different from the standard approach, which is to focus on the specific topic of the meeting.

For example, if the meeting is a quarterly review of business operations or a strategy session, we might start by looking at asset performance with various derivatives of cash flow and cost. At the field level we might begin with a discussion on a specific operational problem.

Instead, consider starting each meeting, regardless of the meeting topic, with a discussion on human capital. I am not suggesting reviewing the complete "Human Capital Metrics Tables" or our AI-generated human performance analysis.

At the start of each meeting, consider posing a question that aligns with our people-first approach.

*"Do we have the right people in place to be successful?"*

This serves as a reminder of our focus on human capital and sets the tone for the meeting. Whatever the topic of the meeting is, do we have the right people in place? If we are unsure, what do we need to do to get the right people? Once that is addressed, move on.

Discussing individual people is always emotional. I do not suggest discussing firing people at the start of every meeting. Maybe we need to add someone or move people around. We may need to put our strongest people in place for a particular operation to reduce risk and cost. We may need to bring in an expert to help with a troubled well or a new cost reduction tactic we are considering.

> **Leveraging the experience of others is one of the best actions to reduce risk and cost.**

When I attempt something that I have never done before, I will search to the ends of the Earth to find someone who has already done it successfully. There is too much risk and money to take chances, roll the dice, and hope for the best.

In some instances, strategic personnel modifications may be necessary, such as switching vendors or individuals within a vendor or releasing a consultant. These decisions, similar to those made by a professional sports team, are made to enhance our performance. However, the stakes in our game, the shale game, are much higher and far more serious than in any sport.

## Emotions Impact Cash Flow

Making personnel decisions to create the right workforce to maximize cash flow with cost reduction is not easy. It is common to change multiple service providers, let good employees go, and release leadership personnel who are not able to deliver a safe, low-cost operation.

Personnel decisions, of any kind, are emotional for all those involved, especially when we know the people on a family level. Creating consistent positive free cash flow is not always fun. Hard decisions must be made. Sometimes that includes letting friends go. Unfortunately, friendships can get damaged during the process and sometimes friends will be lost.

If emotions are controlling operational decisions, it will be difficult to make good cash flow choices. Money and personnel are emotional subjects, especially when critiquing someone's operational costs and how they need to be reduced. Some folks do not handle it well. I have had many people take it personally and become emotional, often leading to anger and aggression. Creating a workforce that respects cash flow and low-cost operations requires hard choices.

Although most of this book is focused on detailed technical aspects of cost reduction in oil and gas operations, the human element provides the foundation and is fundamental to success. If people do not enjoy coming to work, their time at work, and working towards a common goal, it will be hard to press ahead, especially when things get challenging.

The power of emotion can work against us, but it can also work for us. An inspired workforce determined to be successful will work far longer and harder at its goals when everyone else has given up, including the competition. Building the right team provides the foundation for everything we will discuss beyond this point.

# 2

# SHALE ECONOMICS

—

**M**aximizing the cash flowing into our business while minimizing the cash flowing out is our fundamental economic objective. However, shale has many challenges that prevent it from occurring. Our strategy to deal with them is to force drilling, completion, and production operations to be exceptionally safe, efficient, and predictable, which ultimately drives down costs and enhances economics.

Consistently delivering costs below current levels requires a tremendous amount of effort. If our team does not know why we are doing what we are doing, they will not come to the realization that our strategy must succeed for our company to succeed. The "why" is very important.

To become passionate about something and eventually obsessed in a productive way, we need to understand the underlying reasons. This is especially true for challenging goals because once things get hard, without a deep understanding of why we need to keep pushing forward, people will become discouraged and give up.

When things become difficult in life, most give up and move on to something else. However, when individuals independently conclude that they must achieve a certain goal, they will not give up easily.

In the oil and gas industry, many people have worked for a company, as an employee or contractor, that experienced financial difficulties, gone through bankruptcy, or was dissolved. During discussions and interviews, when asked why they think the company failed, most blame management.

When discussing specific problems they had on various wells and the associated cost overruns, the same folks who blame management for the economic failure, rarely make the connection to the engineering, operations, and field decisions that contributed to the high-cost operation and incidents that slowly eroded their company's financial future.

## Cash Flow Is Vital

It's cliché, but cash is king in business. Arguably, the number one rule in business is: Do not run out of cash.[1] In the field, on the operations side of the business, we often do not concern ourselves with where the cash to drill, complete, and produce wells is coming from and what happens when we exceed AFEs or have a major operational issue requiring unplanned expenditures.

In the field, and office to a certain extent, the mindset is that undesirable events and cost overruns are just part of the business. Additionally, we rarely ponder what happens when a well underperforms projections, or is a dog producer, due to an unexpected geologic anomaly or operational mistake.

All of these dynamics impact cash flow. Most operations personnel do not worry about how undesirable daily occurrences impact economics. People shrug it off and continue with business as usual. Cash flow is given little attention until there is a funding problem that directly impacts them, such as when or if they will get paid.

Every dollar spent that does not have to be spent puts us closer to the day we could run out of cash. By running a low-cost operation and continuously looking for new ways to reduce costs, we are taking a proactive approach to managing cash flow. The more money we spend on each well the more risk we take. There is no guarantee we will see that dollar returned in the form of production over many uncertain days, months, and years.

## Free Cash Flow

The primary objective of an oil and gas company is to maximize value for its stakeholders (i.e., owners, investors, shareholders, employees, bondholders, managers, executives, mineral owners, working interest owners, partners, landowners, customers, communities, and all other people and entities with an interest or concern in the business) while maintaining excellent environmental and social responsibility.

Each stakeholder views the value of the business slightly differently, forecasting future value creation and financial strength using a variety of methods and metrics. One of my favorite metrics to determine financial strength and profitability is free cash flow (FCF).

Cash flow is the amount of cash moving in and out of the business. We can find this number at the bottom of the Cash Flow Statement, usually listed as "Cash and Cash Equivalents." This number includes the net cash flow amounts for operating, investing, and financing activities.

In isolation, this number is not a good indicator of the strength of a business because it encompasses cash flow from all sources, including cash flow from issuing bonds, borrowing debt from a bank, or issuing equity. None of these things are necessarily bad, unless the company cannot cover the cost of capital and create value for stakeholders. However, the cash flow number at the bottom of the cash flow statement does not remove these aspects to get a clearer picture of whether the company is creating value by generating surplus cash.

Free cash flow (FCF) is a better but not perfect metric because it indicates the cash generated from operations net of operating expenses (OPEX) minus capital expenditures (CAPEX), including drilling and completions costs. There are multiple ways to calculate FCF, and various terms used. Different operating companies and analysts use different variables depending on the specific situation or preference. The objective is to see whether a company is generating cash flow above the cost of operations. The next page lists a variety of formulas to calculate FCF.

With these formulas, we can see if a company can fund its operations from production or if it's selling assets, using financing activities, or some other means.

# Standard Free Cash Flow Formulas

FCF = Cash Generated from Operations after Cash Spent on Costs

FCF = Cash Flow from Operations (CFO) – Dividends – CAPEX

FCF = EBIT * (1–Tax Rate) + D&A – $\Delta$Working Capital – CAPEX

FCF = Adjusted EBITDA + Non-cash items – CAPEX

FCF = EBITDA – Interest Expense – Dividends – CAPEX

FCFF = CFO +(Interest Expense *(1 – Tax Rate)) – CAPEX

FCFE = FCFF – (Interest Expense*(1 – Tax Rate)) +Net Debt Issued

*These formulas are based on multiple sources. See Sources, Chapter 2, note 2, infra*

Although the above formulas are standard, they do not resonate with our ops team, responsible for maximizing FCF. Therefore, when selecting metrics to measure our performance, consider reframing with variables that pack a more powerful punch.

# Trench Warfare Free Cash Flow Formula

**FCF = (Net Production × O&G Prices) – Field Tickets**

Our team knows production, commodity prices, and field tickets—three critical FCF variables. We cannot change oil and gas prices, but we can change production and the field tickets. In terms of cost reduction, the tickets are critical, which is what the Trench Warfare FCF Formula is designed to illustrate. Everything we are doing to deliver massive cost reductions must be realized in the field tickets and final invoices.

As hard as we work to reduce costs, we must work just as hard to ensure our efforts are reflected on the invoices. Field ticket and invoice reduction is an objective we will attack from multiple angles.

Many companies do not have positive FCF because they invest to grow production by drilling a lot of expensive wells. Thus, drilling and completion costs are larger and growing faster than cash flow, with the hope that, eventually, production from the wells will generate more cash than the total cost to run the development program. At this point, the company can fund growth from production and generate positive FCF. If a company cannot do this, it will eventually run out of cash once it exhausts its ability to raise capital.

# Oil and Gas Company Valuation

*This simplified idealized valuation flow diagram is for a company whose strategy is primarily focused on production growth through drilling and completion activities.*

**OPEX**  = Operating Expenditures = Production Costs
**CAPEX** = Capital Expenditures = Drilling & Completion Costs

---

(Commodity Price – **OPEX**) X (Production + Growth) = CFO

↓   *Drilling for growth* ↗

(CFO) **Cash Flow from Operations** – **CAPEX** = FCF

↓

(FCF) **Free Cash Flow** + Risk = NAV

↓

(NAV) [Net Asset Value – Net Debt] / Number of Shares

↓

**($ / Share) Stock Price**

---

*This valuation model was constructed referencing multiple sources. See Sources, Chapter 2, note 3, infra*

Notice how significant CAPEX and OPEX are in determining FCF. Operations personnel rarely make the connection between their performance and the cash flow of the company they work for. Perhaps, more than any other industry, due to the capital intensity of oil and gas operations, a decision by one employee or vendor in a remote area of the business can determine whether a company is FCF positive or negative, thus determining whether it will continue to exist.

Furthermore, seemingly minor management, geologic, or engineering decisions can put a company on the path to bankruptcy before the fiscal quarter's first well is drilled. Understanding these risks will help us stay diligent, especially when things get tough, and people want to give up on our cost reduction goals. When everyone understands why we need to work hard to push costs down, it helps keep the team energized and moving towards our targets.

At the beginning of a cost reduction effort, people often ask why we are doing what we are doing, especially when commodity prices are perceived as high or well results are robust. Explaining why we must control costs and get them as low as possible has helped achieve targets by enabling people to become believers in the cause as they understand the economic challenges at the asset and the corporate level.

Since we cannot know everyone's economic challenges with shale development, currently and in the future, we will review the most common that companies face. We will review each one because the more reasons we have, the more dedicated we will be.

# 27 Reasons for Shale Cost Reduction

## 1. Commodity Markets

Unfavorable commodity prices combined with high and unpredictable well costs are among the most significant factors motivating cost reduction. Oil and natural gas prices are unstable, impacted by various factors that no single company or government controls.

We are price takers in the oil and gas business. The market sets the price, and that is what we get. Unlike other industries, we cannot set the price we sell our products. For example, if we were manufacturing vehicles or electronics, each producer could exercise some control over the sales price. Oil and gas companies must take the price offered for their product or shut-in production, which is usually not a viable long-term option. Ultimately, prices are set by local and global markets, a function of multiple factors.

### Major Factors Impacting Oil and Natural Gas Prices

- **Supply:** primarily influenced by OPEC+ and U.S. drilling
- **Storage:** available space, current inventory, imbalance
- **Demand:** primarily dependent on economic activity
- **Reports:** economic outlook, consumption, inventory, news
- **Governments:** policies, laws, regs, taxes, interest rates
- **Geopolitics:** power struggles, relations, behavior
- **Uncertainty:** crisis, conflict, war, unrest, strikes, terrorism
- **Market:** value, risk, correlation, speculation, manipulation
- **Derivatives:** futures, swaps, options, shorting
- **Weather:** storms, hurricanes, seasons, extreme temps

Since so many unpredictable factors impact oil and natural gas prices, which we cannot control, we must fiercely control what can be controlled, and that is cost. None of the major factors impacting oil and gas prices can be controlled by a single company for an extended period.

However, capital efficiency, risk management, low-cost operations, and cost reduction are controllable. Since we cannot control the price we sell our product for, we must control the cost, and it must be done aggressively.

## 2. OPEC

The 12 member countries of the Organization of Petroleum Exporting Countries (OPEC) account for about 40% of global oil production and 80% of the world's proven oil reserves.[4] As a result, OPEC wields significant control over global oil markets. OPEC's official mission is to "coordinate and unify the petroleum policies of its Member Countries and ensure the stabilization of oil markets to secure an efficient, economical, and regular supply of petroleum to consumers, a steady income to producers, and a fair return on capital for those investing in the petroleum industry."[5]

Although the official mission of OPEC appears to align with the economic interests of American companies, shale is a direct competitor to OPEC and has taken significant market share. As a result, OPEC has taken actions in the past to flood the market with oil to force American producers to cut production.

OPEC often works with non-member countries, including Russia, referred to as OPEC+, to achieve mutually beneficial goals.[6] This can lead to agreements, disagreements, and all-out price wars. Since the United States does not directly work with OPEC+, the group has few options to get American producers to help balance the market other than to flood the market with oil, forcing global oil prices to decline, thereby incentivizing American producers to make cuts.

This approach has damaged American companies, many of which have been driven out of business. If American companies worked with OPEC, it could be argued that rebalancing oil markets would be far less destructive to the United States and American energy companies. However, this is currently not the case. Therefore, the best defense to undesirable OPEC actions is to significantly reduce costs.

## 3. Hedge Position

An unfavorable hedge position is a reason to reduce costs. Since hedging locks in price for a specific volume and time, if prices move unfavorably, the producer is protected. I firmly believe in hedging because if we decide to drill at a particular commodity price, it pays to honor that decision with hedges to protect downside risk. However, if we hedge in a rising commodity market, with services costs also rising, we will have no choice but to reduce costs to maintain margins.

Reserve-based lending (RBL) and other forms of debt often contain covenants requiring certain hedge levels. If a company is actively drilling and does not hedge, should prices fall, there is no choice but to reduce all costs.

With so much risk in this business, if there is an option to limit downside risk for modest cost, it should be considered. If we do not hedge, we are speculators. With extensive due diligence, this can be profitable, but it is an additional risk in an already risky business. We don't have to hedge 100% of our production, and it's not always possible. However, if we are operating with significant debt, it is prudent to hedge enough production to ensure the debt can continue to be serviced in case commodities prices decline.

If the decision is made to sign a rig contact for a certain number of wells or for a period of time, and that decision is made based on a commodity price forecast using the current forward curve, then to honor that decision, it is good risk management to hedge and lock in a stream of cash flows honoring our economic decision. This is especially true for shale wells since the first several years of production are critical for the economics to work.

## 4. Shale Production Profile

Shale wells generally exhibit a production profile with a high initial decline. An 80% decline in the first year is not uncommon. Initial production (IP) rates are often misleading. This is why referencing short-term rates, 24-hour and 30-day IPs, should be avoided.

Short-term IPs are not an accurate portrayal of what is happening and can demotivate a cost reduction program. People who do not understand the high decline production profile often reference success or failure based on 24-hour and 30-day rates. Once that number makes its way to the field, people do quick math and carry it forward indefinitely, concluding the well paid out in 3-months.

Many team members who are not watching the decline, use the high initial production rates to ease off cost control, cost reduction initiatives, and invoice diligence. For example, let's say a few wells hit +1,000 barrels of oil per day (bopd). Management starts referencing 1,000 bopd in conversations, meetings, and presentations. All the field guys hear that and think there is no need to reduce costs, or they think the oil company is extremely rich and greedy.

When service providers hear how "successful" the wells are, they hit us with inflated tickets and increased prices. If a service provider is on the fence about a specific charge and hears about how big our wells are, rest assured that we will get hit with every charge imaginable because we can afford it.

When everyone starts referencing +1,000 bopd rates, everyone's mental math becomes based on 1,000 bopd for all the wells. They think, "Why should we be more efficient when the wells are making a thousand barrels per day?"

Then, people start taking what they think they deserve. Overcharge here, add extra there, and it adds up. Little do they know, +1,000 bopd only lasted a few days or months before the wells quickly declined. Short-term IPs are vanity metrics and do not provide critical information on the decline. It is the detailed long-term dynamics that determine success or failure.

## 5. Underperforming Wells

There are many potential reasons for wells to underperform production expectations. One of our best weapons to combat below type curve performance is to reduce costs.

# Type Curve Underperformance Causes

| Failure Cause | | Key Points |
|---|---|---|
| Projection Model | Data Availability | Delineation and appraisal phase did not have sufficient producing well count and/or production time data for accurate type curve development. |
| | Aggressive Forecasting | Optimistic production projections based on actuals. Unfavorable wells were excluded from forecast. |
| | Data Lumping | Multiple type curves were not constructed and segmented by area, geologic variables, volumetrics, target interval, and frac designs. |
| Geologic Model | Rock Variability | Reservoir characteristics (porosity, perm, thickness, pressure, etc.) are more variable than originally thought. |
| | Oversimplified Mapping | Geologic analysis was not thorough and lacked attention to detail. |
| | Significant Faulting | Faults impacted production (high water production, difficulty staying in zone, stolen frac energy). |
| Development Model | Engineering Design | Ops techniques not optimized based on subsurface characteristics. One size fits all approach impacts results. |
| | Well Spacing | Parent well depletion, communication, well-bashing, all impact new wells and legacy production profiles. |
| | Product Variation | Changing GOR's, unfavorable oil or gas (gravity, $H_2S$, $CO_2$, BTU, NGLs) impact forecasts and economics. |
| | Production Downtime | Well issues, lift failures, frac hits, pipe shut-ins contribute to downtime impacting actual production profiles. |

When looking at production on a well-by-well basis, the Pareto Principle often applies. The Pareto Principle, also known as the 80/20 rule, implies 80% of the production comes from 20% of the wells.[7] In other words, 20% of the wells carry most of the production, making up for the underperformers.

The problem is that all the wells are expensive, so we get no relief on the cost side of the equation. A few bad wells could put the company's cash flow in jeopardy. For example, if well costs are $10.0 million/well, an underperforming multi-well pad of 10 to 20 wells could put the company in financial peril.

However, if well costs are reduced to $3.0 million per well, even if production performance is below type curve, the wells have a much better chance of paying out and will not impact the company as a going concern. If well costs are low, a string of underperforming wells can be financially absorbed. Drilling a well is like placing a bet. The higher the cost per well, the more financial exposure to the entire company each time we drill. The key to long-term success in any business is staying in the game—a strategy that low costs enable.

## 6. Time-Lag

One of the most difficult challenges in any business is identifying problems quickly enough to prevent undesirable results that cannot be rectified. What makes shale oil and gas development especially challenging is that, unless we are extremely diligent, by the time we realize there is a problem, significant capital has already been deployed.

A time lag exists between acquiring a land position, permitting, drilling, completions, and having enough production data to determine if the unit is economical. I call this the "Clarity Gap" or time until economics are clear. The table below showcases the clarity gap in days from assembling a 1280-acre drilling and spacing unit (DSU) to obtaining enough data to determine the production profile. It is preferable to have 90 days of production to make an informed decision, but waiting that long is not always possible or advisable.

## Clarity Gap

| Well Phase | Time in Days | | |
|---|---|---|---|
| | 1-well pad | 3-well pad | 5-well pad |
| Land, Legal, Regulatory | 100 | 100 | 100 |
| Pre-Spud Operations | 15 | 15 | 15 |
| **Drilling Operations** | **20** | **50** | **75** |
| Mob, Setup, Standby | 10 | 10 | 10 |
| **Completion Operations** | **15** | **30** | **50** |
| Mob, Setup, Standby | 5 | 5 | 5 |
| Flowback | 10 | 10 | 10 |
| Production | 90 | 90 | 90 |
| **Spud to First Sales** | **50 days** | **95 days** | **140 days** |
| **Total Time** | **265 days** | **310 days** | **355 days** |

*Based on 10,000' TVD with 10,000' lateral, 50 stage PnP completion with simul-frac, millout, tubing and gas lift install. Land, Legal, Regulatory days include time to put together multi-section position, file orders, hold hearings, survey and stake location, file State/Federal permits. Pre-Spud Ops includes days to build pad, set conductor, drill surface, wait on big rig. Drilling includes mob and spud to rig release. Completion includes prep, frac, millout, tubing install, facility construction. Flowback includes days for well to clean up. Production includes days required to determine production profile. Total days assumes midstream gathering system is being constructed simultaneously.*

As the table shows, multi-well development increases the clarity gap and compounds capital at risk, potentially placing our company on the road to insolvency, should a relatively significant amount of capital be deployed in areas that turn out to be uneconomic or there is an operational problem.

When running multiple rigs in full-scale development mode, if anything is off, even slightly, due to the scale, by the time we realize there is a problem, it may already be too late.

To illustrate the point, let's assume the development plan is to scale operations in an area by running ten rigs, down space core areas, and step out into expansion areas with respectable type curves based on test wells. Assuming $8.0MM cost/well, by the end of the first year we will spend almost $1.0 billion. If something is off, let's say child-parent depletion, expansion area geology is inconsistent, operational problems occur, or any other issue, a significant amount of capital could be lost.

## Development Capital Risk Over Time

| Phase | Q1 | Q2 | Q3 | Q4 |
|---|---|---|---|---|
| Drilling | $120 MM | $120 MM | $120 MM | $120 MM |
| Completion | $80 MM | $120 MM | $120 MM | $120 MM |
| Quarterly Risk | $200 MM | $240 MM | $240 MM | $ 240MM |
| Cumulative Capital Risk | $200 MM | $440 MM | $680 MM | $920 MM |
| Clarity Gap -------------------------------------- | | | | |

*Based on running 10 rigs, drilling 1 well /month/ rig, $4.0MM drilling cost, $4.0MM completion cost. Total risk is cumulative capital expenditures for the first year.*

Based on this example, due to the time lag between the development capital outlay and having enough production data to determine whether the drilling program continues to deliver on its economic targets, we could have $500MM deployed before realizing there is a problem.

If our program starts to have challenges due to spacing issues or inconsistent rock quality, and ten rigs are running with a completed well cost (CWC) of $8.0MM, a serious amount of money could be lost should things not turn out as planned. Wells drilled in the first quarter (Q1), January to March would not be online until the May to June period (Q2). By that time, we already drilled 50 to 60 wells.

The combination of the capital intensity during drilling and the clarity gap can be financially deadly. This is why we must get costs as low as possible.

---

**When costs are high, time is our enemy.**

---

Low costs allow us to stay in the game through the rough patches and uncertainty, providing time to create stakeholder value, ride out the market cycle, or just get lucky.

## Tactics to Deal with the Clarity Gap

In addition to reducing costs to reduce exposure, there are several tactics to consider to further reduce cash flow risk during full-scale shale development. First, look for correlations between gas shows during drilling and long-term production performance. In many horizontal plays, gas shows are a good indicator of strong production, and a loose correlation can be made to predict production based on gas shows during drilling. Second, look for geologic surprises or concerns with the geology relative to our GeoProg expectations during drilling. Third, look at the cuttings to see if they match expectations.

If the development plan is to drill multiple wells in a DSU, but the first well is concerning, consider changing the plan immediately—we must react. There have been numerous situations in which a team became concerned after the first well, but nothing was done. No action was taken.

All the wells were drilled as planned, only to realize later that it was a huge mistake. On the first well in the DSU, if the gas shows do not look right, a meeting should be called to discuss concerns and determine the best course of action.

## 7. Capital Structure

Companies utilize debt and equity to finance operations and acquisitions. The amount and type of debt and equity is the capital structure. Super Majors and conventional producers are generally self-financed entities. However, shale operators are often highly leveraged, funding growth with high debt levels.

While in business school at Columbia University, I took a class called "Theory and Policy of Modern Finance," taught by Joel Stern—the creator of the term Free Cash Flow (FCF).[8] One of the learnings that stuck with me was based on a question Joel posed to us on the first day of class.

He asked everyone individually, "What is more expensive, debt or equity?"[9] He went around the entire class, asking each student that question. Almost everyone said equity is more expensive. After everyone answered the question, he ended the class for the day without giving his opinion. At the end of the first week, we circled back to that question, and he explained that although it appears equity is more expensive, it is debt that is more expensive because of the risk to the company as a going concern.

Adding debt increases the risk of bankruptcy due to an unforeseen event, which is more common than management usually thinks. In his experience as a banker and consultant, management is not good at predicting future cash flow. Unforeseen events occur more often than management predicts, making debt more expensive than equity.[10]

On paper, debt is cheaper than equity because it is cheaper to pay interest, usually around 3% to 5%, than to give up more profits from the company, usually at a cost of around 8% to 10%. However, as the investment risk increases, the cost of both debt and equity increases, sometimes dramatically.

In exchange for a higher expected return, equity holders are subordinate to debt holders, and debt holders have a contractual obligation to receive interest payments and a return of principal when the bond matures. Where debt gets more expense is if the company is unable to make an interest payment, meet debt covenants, refinance, or repay the principal. The bondholders can then obtain ownership of the company and wipe out the equity holders.

Therefore, one of the best methods to reduce default risk is to reduce costs and maximize FCF. The higher the free cash flow of the company, the lower the risk that the company will not be able to service its debt.

# Oil and Gas Company Capital Structure

| | Liabilities & Equity | Key Points |
|---|---|---|
| **Oil and Gas Assets** | **Short-term Borrowings** | • Debt due within one year<br>• Current liabilities |
| | **Reserve-Based Lending (RBL) Revolver** | • Revolving credit facility or term load, redetermined semi-annually<br>• RBL is a source of capital for drilling and completion operations<br>• Cost is a key factor in establishing the RBL borrowing base |
| | **Senior Secured Debt** | • Prioritized for repayment<br>• Secured by company assets<br>• Not making interest payments on debt usually triggers a default<br>• Maximizing cash flow with cost reduction will help ensure debt can be continuously serviced |
| | **Senior Unsecured Debt** | • No specific collateral<br>• Higher risk and interest rates |
| | **Subordinated (Junior) Debt** | • Unsecured debt below senior debt<br>• High risk, high yield, mezzanine |
| | **Mezzanine Debt** | • Highest risk form of debt often with highest returns<br>• Frequently include Warrants, Payment-in-Kind (PIK) interest |
| | **Hybrids** | • Securities with debt and equity characteristics: convertible bonds, preferred, in-kind toggle notes |
| | **Preferred Equity** | • Fixed dividend, redemption value<br>• Priority over common<br>• Limited rights |
| | **Common Equity** | • Ownership, unlimited upside<br>• Voting rights<br>• Variable or no dividend |

*This table was constructed referencing multiple sources. See Sources, Chapter 2, note 11, infra*

## 8. RBL Borrowing Base Redeterminations

Shale companies depend on reserve-based lending (RBL) as a source of working capital.[12] As a result of the number of oil and gas bankruptcies, banks are more conservative with the size, withdrawal amounts, usage of funds, and other covenants for RBLs. Due to risk, banks will reduce the borrowing base.

The borrowing base is the loan amount based on the value placed on the collateral. This is the value of proved reserves.

Part of the RBL underwriting process is an engineering reserve report, prepared twice a year. According to the Office of the Comptroller of the Currency (OCC), which regulates all national banks, the engineering report should address four critical concerns: Pricing, Costs, Discount Rate, and Timing. [13]

Pricing is based on a price deck discounted to NYMEX futures, industry forecasts, and pricing the operator receives. Discount rate is based on the required rate of return based on risk. Timing refers to a requirement that the reports should be based on data that is not older than six months. Cost is the one factor that the operator has the most control over and can be critical during an RBL borrowing base analysis.

According to the OCC:

> *"The bank's engineer reviews the reserve report for estimates of OPEX and expected ultimate recovery of reserves, production rates, and CAPEX needed to convert reserves into the PDP category. The bank's engineer makes technical adjustments based on his or her professional judgment. Examiners should assess whether adjustments are well supported and documented."* [14]

Additionally, the OCC states:

> *"Cost assumptions ... should be realistic and fully supported. Costs affect economic life of reserves primarily in two ways: development costs and production costs.*
>
> *Production costs are a key focus in underwriting because the borrowing base is based primarily on PDP reserves. Production costs include lifting costs or lease operating expenses, which include operating and maintenance expenditures for materials, supplies, fuel, insurance, maintenance, and repairs. Additional production costs include property and severance taxes.*
>
> *If there are plans for further development, engineering reports may include development costs, or CAPEX, for PDNP and PUD properties as well.*
>
> *CAPEX may include roads, utilities, drilling pads, site facilities, development wells, wellheads, well casing, and pipe and well equipment. To a lesser extent, CAPEX may include workover costs for PDP wells."* [15]

As we can see, even the banks want us to reduce costs. They want to lend money. That is their business. They do not want to reduce the borrowing base. However, sometimes they must, and the high-risk, high-cost operator is not typically viewed as a prudent steward of capital.

Banks lend money to all types of oil and gas companies. They see everyone's costs. If someone can significantly reduce their CAPEX and OPEX, that looks very favorable in the engineer reports which help determine the loan size, allowed withdrawal amounts, usage of funds, and other covenants.

# 9. Debt Covenants

Covenants are requirements lenders place in agreements with borrowers to align interests, address concerns, and protect the lender. They are financial and non-financial, structured to reward lenders and penalize operators. Turmoil in the industry has resulted in an increase in new covenants and a tightening of existing ones. Maximizing cash flow with cost reduction addresses the most common covenants.

## Shale Debt Covenants

➤ Operator must provide projected cash flows, CAPEX, OPEX and production forecasts, LOE reports, budgets, reserve reports, development plans, and financial statements
➤ Debt to EBITDAX < 4.0x (leverage ratio)
➤ EBITDA to Interest Expense > 2.5x (coverage ratio)
➤ Current Assets to Current Liabilities > 1.0x (current ratio)
➤ Total Debt to Cap Restrictions, Subordinated Debt Limits
➤ Minimum Liquidity Obligations, Anti-cash Hoarding Provisions
➤ Hedging Requirements and Restrictions
➤ Lien restrictions for Taxes, Vendors, and Non-operators
➤ Limits on Dividends, Sales, Acquisitions, Stock Purchases, Investments, Personnel Expenses, and Staffing
➤ Restrictions on Transactions with Affiliates
➤ Maintenance of Insurance Requirements
➤ Report SEC Inquires, Payment of Taxes Requirement
➤ Compliance with Laws, Rules, Regulations and Orders
➤ Governmental Notices, Citations, Summons Notification
➤ Report Damage to Oil and Gas Properties > $100,000
➤ Environmental Matters, Notices, Violations, Allegations
➤ Take-or-Pay or Other Prepayment Restrictions
➤ Redemption Notice Deadlines, Redemption Provisions

*This list was constructed referencing multiple sources. See Sources, Chapter 2, note 16, infra*

As the energy business environment evolves, specific lenders may continue to toughen terms. Oppressive covenants make certain shale business strategies unviable. This dynamic negatively impacts prospects by hampering growth and opportunities to create long-term value.

Understanding how debt covenants impact one's value creation plans is essential. Regardless of the situation, maximizing cash flow with cost reduction addresses most covenant concerns, as increasing free cash flow increases options for all scenarios. Reducing costs is welcome music to the ears of the CFO and other executives responsible for managing the capital structure, financial risks, and obligations in an ever-changing economic environment.

## 10. Debt Maturities

As debt matures, the principal must be repaid to investors. Usually, oil and gas companies issue new bonds with a longer time to maturity and better terms, using the capital to pay down current outstanding debts. Favorable debt markets or improved credit ratings can incentivize companies to refinance or restructure their debt.

However, many shale companies are forced to refinance current debt at higher interest rates because they cannot meet their current debt obligations or do not have the capital to repay the principal. In these situations, there are usually fewer interested investors, and the investors that are interested consider the investment higher risk. Maximizing cash flow with cost reduction helps make an investment more attractive to debt investors because it lowers the risk the borrower will be unable to meet debt obligations. This in turn can reduce the cost of capital, providing more opportunities for a company to finance its operations.

## 11. Access to Capital

It is becoming more difficult for oil and gas companies to raise capital primarily due to shale wells producing at lower rates than initial forecasts. Banks are seeing actual production come in under type curve. Lower-than-expected production performance combined with higher-than-expected costs does not give investors confidence.

Additionally, environmental activists are putting pressure on financial institutions to limit or even eliminate investments in carbon-based natural resources. Therefore, companies must depend on their own internal ability to generate cash for growth. Reducing costs is a great way to create value and free up capital in an environment in which access to capital markets is constrained.

## 12. Bankruptcy

The last thing an owner, investor, employee, vendor, or any stakeholder wants is the operating company to go through bankruptcy. As stressful as events leading up to a potential bankruptcy are, the process is full of significant amounts of work and legal fees, generating hundreds of painful actions.

If nothing else motivates the team to be a low-cost operator, continuously looking for ways to drive costs out of the operation, I hope avoiding the pain and suffering of bankruptcy is that factor.

## 13. Acquisitions

Asset deals and corporate acquisitions are part of our value-creation strategy. It is common for a small oil and gas company to create value by acquiring assets in an unproven or marginal area, showcasing that they are economical, and then selling to a larger entity for full-scale development. Depending on the purchase price, this is usually a low-risk strategy for the buyer; however, the larger the acquisition, the more risk exposure there is.

A sizable corporate acquisition in the energy industry can quickly get into the billions of dollars. Valuations are based on hundreds if not thousands of assumptions, as the final purchase price is based on current and future projections.

All valuations are incorrect to a certain degree. The question is, by how much? If a critical assumption is incorrect, it is easy to overpay. When this happens, we must take immediate action to create additional value to neutralize the valuation mistake or unexpected circumstances.

Extreme cost reduction is a proven strategy to liberate cash flow to fix an acquisition. For example, if our acquisition was underwritten based on a development plan with $10 million well costs and we can get costs down to $5.0 million, the value created could offset undesirable issues that were only discovered after the deal closed.

It is common to find financial modeling mistakes, incorrect assumptions, and costly operational problems with an acquisition after several months of operating the asset. Acquisition surprises are almost always negative.

# 14. Land Costs

During the shale land rush, acreage costs were bid up to extreme levels. These expenses are often not accounted for when running economics. However, they were paid for and continue to be paid for when leasing new areas. When bidding against a competitor, the price goes up. If the acreage is necessary, we have no choice but to pay more than what we wanted to pay and more than what was planned. These costs need to be offset for the economics to work. That's where cost reduction comes to the rescue.

# 15. Asset Quality and Areal Extent

As shale oil and gas production matures, high-quality areas become fully developed. Therefore, lower reservoir quality areas must be drilled to maintain or grow production. Economic extents are determined during this process as technological limits are pushed. Larger, more expensive fracture treatments are often utilized to maintain production performance. However, they do not always live up to expectations.

Lower rock quality and higher costs are not an attractive value creation proposition. Therefore, to counteract the financial impact of degrading asset quality, cost reduction initiatives should be implemented. By significantly reducing well costs, the negative economic impact of non-core acreage can potentially be alleviated. Financial success depends mainly on the degree of cost reduction that can be consistently achieved.

For example, reducing a $10MM well cost to a $3.0MM well costs could, depending on the play, transform non-core acreage into the most economically attractive acreage for development. Of course, there are many variables to consider, but a 70% cost reduction achievement would be a game changer. Although this is not easy to accomplish, it is possible with a focused, dedicated team of professionals with the time and determination to realize a specific cost reduction objective.

## 16. Changing Reservoir Fluid Composition

The decision to drill is driven by economics, which is heavily based on product composition. Key aspects include oil volumes, oil gravity, water volumes, water salinity, natural gas volumes, BTU content, natural gas liquid (NGL) volumes, NGL composition (ethane, propane, butanes, pentanes), and non-hydrocarbon gases including carbon dioxide, hydrogen sulfide, nitrogen, and helium. Once production starts, static data in the laboratory becomes dynamic data in the real world. If elements change unexpectedly, it will impact economics.

Shale oil wells exhibit increasing gas-oil ratios (GOR) over time. As a well is produced, reservoir pressures decline. Nearly all oil reservoirs hold natural gas in the oil at reservoir pressure. Bubble point pressure is the pressure at which gas first comes out of the oil solution and forms bubbles. Once reservoir pressure drops below bubble point pressure, gas will separate from the oil. This usually impacts economics negatively and can be unexpected.

Shale condensate wells can exhibit condensate banking. Condensation banking occurs when reservoir pressure drops below dew point pressure, and condensate starts accumulating around the wellbore. As reservoir pressure declines due to production, a liquid wall "bank" forms in the near-wellbore region. This reduces hydrocarbon production or the loss of heavy components in the production stream. It is hard to combat rising GORs or condensate banking's impact on economics. However, reducing costs can offset the impact and give us another motivating reason to protect our economics.

## 17. Midstream Agreements

The oil industry is divided into three sectors: upstream, midstream, and downstream. The segment most often misunderstood is the midstream sector, which is focused on the transportation of produced fluids.

Transportation by pipelines, trucks, marine vessels, and rail cars are most common. Each method has an associated cost addressed within long-term contracts. Due to the uncertainty inherent in long-term forecasting, it is common for these contracts to become highly unfavorable to the producer.

Undesirable midstream agreements can make an otherwise profitable asset uneconomic due to the associated midstream burdens. This is especially true with natural gas. Many long-term gas agreements were put in place under assumptions that did not pan out. These agreements contain a plethora of fees frequently referred to as "contracts that kill" through "death by a thousand cuts," implying many small burdens bleed an operator to death. Shrewd midstream contract architects and negotiators are known to take advantage of eager producers looking for a midstream partner to develop an asset.

# Natural Gas Midstream Burdens

| Burden | Description |
|---|---|
| Gathering Fees | Fee to collect gas from the pad using small diameter pipelines which move gas to the next destination, typically a larger diameter mainline pipeline. |
| Compression Fees | Fee to compress gas to transmission pressures to enter a long-distance interstate transmission pipeline system. |
| Processing Fees | Fees to prepare gas for sale including separating natural gas into its components. |
| Metering and Measurement | Fees to install, maintain, operate, and test gas measurement devices. |
| Gathering System Fuel | Quantity of gas deducted to power and/or operate processes and equipment across the gathering system. |
| Compression Fuel | Gas utilized as fuel for compression. |
| Plant Fuel | Gas utilized to power / operate equipment and processes in the processing plant. |
| Transportation Fuel | Gas utilized to power and/or operate equipment and processes involved in the transportation of natural gas. |
| Fee Escalations | Fee increases based on Consumer Price Index (CPI) and other statistical estimates. |
| Midstream Shut-ins | Loss of cash flow due to midstream shut-ins or downtime for various reasons including maintenance, upgrades, unplanned repairs, problems, weather, offtake capacity, storage capacity, market conditions, and force majeure. |

# Natural Gas Midstream Burdens continued...

| Burden | Description |
|---|---|
| High Line Pressure | Subdued production due to high line pressure impacting well performance. |
| Ethane Handling | Cost or lost revenue due to ethane handling (rejection or recovery). |
| Bypass Gas | Lost revenue due to gas that is not processed (bypassed around the plant). |
| Lost & Unaccounted For Gas | Gas lost after custody transfer due to leakage, venting, flaring, mishaps, theft, line pack, measurement inaccuracies, timing, midstream use, non-metered consumption, composition corrections, accounting errors. |
| Conditioning | Fees for nonconforming gas according to gas specifications required by contract. |
| Environmental | Fees to address future environmental laws concerning midstream operations (modify facilities, emissions, regulatory fees). |
| Minimum Volume Commitments (MVCs) | Commitments to deliver a minimum volume during a specific period. If the producer does not meet the MVC, a fee must be paid. MVCs can influence a producer to act irrationally to avoid penalties. |
| Percentage-of-Proceeds (POP) Provisions | A percentage of proceeds from selling the residue gas and/or NGLs goes to the midstream company. |
| Unfavorable Royalty Terms or Gross Proceeds Rulings | Gross royalty clauses or court rulings may not allow operators to deduct post-production costs. Therefore, operators may have to carry 100% of the midstream burden including all midstream costs to process mineral owner gas and NGLs. |

## 18. G&A

I once worked with an executive who converted all dollar amounts into units of General and Administrative (G&A) expenses. His mental model of the world was calibrated based on G&A expenses. He thought of everything in terms of G&A.

This man despised G&A, so to constantly remind everyone of the expense, he would convert all dollar amounts into units of G&A, often months and days of G&A. For example, if a well produced 1500 bopd or something cost several million, he would convert it into days of G&A and then tell you how many days of G&A that was.

Let's say annual G&A is $365 million, and we are running 20 rigs drilling two wells per month per rig. If we were to find a way to save $25,000/well, he would convert that into days of G&A. You would say, we can save $25,000/well, and he would say, that's about 12 days of G&A. He became so obsessed with reducing G&A that he started converting everything into minutes of G&A.

General and Administrative expenses can consume a significant amount of cash flow directly and indirectly. When entities reorganize, restructure, downsize, merge, or engage in actions to optimize business expenditures, G&A is usually addressed. High-cost G&A items are focus points during cost reduction efforts. High-cost, low-valve personnel are a burden on cash flow that can easily be lowered, most commonly by reducing headcount.

Although it is easy to cut expenses by leaning out the workforce, personnel reductions are never a pleasant or fun process. Nobody likes to do it, but when done correctly, it is an effective tool to optimize an organization.

---

**Maximizing cash flow with cost reduction strengthens job security for everyone.**

---

High operating costs put pressure on management to reduce the workforce. Low-valve individuals not interested in making changes to reduce business costs are an easy target during overhead reduction efforts.

This fact should motivate the workforce to maximize cash flow with CAPEX and OPEX reduction efforts. A low-cost operator generating positive cash flow can avoid painful downsizing and restructuring events, limiting layoffs and other undesirable occurrences.

## 19. Legal Situations

Legal situations are a significant cash flow and intellectual drain. Disputes involving land, operatorship, permitting, liens, invoicing, and operations consume significant amounts of resources. Undesirable legal situations command workforce attention and mindshare, which is a major distraction to effective operations. Therefore, legal issues damage cash flow directly and indirectly. The associated costs are usually unexpected and drag on for years.

Large amounts of capital flowing across an industry can attract lawsuits for a variety of reasons. The perception is that oil and natural gas companies are wealthy entities swimming in surplus cash. Those of us who understand the financial aspect of the industry know that this is not true. Reducing operational risk and costs help prevent and counteract these unexpected expenses impacting free cash flow.

## 20. Production in Paying Quantities

A critical but often overlooked reason to reduce lease operating expenses (LOE) is to maintain production in paying quantities (PPQ). When an oil and gas lease enters the secondary term, production must continue in paying quantities to keep the lease valid. Defining PPQ status is based on revenue minus expenses. If PPQ is challenged by mineral owners or competitors, the matter often ends up in a court of law, with a focus on the details of our lease operating expenses.

To show the lowest possible LOE, many operators exclude certain expenses from the calculation, a strategy that has worked in the past. However, the items required to be included in LOE vary on a state-by-state basis. For instance, in some areas, insurance, if directly attributable to the well, must be included when determining PPQ.[17]

These developments are a strong incentive to significantly reduce OPEX to maintain valid leases, especially on marginal wells. A prudent operator must demonstrate a good faith effort to reduce expenses to maintain production in paying quantities for the benefit of all interested stakeholders.

## 21. Asset Retirement Obligations

Asset retirement obligations (AROs), more commonly referred to as plugging liabilities, encompass the costs to plug wells and reclaim the surface location to its original condition. Once a well is drilled or acquired, a liability is established in the form of a future plugging obligation. ARO liabilities are estimates based on the present value of the projected plugging expense.

On a low-rate or marginal producer, if a problem occurs during routine daily operations, a producing asset can quickly and unexpectedly become a liability. Many low-rate producers are one problem away from becoming a plugging expense. During plugging or reclamation, operational costs can quickly increase beyond expectations. These are real risks and expenses that need to be accounted for.

There are state and federal guidelines on plugging and abandonment (P&A) requirements with differences between regions. In general, inactive wells are required to be plugged within a year after operations cease. Defining a well as inactive can be subjective. However, during extended periods of challenging economics, more wells become neglected or shut-in and the number of wells classified as inactive increases.

Additionally, it is common for potential acquisitions to be heavily burdened or fail due to ARO liabilities from inactive wells attached to producing asset packages, as current and future plugging expenses damage cash flows and valuations.

This dynamic further expands the P&A problem as assets fail to change hands to new owners that may revitalize the inactive wells or address the P&A obligation. As a result, ARO liabilities should be a motivator to reduce costs in order to return inactive wells to production (RTP) or cover plugging expenses.

## 22. Non-Operating Working Interest Owners

An operator is responsible for the prudent use of capital on behalf of its non-operating partners. As a result, it is not uncommon for an operator to be questioned about CAPEX and OPEX by non-op working interest owners. If we are not controlling our costs, we will hear about it from the non-op partners. If issues are not addressed, depending on contractual obligations in certain situations, an operator who is not reducing costs could lose operatorship and be replaced.

## 23. Competing Energy Sources

Coal, nuclear, geothermal, wind, solar, biomass, and hydroelectric compete with oil and natural gas for market share. Significant capital and human resources are currently focused on alternative energy cost reduction, making them more attractive to consumers. With all that is happening within energy markets, oil and natural gas producers cannot sit still when it comes to cost reduction. There is too much competition from all forms of energy.

Hydrocarbons have enjoyed a favorable legacy position when it comes to energy economics compared to other sources. However, advancements within the alternative energy sector are making them more attractive relative to fossil fuels. The smartest people in the world are working hard at reducing the cost of competing energy sources. This fierce competition is an excellent motivator for the oil and natural gas industry when it comes to shale cost reduction. Make no mistake: the battle to be the preferred source of global energy is currently underway and cost is at its core.

## 24. Environmental, Social, Corporate Governance

Stakeholders and potential stakeholders are looking at a lot more than traditional accounting metrics to determine future value. Businesses and industries are being analyzed using Environmental, Social, and Corporate Governance (ESG) criteria to determine value and future viability.

Environmental topics receive a lot of attention from the media and politicians, impacting the public's perception of natural resources industries, especially the energy industry. The spin against oil and natural gas is negative, and we, as an industry, have not done a good job showcasing the value we deliver to society. Additionally, there are social concerns regarding climate change and sustainability with industries that emit carbon dioxide and produce nonrenewable resources.

Our industry can address ESG concerns; however, it will cost money. There are many actions we can take to be carbon neutral or even carbon-negative, meaning we, as an industry, remove more $CO_2$ from the atmosphere than we emit.

I am 100% convinced the oil and gas industry can be carbon-negative, but it will not be free. It will cost money, which is much easier to absorb if our base operating costs are extremely low and we are generating massive free cash flow.

A robust cost reduction strategy is an excellent plan to address ESG concerns because lower costs provide the resources to address the ESG factors stakeholders value.

## 25. Government Regulations

History suggests that governmental regulations expand in almost all sectors over time. Energy, particularly the oil and gas industry, is a highly politically charged subject, resulting in additional attention and increased regulation with cost implications. In general, increased regulations result in increased costs. To neutralize increased regulation, operational cost reduction must be employed.

It is not unheard of for regulations to be used as a political weapon to achieve an objective, often resulting in economic damage to an industry. For example, although it is unlikely a specific technology, like fracturing, will be outright banned across the United States, it could be regulated to the point that it is uneconomic, effectively using cost as a weapon to destroy the American oil and gas industry. At that point, our only choice would be to deploy massive cost reduction efforts.

Furthermore, it is easy for a government, group of politicians, or radicalized activists to cancel an industry that is financially weak. If an industry has cash flow, it has the resources to fight back.

**Cash flow is power.**

Cost reduction provides our industry with the cash and the credit to fight for survival in a world trying to eliminate American energy. Politics is highly emotional and a great motivator to get the team behind a technically challenging cost reduction effort. Consider utilizing a political angle to motivate the team to put in extra effort to fight back and maximize free cash flow with cost reduction.

## 26. Black Swans and Gray Rhinos

Oil and gas markets are influenced by almost every notable occurrence across the globe. The most thought-out business strategy or development plan could be thrown off course, rendered useless, or effectively dismantled due to an unexpected event halfway across the world.

Rare, unexpected events with significant negative consequences are often referred to as black swans. The term was popularized by Nassim Taleb in his book "The Black Swan: The Impact of the Highly Improbable."[18]

Although black swan events are considered rare and unpredictable (e.g., World Wars, Boxing Day Tsunami, Global Pandemic), due to the interconnected nature of energy markets, primarily crude oil markets, any significant unexpected event impacts the oil and gas industry. In recent years, there seems to be at least one significant black swan event per year that impacts the energy industry.

While black swans are considered unpredictable rare events, "gray rhinos" are considered "high probability, high impact"[19] hazards that are ignored. The term was coined by Michele Wucker in her book "The Gray Rhino: How to Recognize and Act on the Obvious Dangers We Ignore."[20]

A rhino in the wild is an obvious threat and considered one of the deadliest animals. As obvious a threat a huge rhino is, some people ignore the rhino and hope it walks away. As convenient as it is to ignore a massive beast, we can't. It might trample us to death, which is no fun.

# Black Swans and Gray Rhinos in the Oil Industry

| Event | Oil & Gas Industry Impact |
|---|---|
| **Economic Crisis** | Demand destruction, commodity price collapse |
| **New Regulations** | Cut locations / development, increased costs |
| **Global Pandemic** | Loss of life, demand destruction, price collapse |
| **Weather Phenomena** | Down production, asset damage, increased costs |
| **Natural Disaster** | Loss of life, assets, production, increased costs |
| **Tech Innovation** | Loss of market share to lower cost and/or new energy sources |
| **Market Failures** | Detached commodity prices, liquidity problems |
| **Industrial Catastrophe** | Cost, negative PR, increased regulation |
| **Loss of Key Personnel** | Operational disruption, knowledge loss |
| **New Taxes** | Financial impact, investment deterrent |
| **Supply Chain Disruption** | Increased costs, delays, NPT, operations impact |
| **Unexpected Competitor** | Loss of growth opportunities, development plan issues |
| **Lawsuit / Legal Event** | Financial impact, distraction, leadership focus |
| **Civil Unrest** | Property damage, production disruption |

# Black Swans and Gray Rhinos continued...

| Event | Oil & Gas Industry Impact |
|---|---|
| Geologic Bust | Uneconomic asset, stressed team, investment loss |
| Capital Constraints | Difficulty raising capital, refinancing, funding growth |
| Workplace Accident | Loss of life, associated costs, fines, lawsuits |
| Operational Problems and Incidents | Increased costs, equipment loss, asset loss, lawsuits |
| Cyberattack | Business damage, cost, operations disruptions |
| War or Terrorist Attack | Human/financial/oil and gas price impact, damage |
| Public Perception | Valuation, access to capital, industry survival |
| Solar Storm | Equipment damage, NPT, production impact |
| Political Power Grab | Technology ban, indefinite moratorium |
| Activist Shareholder | Distraction, forced actions, short-term agenda |
| Subsurface Surprise | Uneconomic asset, out of zone, water wells, financial impact |
| Financial Error | Unexpected fees, fines, make-whole penalties |
| Development Mistake | Asset underperformance, value destruction |
| Reservoir Misanalysis | Asset value loss, development plan disruption |

## Black Swans and Gray Rhinos continued...

| Event | Oil & Gas Industry Impact |
|---|---|
| **Manufacturing Defects** | Multiple well failures, accidents, financial impact |
| **Oil Storage Constraints** | Negative prices, environmental risks, production shut-ins |
| **Takeaway Bottleneck** | Widening spreads, flaring, production shutdowns |
| **Population Decline** | Energy demand decline, price collapse |
| **Raw Material Shortage** | Increased costs, production growth impact |
| **Trade War** | Materials costs, destabilized supply chains |
| **Power Grid Failure** | Human life impact, loss of public trust |
| **Nuclear War** | Major global impact |
| **Electromagnetic Pulse** (natural or artificial) | Damaged equipment. Loss of production |
| **Water Shortage** | Human impact, operational impact, increased cost |
| **Draconian Policies** | Unviable assets, bans |
| **Stagnation** | Sluggish demand, price decline |
| **Climate Change Events** | Human impact, infrastructure damage, regulation |
| **Magnetic Field Reversal** | Directional drilling challenges, mass extinctions |
| **Unknown Unknowns** | Assume it's not good |

One of the best ways to prepare for any of these events is to have a risk management plan in place and be proactive. Reduce costs, build capital reserves, and hedge. Our plan will manage black swan and gray rhino risk by maximizing FCF. Lean operations that deliver strong cash flows are a powerful combination that can withstand hard-to-predict events.

## 27. Inflation

Costs creep up for many reasons beyond the standard measures of inflation. Oil and gas prices may or may not reflect the same level of inflation as services and materials. Therefore, to offset it, we must have cost reduction efforts in place.

In addition to inflation, vendors are motivated to increase their revenue, and have many tools to skillfully do this without attracting attention. Just as we work to maximize FCF, so do service providers, which are motivated to extract as much capital from the operator as possible without jeopardizing the relationship, in most cases, but not all.

> **If we are not aggressively pushing costs down,**
> **the shale machine will push them up,**
> **eventually consuming all our capital.**

Some service providers will sacrifice the client to maximize their cash flow. These service providers are notorious for slowly bleeding operators to financial death, then kicking them when they are down, extracting every last cent before moving on to their next victim.

Inflation and aggressive service providers, in all forms, must be incorporated into financial projections. When performing economic modeling, most financial professionals model healthy commodity price appreciation if warranted by the forward curve and the company's commodity price forecast. A lot of resources and smart people are dedicated to making predictions about future commodity prices. Many strategic, non-reversible decisions are made based on these commodity forecasts. However, what is rarely included in these forecasts is what service costs are going to do as a result.

For example, if our company is projecting crude oil to increase from $50/bbl to $100/bbl over a 3-year period, this projection is entered into the financial models without addressing what it will do to CAPEX and OPEX.

It is common to put very little thought or work into what service costs and material prices are going to do as a result of rising commodity prices. The default assumption is to model in a minor 3% to 4% increase due to inflation. It is almost always too conservative.

If commodity prices increase 50%, costs will go up at least 10% to 20% across the board. Of course, nobody wants to model that. Therefore, assumptions often include projecting operational efficiency and technical breakthroughs that have not yet occurred to offset the cost appreciation.

This is rarely modeled and often just implied by holding costs constant or reducing costs over time to make the economics work. This dynamic is another reason cost reduction must occur within every company involved in shale development.

# Shale Operations Demand Cost Reduction

Shale economics and the "27 Reasons for Cost Reduction" provide the foundation for why we must aggressively reduce costs. Without a robust ongoing cost reduction strategy, generating positive free cash flow from shale operations over extended periods is difficult, if not impossible. There are too many things pushing up costs.

Due to the nature of the business, maximizing cash flow from shale operations requires ongoing cost reduction, or we will get overwhelmed by problems from the shale machine, which can consume all our capital.

**Shale Economics do not work without Cost Reduction. Cost Reduction does not work without Risk Management. Risk Management does not work without Good People.**

Due to the operational intensity of shale development, our economic success is closely tied to the quality of our people and our ability to manage risk. Encourage a corporate culture that looks for problems because that is how we manage risk and where we find hidden cost reduction gold.

Additionally, significant FCF provides the ability to take advantage of growth opportunities. If we are the lowest-cost operator, our competitor's assets are more valuable under our control. Free cash flow enables acquisitions, which can provide an opportunity to reach our 100X value creation objectives.

# 3

## SYSTEMIZE & AUTOMATE

—

Preventing costly mistakes is one of the most important actions we can deliver to shareholders—whether that's mitigating inaccurate subsurface interpretations, eliminating operational errors, stopping undesirable situations, or averting bad executive-level strategic decisions. Developing systems to avoid mistakes helps us establish a stable foundation from which to maximize free cash flow.

Having the lowest costs per well or per barrel means little if we have trainwrecks every other week. To reduce operational risk and prevent mistakes, our plan is to systemize and automate everything. Then, we will integrate artificial intelligence to the limit of our ability.

Systemizing our business involves structuring procedures and step-by-step processes to reduce risk, improve efficiencies, and deliver consistent results. Automation refers to converting aspects of our business to achieve outcomes with reduced human interaction and intervention. Automation can be implemented with or without computers.

In conjunction with systemizing and automating our operation, we will incorporate artificial intelligence (AI) to help eliminate mistakes. AI is any system that enables a computer to replicate a human action.

There are different levels of AI. Initially, we will focus on classical AI systems. These systems use simple algorithms to compile information and perform analysis. Algorithms are a set of rules or instructions to achieve a goal. I prefer not to utilize an AI black box system, in which we do not know how the AI makes its decisions and recommendations.

Although AI will not make mistakes that humans would make, it will make mistakes that humans would never make. When expanding this technology, including automated systems and artificial intelligence, we cannot have the mental framework that we are replacing good human judgment and diligent human intervention with a computer. AI in the oil and gas industry is not designed to replace humans; it can't. AI is an advisor, working with humans to help eliminate mistakes.

We have all seen automatic driller systems spiral out of control, frac electronic kick-out systems fail to shut down horsepower, and production SCADA systems send erroneous alarms or stop working without notification. This does not mean they should not be utilized. We need to know what it looks like when the tech fails so we can recognize it and take action. Studying digital technology disasters from other industries will help us avoid a similar fate.

# Digital Disasters Applicable to Shale Operations

| Event | Software Failure Details |
|---|---|
| **Zillow AI Disaster** | Predictive analytics AI algorithm caused Zillow to buy +18,000 homes at inflated prices, +$500MM losses, 25% laid off. |
| **Boeing 737 MAX Crashes** | Flight control software (MCAS) put planes into a nosedive due to a single faulty sensor that sent bad data. 346 people died. |
| **Knight Capital Automated Trading Algorithm Fiasco** | Defective trading algorithm lost $440MM in 45 minutes. Caused by coding errors, poor version control, no written procedures, missed alerts, no circuit breakers, and no emergency kill switch. It took 17-years to create Knight Capital, and 45-minutes to destroy it. |
| **2003 Blackout (Worst in U.S. history)** | On a hot summer day, power lines sagged onto trees, starting a domino effect impacting 508 generating stations and 256 power plants including nine nuclear plants. SCADA alarms failed on the systems; operators did not know as SCADA continued working but would not send out the alarms, so operators did not think there was a problem. Therefore, operators did not act fast enough to prevent the blackout from expanding. Cost $7.0 to $10.0 billion, and 100 deaths. |
| **Nike Software Disaster** | Nike spent $400MM implementing a new ERP system to enhance its supply chain. The software had bugs, difficult to implement, made poor forecasts causing the wrong sneakers to be produced. This caused a 20% stock drop, and at least a $100MM lost. |

# Digital Disasters Applicable to Shale continued...

| Event | Software Failure Details |
|---|---|
| **NASA Mars Climate Orbiter Failure** | After traveling to Mars, the Orbiter crashed because ground software used English units, but onboard software used Metric units. Engineers did not catch it. |
| **FoxMeyer Software Caused Bankruptcy** | 4th largest drug wholesaler implemented a popular ERP software system to increase efficiency and cut costs, but it could not handle transaction volume, forcing FoxMeyer into bankruptcy. |
| **Soviet Early Warning System Malfunction Indicating U.S. Nuclear Attack** | In 1983, Soviet systems signaled a U.S. nuclear strike. Protocol was to retaliate but officer Stanislav Petrov did not. He knew the system in detail and thought it was a computer error due to tech newness, unrealistic AI confidence level, low number of missiles, and ground radar that did not confirm the attack. His actions prevented WWIII. |
| **U.S. $0.46-cent Computer Chip Indicates Soviet Nuclear Attack** | In 1980, U.S. systems indicated the Soviets launched a massive nuclear strike. U.S. computer systems showed a random number of attacking missiles. Comparing signals from several sources, it was determined to most likely be a false alarm. Later it was found a $0.46-cent computer chip "simply wore out,"[2] causing the incident. |
| **NASA Mariner Software Typo** | A typo (of just one symbol) in the launch vehicle's guidance software caused it to veer off course during launch. |

*This table was constructed referencing multiple sources. See Sources, Chapter 3, note 1, infra*

# Digital Disaster Lessons for Shale Operations

- People working in AI have knowledge depth on specific aspects of it. However, they often do not understand or appreciate interactions and impacts on the entire operation.
- AI can elevate risk by increasing human complacency.
  - e.g., people asleep at the wheel in cars on autopilot.[3]
- Establish written checklists, processes, and procedures addressing digital tech and artificial intelligence systems.
- Understand weather's impact on digital technology. Weather often contributes to digital failures, including the worst blackout in U.S. history and the Soviet nuclear missile attack. The satellite software system interpreted sunlight reflecting on high-altitude clouds above North Dakota as a U.S. missile launch.[4]
- Routinely confirm digital sensors match analog gauges, straps, and reality. Establish a data quality checklist.
- Have a digital system health check process.
- Require redundancy on critical sensors and SCADA equipment. Lack of redundancy has contributed to many digital disasters, including the 737 Max crashes.[5]
  - Redundancy prevented WWIII during multiple false alarms on both the U.S. and Soviet side when digital tech indicated attacks were determined to be incorrect.[6]
- Test software compatibility before implementation.
- Incorporate independent software reviews.
- Familiarize the team with the control systems.
- Ensure the team understands AI bias in all its forms.

- Many digital systems are based on the future resembling the past, which is often not the case. This played a role in Zillow's home-buying fiasco, as the machine learning systems couldn't predict unique housing market changes.[7]
- Confirm the units are correct on all systems and coding.
  - o Oil and gas data is often expressed in different units.
- Identify digital drift introduced into the sensors.
- Strengthen operator and tech vendor personnel interaction.
- Minor digital anomalies can reveal big software problems.
- Have a plan if someone pushes the wrong button.
- If any aspect of the system goes down or offline, confirm the system will send an alert that the system is down.
  - o Have a process to alert operators of a system failure independent from the system.
- All systems have bugs. Some are known, but many are not.
- Don't blindly trust digital systems. During the 2003 Blackout, operators in the command center dismissed calls from the front lines warning them about conditions worsening in the field. They were ignored because the SCADA system did not indicate anything was wrong.[8]
- Be on the lookout for individuals who do not embrace technology because they are afraid of losing their job. In the FoxMeyer ERP implementation bankruptcy, angry warehouse workers who thought they would be replaced by automation performed massive sabotage.[9]
- Do not allow digital implementation to be handled by trainees or to be used as a training ground.
- Minor errors in digital tech can have a cascading effect.

There is a significant opportunity for the shale industry to leverage AI to create value. In general, oil and gas companies trade at lower multiples compared to tech and SaaS companies for several reasons, including higher CAPEX and higher risks. Digital tech, on the other hand, is easily scalable at zero marginal cost, has lower risk, can be quickly updated, and is more appealing to investors.

If we can leverage AI to reduce cost and risk, our industry will become stronger, more financially successful, and more appealing to the investment community. This supports earnings growth, multiple expansion, and increased valuations, encouraging further investment.

Our cost reduction system addresses this opportunity, leveraging four essential components that work together to achieve results. These components include implementing systems and automating them, setting a specific cost reduction target, a time frame to accomplish the goal, and an honest dedication to achieving success. All future cost reduction actions are based on these four fundamental components.

## Essential Cost Reduction Components

1. Systemize and Automate to the Limit.
2. Establish a Numeric Cost Reduction Target.
3. Set a Time Frame to Accomplish the Goal.
4. Make a Selfless Commitment to Succeed.

These four components form the foundation of our cost reduction system and shape the detailed technical action plan to achieve the chosen cost goals within the selected time frame.

# ESSENTIAL COMPONENT #1

## Systemize and Automate to the Limit

Employing systems and subsystems across our operation will add value to our cost reduction efforts and our company by providing a structure from which to work. Systems provide an organized framework on how something should be done, which is particularly helpful in the oil and gas business due to the associated risk and uncertainty in dealing with the ever-changing surface and subsurface environment.

Our cost reduction program is structured as a system with four interconnected components. To enhance the operation and consistently reduce costs, we seek opportunities to put as many systems in place as possible, systemizing to the limit. This can be done with detailed written procedures, processes, checklists, hardware, software, artificial intelligence, risk management, and a variety of other tools we will incorporate into our operation through our action items.

Once systems are in place, look for opportunities to automate them. For example, cost reduction systems could be turned on during specific periods throughout the year in response to actual cost overruns, to address underperforming economics, or in response to market conditions.

One method to automate cost reduction would be to establish a clear IF-THEN RULE in which cost reduction systems automatically turn on for a specific asset if actual costs exceed AFE by a certain amount within a certain period, let's say a cost overrun of 5% or more on two or more wells within a 3-month period. This automated system could be implemented at the corporate level, requiring each asset division to take specific actions.

The goal is to put automated systems in place to force the shale machine to acknowledge and react to what is happening within the business before it becomes a serious problem. During large-scale horizontal operations, costs can very quickly get out of hand. There must be an automated system to stop this from happening. A system that forces the team to take action and be aggressive. In the shale game, anything that can go wrong will go wrong and happen across all the wells.

The best teams can navigate the big-picture strategy and day-to-day operations simultaneously without losing focus on either one. If we do not address day-to-day operations, we will have problems, but if we lose focus on the big-picture strategy, we can successfully run the company right off a cliff. Automated systems increase bandwidth and help the team navigate the shale machine through this dynamic.

Regarding cost reduction, we are going to be making changes. Change carries risk. Sometimes, making changes can introduce new risks; other times, it can uncover risks that have always existed. In many organizations, when an operation is working, the team hesitates to change anything for fear of inadvertently messing things up. This is a valid concern that will paralyze progress. However, it can be the right approach for a period of time, but not forever. To address this dynamic, we will systemize as much as possible. Employing systems throughout our operation can effectively address the risk inherent in change.

---
## ACTION
### Build Systems
---

Humans love risk. However, undesirable situations often occur when risk is not properly managed. There is a precedent for organizations to get blindsided by situations in which executive leadership was not fully aware or did not understand the potential consequences, resulting in unthinkable events and tragedies. Studying other companies where this occurred will help us avoid a similar situation.

## CEO Departures due to Risk Management Dynamics

| Company | CEO | Event | Result |
|---------|-----|-------|--------|
| BP | Tony Hayward | Macondo accident | CEO steps down |
| Boeing | Dennis Muilenburg | 737 MAX crashes | CEO fired |
| Massey Energy | Don Blankenship | WV mine explosion | CEO steps down, then federal prison |
| Suncor Energy | Mark Little | Multiple fatalities | CEO steps down |
| VW | Martin Winterkorn | Emissions scandal | CEO resigns, Fugitive in the U.S. |
| PG&E | Geisha Williams | Multiple wildfires | CEO steps down |
| Equifax | Richard Smith | Massive data breach | CEO sudden retirement |
| Philips | Frans van Houten | Product recall | CEO steps down |
| Silicon Valley | Gregory Becker | Bank collapse | CEO fired |
| Rio Tinto | Tom Albanese | Bad acquisitions | CEO resigned |

*This table was constructed referencing multiple sources. See Sources, Chapter 3, note 10, infra*

In order to prevent these types of situations, our action plan is to put robust systems in place. If better decisions were made within the teams at BP or Boeing could these situations have been avoided? Absolutely.

It only takes one influential person to prevent these types of events. However, the reverse is also true because it may only take one person to cause these situations to occur.

With solid risk management systems, we can prevent one person from making a catastrophic decision that puts the entire enterprise in jeopardy while simultaneously enabling all personnel to take action to prevent undesirable situations, protecting the enterprise and its leaders. We do not want to be in a situation in which an individual or group unknowingly makes a high-risk decision on location or in the office that puts the entire company in jeopardy. Unfortunately, this happens more often than you think.

## Decisions which caused Executive Terminations

- Weak operations risk management at the field level.
- Asset development decisions and resulting performance.
- Signing undesirable long-term drilling or frac contracts.
- Ineffective cost control relative to peer group.
- Value destructive acquisitions.
- Unsuccessful capital allocation decisions.
- Inappropriate relationships or ethical misconduct.
- Executing unfavorable midstream contracts.
- Decisions by employees or contractors that destroy value.
- Inadequate vetting process for employees or contractors.

## ——————— ACTION ———————
## Operational Issues List
### (aka "The Pinks")

I enjoy studying great engineers, executives, and entrepreneurs. One of my favorite engineers exemplifying the concept of responsibility is Hyman Rickover. Admiral Rickover built Nautilus in 1952, the world's first nuclear powered submarine.[11] He then oversaw the construction of the first full-scale commercial nuclear power plant, the Shippingport Atomic Power Station, which reached criticality, a self-sustaining controlled fission chain reaction, in 1957.[12]

Rickover's operating prowess, methods, and assignment of responsibility are credited with achieving zero nuclear reactor accidents on nuclear-powered submarines.[13] Rickover believed deeply in personal responsibility. You could say his motto, which he used to build a culture of safety, was "I am personally responsible."[14] U.S. President Jimmy Carter said Rickover was:

*"The greatest engineer who ever lived on Earth."* [15]

Rickover believed in detailed oversight of all issues across his operations, which involved thousands of workers, contractors, and subcontractors. His preferred tool for accomplishing this was called "The Pinks." The name refers to pink carbon copies of memos he required people to write to him every week about any issues or problems in their area of responsibility.[16] This was his way to avoid bureaucracy and quickly get to the heart of any potential issues before they became big problems. If you did not report your issues or tried to hide something, there was a good chance you would be fired.

With modern technology, we have a lot more tools at our disposal to communicate issues quickly. However, we often do not take advantage of it, or we use modern technology to hide things. As Jeff Bezos, founder of Amazon, has said, "We outlawed PowerPoint presentations at Amazon…and it's probably the smartest thing we ever did."[17] Bezos thinks, "PowerPoint is really designed to persuade, it's kind of a sales tool and internally the last thing you want to do is sell. You are truth-seeking. You're trying to find truth."[18] Bezos prefers six-page memos instead of slideshows.[19]

Admiral Rickover's pink sheets were also memos, but they were not required to be six pages or written in a specific way, although Rickover was known to correct people's grammar and writing errors. Bezos' and Rickover's goal is to understand the issues thoroughly, and well-structured writing provides increased clarity over a modern slideshow "sales-type" approach. Since structured memos take a long time to write and are not ideal when dealing with serious problems in real-time, consider capturing important issues with a written list, which I refer to as the "Operational Issues List."

Throughout the day during drilling, completions, and facility construction, require the wellsite leaders identify any issues, problems, or concerns that could impact operations, no matter how small, and capture them in the daily report or create a separate software module to spotlight them.

If we do not want to deal with software, the best thing to do is:

> **Require each wellsite leader to upload a list containing all issues from the past 24 hours.**

The "Operational Issues List" is sent out each morning for discussion at the morning meeting. Many things happen on location that are not recorded. Daily reports are mostly a cut-and-paste redacted biased version of the truth, regurgitated from well-to-well. They are not unique like "The Pinks," and the daily report timeline does not spotlight concerns.

Each morning, when looking through the reports, we must read between the lines of generic blurbs and identify how much time each process took to try and find issues that occurred or need attention.

From a management perspective, standard daily reports are not ideal for robust, detailed oversight of issues or concerns. The ideal morning report is a list of each issue, concern, anomaly, or problem over the past twenty-four hours. Operations occurring as expected is not a concern, it's the problems or potential problems that need attention.

Plus, management does not have time to search through each well report or time log and find the key issues. That is inefficient, in addition to people intentionally burying things in hopes nobody in the office notices or asks questions.

Consider establishing specific requirements to include in the "Operational Issues List" or, as Admiral Rickover would say, "The Pinks."

# Operational Issues List: *Suggested Requirements*

1. Near-misses, close calls, red flags, good catches.
2. Operational errors, mistakes, fumbles.
3. Cost overruns.
4. Inefficient situations, procedures, vendors, or personnel.
5. Wasteful events (time, materials, human resources).
6. Engineering procedure errors.
7. Engineering procedures that do not reflect reality.
8. Problems getting ahold of engineers or geologists.
9. Geologic uncertainty. Data quality.
10. Problems with artificial intelligence (AI) systems.
11. Failure of operation to be completed as designed / planned.
12. Non-Productive Time (NPT).
13. Concerns about specific vendors or vendor personnel.
14. Any minor or major concerns expressed by vendors.
15. Vendors that do not have their own written procedures.
16. Vendor performance. Lack of experience notification.
17. Unprofessionalism. Vendor not showing up on time.
18. Vendor does not answer phone or return texts promptly.
19. Invoicing issues (overbilling, errors, shenanigans).
20. Invoicing mistakes that need to be shared with the team.
21. Problem invoices. Invoices rejected on location.
22. Equipment problems.
23. Issues with digital tech (sensors, software, harness).
24. Automated alarm issues. Areas that need alarms.
25. Upcoming weather concerns and impacts to the operation.
26. Anomalies, irregularities, unexplainable events.
27. Safety or environmental concerns. SWA events.
28. Any operational concerns.

---------------- **ACTION** ----------------

## Automate Operational Improvement

---

OSHA defines a near-miss as "a potential hazard or incident in which no property was damaged and no personal injury was sustained, but given a slight shift in time or position, damage or injury easily could have occurred."[20] OSHA's definition is narrow. I prefer a broader definition which is:

*Any situation with a potential negative impact.*

This broader definition makes it easier for people to report events without having to think deeply about whether there was potential damage or injury risk, which can be subjective or hard to determine. Plus, when only focused on damage or injury, it stigmatizes reporting, making it less likely to occur. The broader definition attempts to remove the stigma.

If near-misses and other issues are not addressed, they will encourage people to take more risks. This dynamic is the result of outcome bias and normalized deviation. To address near-miss events, the team must be comfortable reporting them when they occur. Additionally, we must identify near-misses as part of our management oversight programs.

Looking at operations from a high level can cause one to judge successful outcomes as a result of successful operations, when in fact, near misses could be occurring. To address this, we need to dig into the details of each operation. It is understandable not to want to engage in this drudgery because it's not a high-level process or software application to deploy across an organization, but that is what is necessary.

The solution is to systematically increase attention to detail by weaving it into the intellectual capital of the company using an automated process. This will be accomplished by automating learnings from operational issues, near-misses, and other valuable situations. When anything occurs out of the ordinary, which happens on every well, the situation must be systematically incorporated into future operations.

The process starts with a detailed written procedure or checklist. After every operation, near-misses, close calls, mistakes, successes, and other notable or valuable situations from our "Operational Issues List" are captured, reviewed, and incorporated into future written procedures for the next well.

One might think that this process is already occurring organically within the organization. This is not the case. Don't believe me? Pick up a procedure from any well you are currently drilling or fracturing today and compare it to a procedure from last month. It will be the same.

Written procedures are cut-and-paste from one well to the next. Due to the remote and rotational nature of the oilfield, valuable knowledge is often forgotten as soon as the next multi-well pad or when a person goes on days off.

Key learnings are not captured and incorporated into the enterprise unless there is a major undesirable event.

This is why the same mistakes happen over and over or small issues get overlooked until they become big problems. Near misses almost always precede major undesirable events.

# Famous Failures Preceded by Near-Misses

| Incident | Unaddressed Near-Misses |
|---|---|
| Union Carbide Bhopal Tragedy | Near-misses were covered up by workers due to mistrust, friction with management, and weak safety culture. |
| NASA Challenger Shuttle Explosion | Multiple O-ring failures linked to cold weather occurred on prior launches that were ignored by managers. |
| BP Deepwater Horizon Spill | Rig leadership agreed that seeing positive pressure during a negative pressure test was common. |
| Chernobyl RBMK Nuclear Meltdown | Similar near-misses deemed "too sensitive"[22] at other RBMK nuclear power reactors were not shared. |
| Toyota Sudden Acceleration | Stuck pedal near-misses were downplayed. Issues unaddressed. |
| Apple iPhone 4 "Antennagate"[23] | Dropped calls design flaws existed for years but were ignored by engineers and management. |
| LZ 129 Hindenburg Airship Disaster | Many accidents and near misses preceded Hindenburg before hydrogen use stopped. |
| NASA Columbia Disintegration | Foam strikes on 80% of launches were not considered near-misses, until the accident. |
| Piper Alpha Oil Platform Explosion | Multiple near-misses occurred in the week leading up to the total loss of the platform. |
| Takata Airbag Flaw and Bankruptcy | Early near-misses of metal fragments injuring drivers from airbags were considered "an anomaly."[24] |

*This table was constructed referencing multiple sources. See Sources, Chapter 3, note 21, infra*

Capitalize on near misses by honoring their discovery. Do not minimize, downplay, hide, ignore, rationalize, or pretend a near-miss did not happen.

Warren Anderson, the CEO of Union Carbide at the time of the Bhopal plant gas leak accident in 1984, which killed thousands and injured over 500,000 people, said, "You wake up in the morning thinking, can it have occurred? Then you know it has and you know it's something you're going to have to struggle with for a long time."[25] The accident haunted Mr. Anderson until the day he died at 92 years old in 2014.[26]

Set a directive across the organization that all near-misses and issues, no matter how small, will be captured, reviewed, and incorporated into procedures and checklists in an automated fashion. Use the Operational Issues List.

No more cut-and-paste procedures. Since we will be making significant changes to maximize cash flow, we need to ensure operations are very strong and everyone knows everything that is happening. Knowledge must be captured, communicated, and shared across the organization laterally and vertically in a timely manner.

The difference between successful and unsuccessful companies is often the free flow of information across the field level to the top of the organization and across divisions.

If there are unmitigated risks and we start to increase efficiencies, unintended consequences could occur due to hidden weaknesses. We must systematically find and eliminate hidden weaknesses, which the following several action items will help address.

--- **ACTION** ---

## Build an Artificial Intelligence Knowledge Bank

(Classical AI system to reduce cost and risk)

If the collective wisdom and totality of experience within our ops team and company are instantaneously accessible during every conceivable situation, a sustainable continuous reduction in cost and risk would be inevitable. The foundation for this system is the knowledge and experience of each team member, our company, and our industry, uploaded into a usable database with a streamlined human-machine interface.

For the next 12 months, if each consultant and engineer captures every item on the "Operations Issues List," paired with its corresponding solution, we will have 365 days of issues and solutions from numerous operations to serve as the initial Knowledge Bank for our AI model. Every day we upload all the past 24-hour problems, solutions, and any other information that we feel is valuable or relevant. The data is then cleaned and labeled for use in our algorithms.

Once built, the AI is configured to provide solutions to daily struggles experienced by our team, identify historic and recent patterns, showcase issues from offset rigs, completion crews, and past events, rank items to raise awareness, analyze pictures to find anomalies, proactively prevent costly mistakes, make cost reduction suggestions and much more, tailored for each operational step. For example, when our wellsite leaders prepare for the day's operations or for any specific task, the system is accessed or automatically populates on the laptop and mobile device with relevant information to review in preparation for the task. The AI is tied to data recorders to send text alerts with a summary and link to prudent information. Or, more simply, before each operation, we click a button for that operation, and the AI-generated dashboard is populated based on knowledge submitted from our company for each specific task.

We can add engineering elements, photos, videos, tips, concerns, and cost reduction items to the dashboard that we want our team to review and acknowledge in preparation for each operational step. AI monitors who is reviewing the Knowledge Bank before executing each task to help address personnel weaknesses and high-risk individuals.

Just recently, a former colleague who works in a drilling leadership position mentioned an incident on a drilling rig in which a million-dollar mistake occurred twice in one month. After the first mistake occurred and they lost the lateral, a detailed email was sent to all the consultants to address the issue related to drilling mud contamination.

Two weeks later, on a different rig with newer consultants, on the first night of their hitch, the same thing happened. That's two wells on two different rigs within a month, having to sidetrack and redrill the lateral: millions in added cost.

I know the company men and engineers, so I heard both sides of the story and asked, "How is information on past experiences and solutions shared?"

The answer was standard industry practice, and that is "by email." During each hitch, they get inundated with emails they must keep track of. Therefore, on days off, it's easy for emails to build up and have critical emails get covered up among many, get overlooked, deleted, or sent to spam.

The team that caused the second incident did not know about the first incident on a different rig. With a better system, this likely would not have occurred.

---

**Cash flow is maximized in the trenches.**
**AI elevates our ability to fight.**

---

Warren Buffet famously said, "My successor will need one other particular strength: the ability to fight off the ABCs of business decay, which are Arrogance, Bureaucracy, and Complacency."[27]

Our artificial intelligence systems are designed to address the ABCs of business decay by acting as an advisor to moderate arrogance-based decisions, avoid bureaucracy, and prevent complacency. It is common for engineers, managers, and field leadership to have strong opinions and confidence, thinking they know everything. While unbeknown to everyone, there is hidden danger. AI is ideal to raise human awareness.

AI systems are not difficult or expensive to construct. It just depends on how elaborate we want to get. On the next page is an outline of how to build a classical AI system. This basic system uses simple algorithms to compile information and perform analysis. A more elaborate statistical AI system with machine learning (ML) to train artificial neural networks would be the next step.

# Building a Classical AI to Reduce Cost and Risk

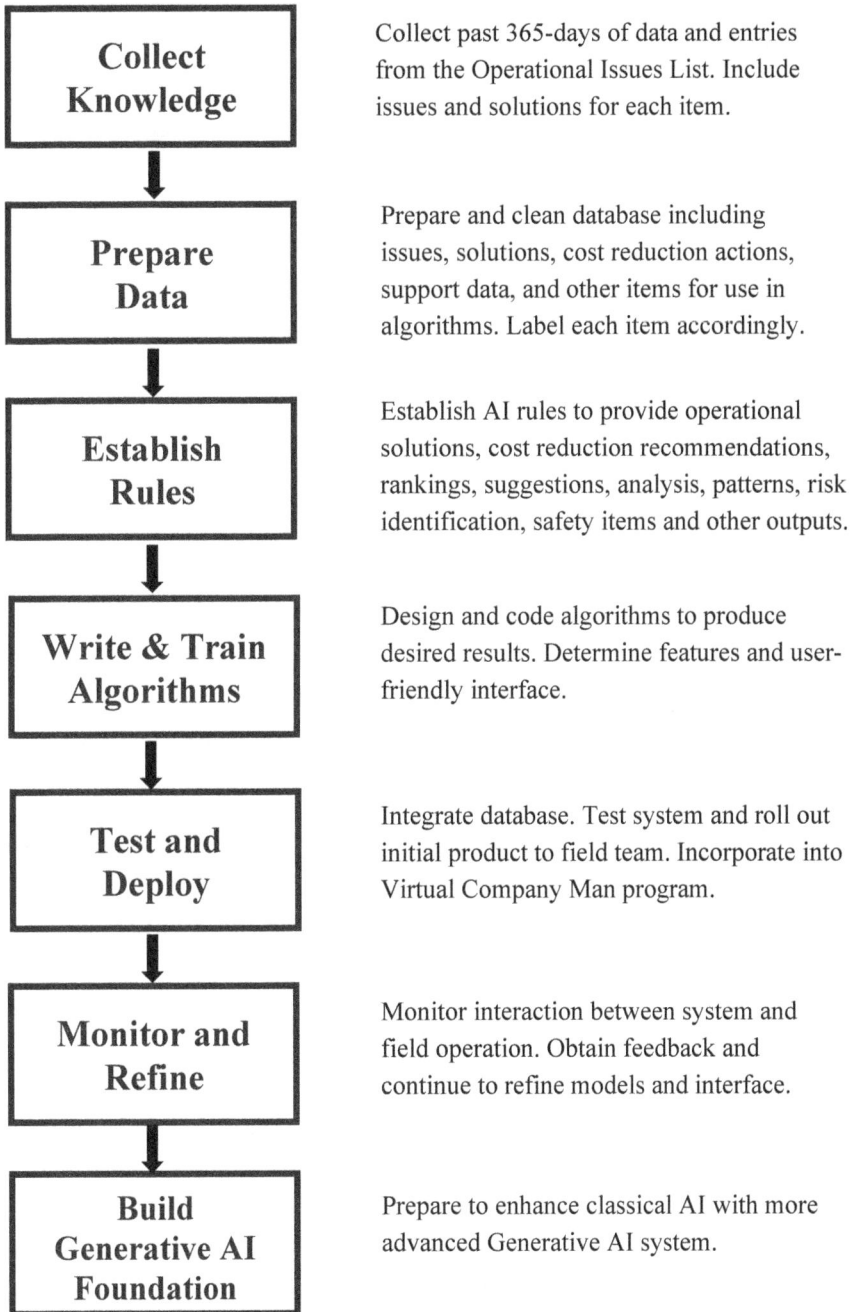

| | |
|---|---|
| **Collect Knowledge** | Collect past 365-days of data and entries from the Operational Issues List. Include issues and solutions for each item. |
| ↓ | |
| **Prepare Data** | Prepare and clean database including issues, solutions, cost reduction actions, support data, and other items for use in algorithms. Label each item accordingly. |
| ↓ | |
| **Establish Rules** | Establish AI rules to provide operational solutions, cost reduction recommendations, rankings, suggestions, analysis, patterns, risk identification, safety items and other outputs. |
| ↓ | |
| **Write & Train Algorithms** | Design and code algorithms to produce desired results. Determine features and user-friendly interface. |
| ↓ | |
| **Test and Deploy** | Integrate database. Test system and roll out initial product to field team. Incorporate into Virtual Company Man program. |
| ↓ | |
| **Monitor and Refine** | Monitor interaction between system and field operation. Obtain feedback and continue to refine models and interface. |
| ↓ | |
| **Build Generative AI Foundation** | Prepare to enhance classical AI with more advanced Generative AI system. |

*This diagram was constructed referencing multiple sources. See Sources, Chapter 3, note 28, infra*

--------------- **ACTION** ---------------
## Daily Visual Report

We perceive reality with five basic senses: sight, smell, hearing, taste, and touch. Our senses allow us to gain intelligence from our environment and make decisions. Each sense is a tool we use to build a picture of what is happening.

Sight is by far our most valuable sense. Unfortunately, our office-based team cannot be on location to visually see essential details and incorporate that intelligence into our decision-making process. This visual operations blindness is a significant weakness that has not been properly addressed by our industry.

Standard industry practice is to monitor operations with daily written reports, sensor-based monitoring systems, and meetings. To take our intelligence, decision-making, and cost reduction to the next level, consider initiating a Daily Visual Report (DVR) for all operations: drilling, completions, and production. The Daily Visual Report is an image record of crucial aspects of operations uploaded to the cloud. Images are captured by our team, labeled, and uploaded to the well file. Cloud technology allows us to use these images in real-time to enhance our intelligence, make more informed decisions, better manage risk, improve safety, and reduce costs.

For example, on wireline operations, we collect 3D images of the plug and guns on each stage before and after running in the hole, label the images, and upload them to the cloud for documentation and analysis using artificial intelligence.

AI is an excellent tool for image evaluation to confirm configurations, detect anomalies, identify problems, or analyze anything we want. We can use imaging checked by AI to verify that the tool is assembled correctly, is not damaged, and that we did not leave anything in the hole, which is an expensive surprise during millout. Additionally, when discussing different aspects of the operation, we can easily pull up the images to visually see what we are talking about. The popular adage is true:

*"A picture is worth a thousand words."* [29]

In many cases, it is hard or impossible to fully describe something in writing; we must see it for ourselves.

The Daily Visual Report also serves as source data for our Knowledge Bank dashboards that our team can reference and from which we train our AI to assist in identifying anomalies, high-risk situations, mistakes, and potential problems before they occur.

For example, anytime there is a problem with a BHA during drilling or completion, everyone always asks if someone took a picture before running it in the hole. Furthermore, when looking back at daily reports over the years, there are so many situations that we wish we had images to accompany the written reports.

Additionally, many office-based team members do not have the opportunity or time to spend on location and have never seen the operations for which they are responsible. This puts office leadership at an extreme disadvantage that our contractors use to their advantage, especially when it comes to billing.

I spend a considerable amount of time on location, maybe too much. As a result, I have seen "how the sausage gets made" and much more. It is not uncommon to hear the following statement repeatedly from vendors writing field tickets:

*"I can put anything I want on the ticket; the oil company has no idea. They don't spend the time out here to even care."*

When you live your life on the front lines of this business, you get to know people on a deep level, and they share a lot. Images from location will help dispute tickets because we can show vendors what was on location, which does not always match the invoices. If we can do that, getting the invoices fixed with minimal pushback is easy.

We then capture this and put it in our Knowledge Bank so our company can benefit during the billing review process before signing the next field ticket. Furthermore, by including images and videos in our Knowledge Bank as a reference, our team can visually see how specific items and tasks should be performed correctly to further reduce risk and cost. It is common for wellsite consultants to oversee tasks and tools from different vendors that they do not have experience with. Referencing images and videos on the task in question provides tremendous value, taking our operation to the next level.

The following page contains a table using completion operations as an example of how instituting a Daily Visual Report can add massive cost reduction value.

# Daily Visual Report: Completion Examples

| Image | Cost Reduction Case |
|---|---|
| **Vendor Items On Location** | Use images during invoice diligence to confirm items are correct before signing field tickets or approving final invoices. |
| **Plug and Guns** | Use 3D images to confirm configuration, detect anomalies, identify if anything was left in hole. Use AI to find issues. |
| **Frac Setup Efficiency** | Capture overhead images to identify if equipment is optimally set for efficiency. |
| **Wellheads Installation** | Confirm wellheads are installed at optimal angles, vendors follow procedures properly. Use images of casing, cutoff, and wellhead internals for damage doc, reference etc. |
| **Frac Stack Installation** | Ensure stacks are setup correctly. Take images after each stage to check for stack movement. Use AI to scan for anomalies. |
| **Backside Risk Management** | Validate backside is open and digital hookup is correct with analog backup. Confirm rig up allows safe pressure release if casing fails or there is a cement issue. |
| **Chemical Accuracy** | Document correct chemicals are on location with amounts to cover job requirements. |
| **Offset Legacy Wells** | Document offset wells before and after frac operations. Protect against legal claims. |
| **Millout BHA** | Show correct configuration with required components. AI scan for issues. Use as reference for fishing operation if needed. |
| **Plug Millout Returns** | Image each plug dump for consistent debris loading, plug parts size, anomalies. Upload for AI analysis of plug parts. |
| **Facility Construction** | Create 3D walkthrough of facility to review construction, identify inefficiencies, bottlenecks, and cost reduction opportunities. |

—————————— **ACTION** ——————————

## Virtual Company Man

Our goal is to create an AI version of the drilling and completion wellsite leader, which I refer to as the Virtual Company Man (VCM). This is not to replace the company man but to provide a super assistant or digital companion, liberating the human consultants and engineers to deliver additional operational value in ways a computer cannot.

We are asking the team to perform at a higher level, particularly during cost reduction efforts. Therefore, we need to provide as much support as we can. The VCM idea is an attempt to assist with some of that support. VCM is not one system but a combination of five systems.

To interface with the VCM, it is suggested personnel have a smartphone and smartwatch. These mobile devices ensure communication with the VCM is streamlined. If the VCM identifies a problem or provides a suggestion, and the team does not receive the information in a timely manner, it is of little value. Mobile technology solves this issue. Below is a list of the minimum recommended VCM components.

**AI: Virtual Company Man** *(system minimums)*

1) Advanced operations alarm system.
2) Knowledge Bank digital advisory with notification system.
3) Camera system with smart detection, sound, and control.
4) Autogenerated written daily report system.
5) Field ticket error detection system.

# 1. Advanced Alarm System

The Virtual Company Man (VCM) must have the ability to notify the team on mobile devices of events, actions, occurrences, and potential issues, similar to what a human company man would do.

Based on our institutional knowledge (i.e., the total knowledge of our company and its employees[30]), we program groups of alarms to notify various team members of specific occurrences. A fully programmed VCM consists of thousands of detailed alarms.

# 2. Knowledge Bank with Notification System

Oil and gas global CAPEX is projected to be +$600 billion in 2024.[31] Each year many wells are drilled, and valuable knowledge is gained in the process. That knowledge must be captured and deposited into our "Knowledge Bank" before it is lost. We discussed aspects of this under the "Build an AI Knowledge Bank" action item.

Ideally, the VCM connects with our Knowledge Bank to provide reminders, suggestions, and solutions throughout the drilling, completion, and production process.

VCM could easily be automated to convey information based on depth or activity. With so many things going on at the wellsite, it is easy to forget a critical step or not remember a new task or process. VCM reminds the field team throughout the operation, conveying intelligence similar to the many smartphone apps that remind us of different things.

If specific events occur, a notification alarm is triggered, similar to a standard alarm system but with more detail to explain the situation with support information based on historical data. For example, if after drilling a certain amount of footage with a BHA, average ROP drops 30% during the past hour and average WOB is stable or increasing, the system sends an alarm notification stating, "There is a 60% chance of damaging the bit beyond repair in the next 30 minutes based on historic data.", as a reminder to take into consideration. These actions leverage our ability to take VCM to the next level based on years of experience from our Knowledge Bank.

---

**AI empowers each team member to operate at the edge of what is known from across our industry.**

---

We take large portions of notable events from operated and non-op wells and code them into the notification system to trigger alerts based on depth, multiple variables, activity, and time, or send them out on a rotating basis systematically throughout the day to continuously feed our team a steady diet of intelligence. Alternatively, or in tandem, we require the team to meet with the VCM before each operational step by clicking a button for that activity to review task-specific knowledge. Engineering elements, photos, videos, tips, concerns, and cost reduction items are included that we want our team to be aware of to address safety, control risk, deliver quality, and reduce costs before each operational step. This will dramatically reduce the risk of a costly mistake.

## 3. Camera System

The oil and gas industry has not historically embraced camera technology for a variety of reasons. However, as costs have declined and technology has improved, the camera system value proposition for oilfield operations has become compelling.

For under $500, a basic solar-powered cellular camera with AI technology can be placed almost anywhere in the United States. We don't need Wi-Fi, electricity, or wiring to install this technology. If entry-level camera technology adds value for our VCM, there are many additional options to customize the system for advanced AI monitoring of oilfield operations.

## 4. Autogenerated Daily Report System

Writing daily reports is time-consuming and not the best use of a wellsite leader's time. This is an ideal task for the VCM. Once the VCM has autogenerated the daily report periodically during each shift, the wellsite leader reviews it and makes changes or adds additional details.

As the technology improves and is customized for each operator, in the future, most actions on location will be automatically captured with digital systems, requiring minimal input from wellsite leaders.

## 5. Field Ticket Error Detection System

Spending time to confirm the field tickets are correct is time well spent. We work hard to be as efficient as possible and need to take every action to ensure we get credit for our cost reduction efforts, reflected in lower field ticket amounts. A good wellsite leader with the skill and diligence to identify ticket errors can add much value during the field ticket reconciliation process on location.

After all our hard cost reduction work, field tickets must be eliminated, or they must be lower than they were before. Otherwise, we are not making progress on our goals. Every service and vendor has nuances when it comes to common errors in the tickets. As previously discussed, every time we find an error, we list it in the Operational Issues List so everyone can see it. Then, we incorporate it into our error detection systems to improve detection accuracy.

Having a digital system auto-check the field tickets before and after they are checked by the wellsite leader can add value as well as reduce the amount of time the wellsite leader may have to spend with the vendors when the tickets are incorrect, which they often are.

Under no circumstances am I suggesting the VCM or AI reconcile the tickets without heavy human involvement. That is a mistake.

Depending on our preferred process, by cataloging all previous invoicing errors, we can provide that data to the wellsite leaders to help them perform the initial check. Then, have the VCM check the tickets to see if something was missed. The final check can occur in the office before payment is approved. On large or complicated tickets, multiple people in the field and office should check the numbers before anything is signed.

—————————— **ACTION** ——————————
## Solutions Book

The Operational Issues List and corresponding solutions are part of the foundation of intelligence that feeds our Knowledge Bank and Virtual Company Man (VCM). Each day, our team compiles a list of all the issues they are dealing with. They also submit solutions for problems. However, as the leadership team, we must discuss the solutions to determine if we are comfortable with how issues are being addressed.

Without capturing all our issues in writing and corresponding solutions, we will not know if we agree with how problems are being solved. Remember, on Deepwater Horizon, the field team said they saw pressure during negative pressure tests all the time. [32] If they had captured similar events from the past and the solutions they used to solve them, there is a high probability that someone would have objected to how the pressure-related issues were being "solved."

There have been plenty of cases throughout my career in which problems were occurring on location, and the team was solving them; however, once I found out about the issues and how they were being solved, I became uncomfortable because I did not agree with how the problems were being addressed.

How problems are solved is essential because we do not want to increase our risk exposure during the process. We cannot stick our heads in the sand and not worry about it because we do not see the problems. If we are not deeply engaged, there is a high probability that problems are happening; our field team is "solving" them but doing so in a manner that sets us up for a catastrophe. Additionally, field solutions can be good, but only a few people may know how to implement them properly. Therefore, we need to document everything in detail.

As we are reviewing the Operational Issues List and proposed solutions, we must collectively discuss and agree to solutions. Approved solutions must be compiled and correlated to the appropriate problems. The collection of approved solutions is what I refer to as the "Solutions Book."

All solutions must be approved by the collective wisdom of the company. Anybody can and must submit operational issues, but the solutions must be discussed and agreed to. Then, they can be added to the Solutions Book for use in our Knowledge Bank.

For example, recently, a large operator was having issues with frac stacks becoming unstable. The field solution to this problem was to tie onto the wing valves on the B-Section because that would add stability to the entire system, which it did. However, due to vibrational issues, the stack continued to move until the connection on the wing valve to wellhead spool failed, causing a massive well control situation that lasted for weeks and cost over ten million dollars to control.

My guess is that the stability problem was not known in the office, or if it was, there was no organized process to document it, analyze it, and solve the problem in a timely manner. Therefore, it might have been something that was brought up nonchalantly in an everyday conversation between the company man and engineer. The company man was fulfilling his obligation to mention it, possibly in a casual conversation, and the engineer ignored it, perhaps because he didn't understand the risks. It is hard to visualize much of this stuff to get the whole picture. Therefore, the company man did what he thought would stabilize the wellhead.

Our Daily Visual Report combined with our Operational Issues List would have put this issue in front of everyone, making it a lot more real. Therefore, it would have been handled appropriately. I have researched many oilfield accident cases, and what is common is that, during discovery, it comes out that someone in the field mentioned to someone in the office about the issue on location, and it got brushed aside or not given attention and priority.

It isn't that it is intentionally ignored; it's just that everyone has things going on that take precedence in their mind, so people kind of forget about the problems on location, do not prioritize them, or do not engage, hoping they go away..............

…

.......

.............

...............

.................

.......................

...........................but they don't.

# ACTION

## Dueling Artificial Intelligence

If there is hesitation with incorporating AI to enhance performance, or developing AI in-house, consider several AI software providers to see whose tech may fit our needs. Speak with different companies and ask them to try their product to see the value. Let them know that they will be in an AI duel with other competitors or our in-house applications.

For example, improving geosteering performance with AI can have a significant impact on total cost and on a unit of recovery basis (CAPEX / EUR) because we are reducing the numerator (cost) while increasing the denominator (oil and gas recovery). All cost reduction actions that reduce cost while increasing estimated ultimate recovery (EUR) or production performance should be explored, even if they are a little intellectually uncomfortable.

AI is suited to enhance geosteering performance because a lot of data must be processed quickly in real-time, 24 hours a day. Uploading decades of experience from our company into the AI will upgrade our decision quality, especially when there is high complexity, faulting, or other uncertainties leading to multiple interpretations. AI can be set up to send alerts whenever an unexpected geologic anomaly is detected or predicted ahead of the bit. This allows the team to focus on other tasks while AI is looking for issues and attempting to predict geologic hazards and provide recommended actions.

Another option to consider is to take two artificial neural networks (ANN), machine-learning models like the human brain, and put them against each other in a Dueling Neural Network (DNN).[33] Applied to geosteering, both AIs would monitor the drilling process. Then, one AI generates a geologic projection and the other AI critiques it. Theoretically, this provides the machines with an environment that more closely resembles human dynamics within the brain.[34] The two systems then duel with each other until they agree.

We would not put this system or any AI system on a live operation until everyone is comfortable based on test data. We train on a simulator using data from the target formation and area of interest. Have it geosteer all previous wells and non-op wells in a simulated environment. As the system is trained, we adjust the AI until it reaches a point where it can perform with a respectable level of accuracy.

Determining what percentage of any given lateral is in the target zone is subjective and depends on a variety of factors. In some areas and formations, it is easy to know we are out of zone, but in other areas and formations it is very difficult.

Some analysis suggests that humans are not very good at geosteering in terms of staying in the best rock. Inspecting outcrops while driving across the United States can help explain why. Our planet is full of small faults. If we have a tight target window and there is a small fault that we do not know about, it is easy to drill out of zone. Then the question is, do we hold course or correct? Seismic helps, but many faults are too small to detect using current technology. This is why there is a lot of hope for AI.

# ESSENTIAL COMPONENT #2

## Establish a Numeric Cost Reduction Target

Cost reduction goals must have a specific target to establish value, maintain accountability, and inspire the team. People want transparency. They want to know the big picture, what we are doing, and why we are doing it. We must understand what is necessary to win and how we perform relative to the target. This is part of the reason sports are so popular. There is a score, and we know who is winning.

Working towards a specific numeric target publicly known within our business is much more powerful than vague, open-ended objectives. For example, implementing a strategy to reduce costs by the maximum amount with a continuous unstructured strategy does not command as much dedication and accountability as implementing a system to reduce costs by 50% or $5,000,000 per well in 45 weeks.

People connect to specific, measurable targets that they can understand. With specific targets, the team will know what is required. Vague strategies using corporate buzzwords often lead to eye-rolling, dismissal, resentment, and a lack of detailed accountability, ultimately impacting cash flow.

Every situation is different, and selecting the best primary numeric variable will depend on what makes the most sense. From a cost reduction and cash flow performance perspective, I have had success by starting with a percentage reduction goal at the executive level and then establishing a specific dollar amount goal on a per-well basis for the team.

For example, let's say current actual well costs for 12,000' lateral length wells are $10,000,000 per well, and we need a 40% reduction in costs to achieve cash flow objectives. Based on this scenario, we would establish our cost reduction target at $4,000,000 per well.

The cost reduction initiative would be based on this specific target, and we would construct everything around this number. Dollar amount numbers are more engaging, easier to understand, and fun to track. Measuring cost reduction success is one of the most enjoyable aspects of each cost reduction campaign.

## Selecting the Exact Cost Reduction Target

Determining the exact cost reduction target will depend on what is desired or required to achieve our economic objectives. In most cases, I suggest a realistic stretch target for the first cost reduction push.

In general, if more than a 25% cost reduction from current levels is required, depending on the situation, consider segmenting the cost reduction program into multiple campaigns with breaks in between each push.

As previously discussed, the exact target depends on a variety of factors, so I am reluctant to generalize this aspect of setting a specific target. However, starting with a 10% to 20% reduction for the initial push cost reduction campaign is a good initial stretch target.

Whatever is decided, it is common for many people to think it is not possible. That is a good sign. When people think it is not possible to achieve the objective, that is a good sign that our target will create value because we are pushing forward. We are taking our shale game to the next level, expanding our technology and engineering abilities beyond what is currently possible. Thinking something is impossible happens before almost every significant breakthrough.

> **It's a normal reaction to think the reduction target is not possible because if it was possible, we would already be doing it.**

It is essential to distinguish hard from impossible. Cost reduction of any amount is hard. For large cost reduction initiatives greater than 20%, consider segmenting the effort into multiple cost reduction campaigns. For example, if we need a 50% reduction in costs, segmenting the cost reduction effort into three distinct campaigns, each defined with a specific amount and time frame, with breaks in between each campaign. Whatever is decided, consider establishing specific targets that people will understand, are memorable, and impactful to the economic objectives.

### Per Foot and EUR Metrics: Cost, Cash Flow, and Risk

The industry has gravitated towards per foot metrics for good reason. They normalize various dynamics and help make relative comparisons between assets, competitors, and technologies. Additionally, CapEx/FT and CapEx/BOE are often utilized in capital allocation decisions and should be incorporated into cost reduction initiatives. In general, on a per foot basis, there is no faster way to improve cost metrics than drilling longer laterals, as long as the target interval is easy to drill and complete.

> **If the rock drills fast and is homogeneous, drilling longer laterals is one of the best ways to improve CapEx / FT and CapEx / BOE.**

If the decision is to run a cost reduction campaign using CapEx/FT as the primary numeric target, there are several things to consider. Longer laterals will reduce CapEx/FT but can increase total well cost and well-level risk, resulting in the opposite impact on cash flow, especially if performance degrades. For example, let's say we are running ten rigs drilling 10,000 FT laterals, 30-day spud to RR, at $7,000,000/well. Over the years, drilling ability improves, and we can now drill 15,000 FT laterals in the same amount of time at $9,000,000/well.

Current Plan Annual CapEx = $840MM at $700/FT
Longer Lateral Annual CapEx = $1,080MM at $600/FT

Cost per foot is reduced by $100/FT (14%), a respectable improvement, but instead of spending $840 Million annually, we end up spending over $1.0 Billion. If the longer laterals do not perform or undesirable situations occur, we can end up reducing free cash flow. Spending a dollar holds much more certainty compared to generating a dollar from projected production. As prime acreage is developed and infill drilling continues, maintaining production performance targets becomes more challenging. Estimated recovery and BOE metrics are uncertain and subject to interpretation compared to increased current incurred costs, which are certain.

The other dynamic is that longer laterals are "risk-on." In general, the longer the lateral, the greater the risk, and higher level of uncertainty. If we start drilling 15,000' to 20,000' laterals and have problems, $9MM wells can easily turn into $12MM wells. We may even lose a few wells due to stuck pipe, casing failure, geologic anomalies, directional survey accuracy, or production issues.

Lost wells severely impact cash flow, especially if they occur during post frac operations because at that point, the majority of our capital has been deployed and is at risk. A comprehensive cost reduction initiative should incorporate a multitude of variables and metrics.

However, as previously discussed, selecting one primary metric will help focus efforts and simplify cost reduction operations. I like using total cost reduction per well with a specific dollar amount as the primary numeric target. However, this is not always the best metric, especially if our company is currently in the process of increasing effective lateral length.

---

**There is no perfect metric. However, if there was one master metric to rule them all, it would be time.**

# ESSENTIAL COMPONENT #3

## Set a Time Frame

Time is humanity's most valuable asset, but we rarely acknowledge its importance or maximize the use of it. Since life takes place in the flow of time, it is easy to overlook or take for granted. In the context of cost reduction, incorporating time is critical.

Due to the nature of how humanity has structured time, it is common to look at goals based on an annual time frame. However, for efficient and focused cost reduction, a year is generally too long and usually results in an unorganized rush toward the end of the year. Additionally, since numerous work-related performance metrics are associated with annualized goals, the cost reduction effort can easily get lost in the crowd or pushed to the back burner.

The other option is to not incorporate any aspect of time into cost reduction efforts and run an open-ended cost reduction operation. In other words, structure a culture where cost reduction is always occurring. This is not a bad option; cost reduction should be a continuous process and part of the culture of an effective business operation.

However, when working to make massive reductions or technological breakthroughs it can add significant value to focus the team on cost reduction, create a sense of urgency, and structure a prioritized effort to deliver breakthrough results.

Based on experience, for the initial push, I have had good

results setting a 10-to-15-week time frame or about one fiscal quarter. Less than ten weeks may not be enough time, and longer than 15 weeks does not get people's attention because there is no sense of urgency. Of course, the optimal time frame must be customized to the specific situation.

For example, if we target a 20% cost reduction on a $7.0MM well, we could structure a 15-week cost reduction campaign targeting $1.4MM per well. If we are targeting more than 20%, consider segmenting the effort into multiple cost reduction campaigns.

For example, if we need a 40% cost reduction on a $10MM well, segment the effort into a three-push program. This allows for breaks to rest and prepare for the next big push. The rest period can be any amount of time we feel is necessary. After 15 weeks of aggressively pushing costs down, consider at least 3 to 5 weeks of rest. This period allows people to take a break, perform analysis, and reflect on the changes made. Additionally, it reduces the risk that something could impact production performance or another aspect of the operation. Below is a potential cost reduction structure to consider.

---

### Cost Reduction Targeting 40% on $10MM/well cost

---

### Target Reduction = $4.0 million per well

| |
|---|
| **Initial Push:** 15 Week Campaign, 15% or $1.5MM/well in 15 weeks |
| **4 Week Rest:** Reflect, Analyze, Recharge and Prep for Second Push |
| **Second Push:** 15 Week Campaign, 15% or $1.5MM/well in 15 weeks |
| **4 Week Rest:** Reflect, Analyze, Recharge and Prep for Third Push |
| **Third Push:** 10 Week Campaign, 15% or $1.5MM/well in 10 weeks |

If you are keeping track of the numbers in our example, you may have noticed that the math does not add up to a $4.0MM per well reduction. It adds up to $4.5MM per well.

> **Costs are never static and are almost always going up, because 99% of people and events are pushing them up.**

Real-world cost reduction campaigns are not exercises in numerical perfection. Cost is almost always creeping up for good reasons and bad. During rest periods, costs will go up, and they are almost always going up directly or indirectly. Since everyone and everything is pushing costs up, we must have systems and people pushing back.

Structuring a system with targets and time frames that we are comfortable with is crucial to delivering results. If the target is $4.0MM/well, we may need to exceed the paper target to get an actual reduction of $4.0MM. Additionally, real-world costs have significant volatility. To cut through the noise and have a material impact on FCF, our effort must be aggressive.

Cost reduction is like a muscle that must be exercised. Since we do not control the price we sell our products for, we must control the cost and be able to do so at will. Therefore, we must have strong cost reduction muscles.

The rest periods between cost reduction sets allow our muscles to recover as we reflect on our efforts to confirm progress, ensure we do not take an action that increases risk, or do something that negatively impacts production.

# ESSENTIAL ACTION #4

## Selfless Commitment to Succeed

---

The single most important action in this book is an individual and team commitment to succeed. If everyone makes a selfless commitment to succeed, I believe the cost reduction targets, whatever they are, can be achieved in a safe and prudent manner, maximizing free cash flow for the company and its stakeholders.

To create value for the company's greater good, personnel must put individual objectives aside. It is tough for humans to put self-interest aside for the greater good. In general, most of the time, people act in their self-interest. People do what is best for them individually.

During cost reduction campaigns, I always ask people to genuinely help and please put personal differences and pride aside. Without honest dedication, the cost reduction initiative will fail. The best plan will fail without a selfless devotion by each person to put the objective above self-interest.

Individual sacrifice for the greater good is not easy. Most people are unwilling to do it, especially when it comes to optimizing something that they have publicly said is impossible—ego and pride are very powerful forces.

During cost reduction campaigns, when we are trying to significantly push costs down, it is inevitable that individuals are going to disagree with specific changes and take things personally. It is common for technical and operational folks to get upset when something they have had ownership of is changed, improved, or optimized, by someone other than themselves. In these situations, I have seen people do unpredictable things to save face, including taking revenge against the company or sabotaging the cost reduction effort.

One option to address this and potentially change a cost reduction resistant culture is to incorporate cost metrics and reduction targets into individual quarterly and yearly performance metrics. With cost and cash flow targets embedded into employee performance targets and tied to recognition and bonus structures, a company can convey a cultural dedication to safely reducing costs.

When incorporating cost and cash flow into individual performance reviews, exercise caution on rewarding ideas over execution. As we have previously discussed, idea generators are good, but being able to execute ideas at scale in the field requires significant ingenuity that often goes unnoticed and unrewarded. At the end of the day, quantifiable results matter most, which is usually the result of a team effort.

Much effort is occurring far away from the office, on a distant location, by people doing everything possible to safely deliver results under difficult conditions. It is easy to forget what we do not see.

The goal is to get the team to embrace a forward-looking strategy. Whatever we are doing now, in the future it will be able to be done more efficiently and for lower cost. That is the nature of engineering and human advancement.

# 4

## RISK MANAGEMENT

—

Risk management provides the foundation for all value creation activities across an oil and gas company, especially cost reduction. Addressing risk properly is the difference between success and failure. Not once or twice, but again and again and again, as operations scale, acquisitions occur, and the number of people we deal with increases. As a result, we become more exposed. Therefore, our focus on risk must expand.

When done correctly, lowering costs lowers risk both financially and physically. Low cost is low risk. High cost is dangerous. A fundamental principle of a prudent large-scale shale developer is the mentality that low cost is low risk. It is a common misconception that running a low-cost operation is somehow unsafe, risky, or more dangerous. This is not true; it's often the opposite. A high-cost, inefficient operator exposes more personnel to more risk over extended periods of time and is more dangerous.

Based on experience combined with researching over 1,000 oil and gas situations for Oilfield Survival Guide, I did not find a relationship between efficient, low-cost operations and safety incidents.[1]

I did find that inefficient operations, non-routine tasks, misjudgments, hasty decisions, lack of knowledge, and extended exposure to high-risk situations can lead to increased incidents and accidents. Additionally, there have been situations in which a high-cost operator exceeded AFE, leading to impulsive decisions to reduce expenses, and resulting in unfortunate consequences.

This could lead the uninformed to conclude that cost reduction caused the incident. However, that would not be true. This is an example of lying by omission. It was a high-cost operator exceeding an already high AFE, which led to financial distress, panic, and ultimately impulsive decisions that the team could not execute safely.

A high-cost team will have trouble safely executing a low-cost strategy without the proper people, planning, discussion, and systems to control risk. One might think we are in the oil and gas business when, in reality, we are in the risk management business.

Warren Buffet, arguably the world's top investor, has said his number one rule is "don't lose [money]."[2] Rule number two is, "don't forget the first rule."[3]

The mindset in the oil and gas business is the same. People who know me will say, "Wait a minute, you always say safety is the number one rule, not money." I do say that all the time, throughout the day on location and in the office. Being safe is equivalent to not losing money. In my mind, it's the same thing. In other words, don't lose money equals safety first.

On location, I always say safety is the most important thing because I do not what someone to misinterpret, "don't lose money." If we are unsafe, high risk, or careless, we will lose money. If being safe means we need to spend more, then that's what we will do because we don't want to lose serious money or make a fatal mistake that will take us out of the game.

Efficient, low-cost operators address cost overruns more effectively as they are not under the same financial pressure as a high-cost operator. Efficient, low-cost operators are less likely to be involved in high-cost, non-routine operations, which inherently carry more risk.

Efficient, low-cost operators do not have personnel exposed to operations for extended periods compared to high-cost operators. For example, if one operator has an average spud-to-sales time of 60 days and another is averaging 120 days, which one is more exposed?

If an efficient low-cost operator can meet growth projections running 10 rigs and a competitor in the same area needs 20 rigs to accomplish the same thing, who is more exposed? There are exceptions, of course, and the details matter, but you get the point.

Time and scale play a significant role in cost and risk. On a well-level basis, efficient, low-cost operators are not on-location exposed to risk as much as high-cost operators. Can safety be sacrificed by rushing through an operation to save money? Of course, but eventually that will be expensive, and a "hurry-up operator" is not efficient or low cost.

## Evolving the Manufacturing Mode Mindset

To maximize cash flow with shale, most operators work towards or are currently in full-scale development mode, often referred to as "manufacturing mode." This mode of operating, large-scale development drilling with multiple wells per pad, often invokes a comparison to traditional manufacturing in some shape or form.

When operating in manufacturing mode, setting up the shale machine with assembly line type systems is often referred to as though drilling, completing, and producing horizontal wells is equivalent to manufacturing cars, consumer products, or some other widget. The problem with this comparison is that it is false.

Dealing directly with a large planet at an extreme level of detail in the subsurface is not manufacturing consumer products on the surface of the Earth in a controlled environment. I have no problem using the term "manufacturing mode" as long as it does not lull the workforce and management into a false sense of security as though we are working in a factory manufacturing oil and natural gas.

Shale operations risk management requires an extreme level of awareness, with all senses on high alert for any abnormalities or variation of even the slightest element. This level of diligence is required on every single well.

The Earth does not care if we are in manufacturing mode. That term can be misleading and negatively impact the culture from a perspective of diligence and attention to detail.

Conveying to the workforce that every well is identical or

should be identical because we are in manufacturing mode can end up increasing risk and costs instead of reducing it. In other words, manufacturing mode can backfire, which it has for many operators and investors.

We previously talked about the time-lag risk (Clarity Gap) in Chapter 2 when engaged in full-scale development mode, but we have not addressed the manufacturing-mode mindset risk. The mindset on location and in the office cannot shift into a mental model in which every well is the same. This is not the way to manage risk in the shale game.

## Every well is different.

Even wells parallel to each other on the same pad are different. It goes even further in that every foot is different within each well. Overburden geology can vary significantly on boreholes 20 FT apart, making the task of drilling down to the source rock unpredictable.

Additionally, there is no such thing as a homogeneous solid block of shale with no variability, especially when expanding operations on a large scale. Furthermore, this aspect of expanded variability will only increase as more wells are drilled and operators are forced out of core areas.

Shale reservoirs contain streaks, faults, depletion, and various rock quality fluctuations that must be addressed to be successful. Furthermore, drilling the vertical and nudge to the target interval contains another source of variability regarding mechanical geologic dynamics.

The other aspects to keep in mind during manufacturing mode are parent-child relationships, frac hits, and depletion. These are factors we generally don't have to deal with during the initial stages of asset development but add risk during full-scale expansion. This risk is normally not included in financial models or risk management dynamics but must be addressed to be successful.

Combined subsurface variability, with surface topography variability, weather variability, and a contractor/subcontractor operational workforce that is constantly changing, with a factory floor moving from location to location, and we have a situation that is very different from traditional manufacturing.

In the shale game, new faces show up on location all the time, adding considerable risk to every operational process. This is different from traditional manufacturing, in which the human workforce is stable, the factory is stationary, work is repetitive, and the entire process occurs in a climate control environment with a high level of transparency and operator management oversight.

The table on the next page compares traditional manufacturing of consumer products to shale manufacturing of oil and gas. Several variables are included to highlight the differences between both industries. Regardless of the differences, in the oil industry, horizontal development at scale is the closest thing to traditional manufacturing, but it is not traditional manufacturing.

# Traditional Manufacturing vs. Shale Manufacturing

| Industry Dynamics | Traditional Manufacturing | Shale Oil & Gas Manufacturing |
|---|---|---|
| **Earth Realm** | Surface | Subsurface and Surface |
| **Visibility** | Highly Visible | Unseeable and unclear |
| **Materials** | Finished | Natural Resource / Raw |
| **Location** | Convenient | Remote, Isolated |
| **Weather** | Protected | Exposed |
| **Climate** | Controlled | Uncontrolled |
| **Factory** | Fixed | Moving |
| **Workforce** | Stable | Highly Variable |
| **Personnel** | Employees | Contractors & Subs |
| **Management** | Onsite | Distant |

To properly manage shale manufacturing dynamics, we must address risk. Increasing attention to detail, proactively identifying risk, and taking action to mitigate risk exposure is critical. However, the most crucial element of shale oil and gas risk management is making good decisions.

# Human Elements, Decision-Making, and Shale

Emotion, stress, pride, office politics, and a host of other aspects related to the human experience hinder good decision making. Oilfield activities are especially susceptible since we deal with problems deep within the Earth. Problems that cannot be visually seen or touched by human hands, ultimately resulting in a higher level of uncertainty.

Furthermore, many challenges involve working on planet-level dynamics. These challenges are highly susceptible to weaknesses inherent in the human condition. Let's review a number of human behavior elements that present challenges to successfully navigating shale operations risk management decision-making. Once we learn about these cognitive biases, we will see them everywhere.

## *Confirmation Bias*

Confirmation bias is the inclination to favor information that supports one's position.[4] This bias distorts the ability to make objective decisions. Due to the uncertainty inherent in the shale industry and the inability to see what is happening in the reservoir, confirmation bias is intertwined with technical analysis, financial valuations, development plans, operational recommendations, and engineering proposals.

Significant amounts of information in our industry are interpreted using various technical methods. If we have a strong opinion about anything in the oil and gas business, it is easy to unconsciously filter information to support our existing beliefs. It is not uncommon for people to search for information that supports their beliefs and ignore information that disproves their beliefs instead of searching for truth.

It is easy to analyze data with a bias to achieve a preferred theoretical result, since finding out the truth often takes a lot of time and money. Confirmation bias negatively impacts objective decision-making, which can result in uneconomic outcomes.

Financial and technical people harbor such strong beliefs on topics that they do not realize how their strong opinions influence the analysis, conclusions, and ultimately, the actions taken based on biased analysis.

If proven incorrect after significant amounts of capital have been spent, the truth can come as a shock. I have seen this play out in shale exploration projects that did not yield producible hydrocarbons, shale step-out areas that were uneconomic, valuations with biased assumptions that resulted in overpaying for assets, drilling and completion engineering designs that failed, optimistic type curves, geologic interpretations, adjusted logs, and adjusted financial metrics. The list is endless.

## *Anchoring*

If we depend too heavily on initial data to make future decisions, even when new information is presented, the bias towards the initial data can be considered an anchoring bias.[5] Initial subsurface information and well results typically set the anchor. Then, a biased interpretation of future information based on the anchor occurs.

Initial well results, if favorable, often set the baseline standard from which to improve, setting the anchor. Typically, a shale development plan is anchored to the initial wells, and those well results are assumed across the acreage position, regardless of the geologic variability or depletion.

There is a popular saying in shale development: "Your first well is your worst well," implying that future wells improve from initial anchor wells. Although this dynamic is often true, in certain areas it is not uncommon for initial wells to be the best wells and for future wells to underperform, think parent-child interference, step-outs, and geologic complexity.

It is common for the best wells to set an anchor from which all future wells are compared or based. This dynamic is present during acquisitions by the sellers based on their initial wells. When future wells do not live up to the anchor, buyers may lose their minds trying to figure it out. Often, people will not let go of the anchor. Once an anchor is psychologically embedded in the mind, it can be hard to let go, especially when a billion-dollar acquisition is based on anchor wells not representative of full field development.

## *Wishful Thinking*

In the shale business, it helps to be optimistic when attempting to get a deal done, drilling in a new area, or trying a new technology. However, we do not want to be unrealistic. Interpreting information as one would wish to be true, as opposed to what the data indicates, is considered wishful thinking.[6]

Excessive optimism often leads to wishful thinking. Embarking on a development program or acquisition, unconsciously based on wishful thinking, can be expensive or outright devastating to our company. However, if we recognize "Shale's Seven Stages of Wishful Thinking," we can hopefully avoid falling into the wishful thinking mind trap.

# Shale's Seven Stages of Wishful Thinking

| # | Stage | Description |
|---|-------|-------------|
| 1 | **Honeymoon** | The development program, acquisition, or financing transaction is initiated. The operation appears to be running smoothly, according to plan. |
| 2 | **Question** | Unexpectedly, there are a few hiccups that raise several questions. Variable performance, hidden costs, mistakes, cash flow fluctuations, or other questionable details emerge. |
| 3 | **Ignorance** | The questions are explained away as anomalies. Although things continue to unfold unfavorably, issues are minimized, dismissed, and ignored. |
| 4 | **Red Flags** | Clear challenges are evident that cannot be ignored. Underperforming wells, type curve issues, high costs, operational errors, flawed acquisition assumptions, modeling mistakes, or financing problems occur. |
| 5 | **The Turn** | Challenges are not able to be addressed quickly, or are not solvable in time. |
| 6 | **Panic** | Problems overwhelm the team, leading to panic. Hail Mary actions are taken. A downward spiral occurs. |
| 7 | **Truth** | The team realizes that unrealistic expectations based on wishful thinking dominated assumptions, decisions, and actions, that led to unfavorable results. |

Cost reduction can help mitigate wishful thinking problems. Initiating a cost reduction program at the first sign of trouble is the key. In Shale's Seven Stages of Wishful Thinking, an aggressive cost reduction action plan at Stage 2 would be most beneficial. The longer we wait, the harder it will be to solve problems before time runs out.

## *Automation Bias*

Considered human labor-saving technology, automation makes a process, procedure, or job occur with little or no human involvement. Automation can lower costs, prevent human error, and reduce the burden on human resources.

However, if relied upon too heavily, automation can cause problems. When recommendations from automated systems are relied upon while disregarding real-world information and good human judgement, this is considered automation bias.[7]

The simplest example occurs when using a mobile phone app for directions without inspecting them. Many people ask for a pin to location and then thoughtlessly take the directions from an app, unquestioningly depending on technology. The automated GPS-generated route is often not the safest or most cost-effective, especially when adjusting for real-world conditions.

Automation has made its way into nearly every aspect of shale operations, including engineering design, supply chain, location construction, drilling, completions, and production. If you're wondering how automation is involved in the supply chain, here is a funny (not funny at the time) story.

When I was on the services side of the business years ago, while working for multiple clients utilizing a variety of drill bits, just as we were finishing our ops program, $10 million of drill bits showed up on our remote location.

When the bits showed up, we were all scratching our heads as to why all these bits arrived just as we finished operations. We thought maybe we had more wells to drill. After looking into the issue, we realized the new automated inventory system noticed we were low on bit inventory and decided to automatically increase our stock to the tune of $10 million.

The supply chain department did not question the system when it ordered the bits because they thought the computer must know. They mindlessly relied on the automated inventory system even though the bit order was excessive and did not make sense. After that incident, several inventory components were taken off the automated reordering system and additional controls were put in place.

A recent example of automation bias occurred when I was looking at our SCADA system on a group of shale wells. I called one of our field team members and mentioned that something might be wrong with our tank level gauges. His response was classic automation bias:

> *"I noticed it looked off, but I did not get an alert,*
> *so I thought it was okay."*

This is very common and a classic example of automation bias. He noticed one of the gauges was stuck because in our water tank we have oil and water gauges that transmit data to our cloud-based automation system. In the water tank, both gauges should read close to the same level. But the water gauge was stuck while the oil gauge was increasing. This can happen if oil goes into the water tank due to an issue. We have alarms set up on the tanks for different scenarios. However, if one of the gauges gets hung up, it may not alarm but is noticeable, which he noticed but did not act because an alarm did not trigger.

He trusted the automation more than his good human judgement. If automation results in personnel not checking things with the human-computer (aka the brain), then automation can be detrimental. It costs us twice, once to install or use and again when personnel stop using human judgement. Automation and other artificial intelligence systems can make people mentally lazy, increasing costs.

## Human Herd Behavior

When humans make decisions strictly based on what others are doing, driven by emotion instead of thinking individually using available information, this is considered human herd behavior.[8] Protests, riots, market bubbles, and shale land grabs are examples in which individual judgment shuts down, and humans follow the behaviors of the mob.

Herd behavior often occurs at times of uncertainty. In engineering and operations, if unsure of the best approach, it is common to go with what others are doing. We leverage the wisdom of the crowd, in which large groups are more intelligent than individuals, under the premise that the chosen method must work if everyone is doing it. Looking at what the majority are doing and replicating it is not a bad tactic. Often, it is a safe approach as long as the situation is analyzed individually and rational actions are taken. However, it is usually not the most economical or efficient approach.

From an operational perspective, there is social pressure to conform to what others are doing. This can be very powerful in the engineering and design area of the business. Every engineer, including myself, has felt significant pressure to conform to what other operators are doing. Due to the capital intensity of the energy industry, there is real risk when doing something different compared to the herd.

From an operations perspective, the thinking is that if you do what everyone else is doing and it does not work, you will not be blamed. You did what everyone else did, so it must be the geology (out of your hands type thinking, must be the rock). Your hands are clean at this point.

But if you do something different and it does not work, then you will be blamed. You're on your own; good luck, you're out of a job or worse. This thinking leads to herd behavior, understandably so.

However, if we are going to reduce costs, we may need to do things differently. Especially if we are interested in outperforming competitors, making competitor acquisitions, consolidating areas of interest, and creating exponential value for shareholders. History suggests companies that create 100X shareholder returns often have a unique approach that differentiates them from the herd.

## *Groupthink*

Groupthink occurs when humans forego personal opinions and accept the beliefs of the group consensus rather than engage in critical analysis, debate, arguments, or disrupt the group's harmony.[9]

If members of the group are fearful of expressing opinions, especially if they disagree with leadership or the consensus, groupthink will occur. Groupthink may also be more likely to occur if there is a super aggressive personality, abrasive manager, or charismatic leader.

The oil industry is full of strong personalities, especially aggressive ones, discouraging people from voicing conflicting opinions. It is not uncommon for diverse views to be diminished or dismissed. Additionally, to avoid disruption and maintain peace in the middle of an operation, it is not uncommon for people to keep their opinions to themselves. To maintain friendly relationships, especially between service providers and operators, conflicting opinions are often held back, leading to groupthink.

There are benefits to groupthink in that it allows large groups to make decisions and take action quickly. Groupthink can enable efficient decision-making. However, this benefit is also a risk if the wrong decisions are made because nobody stepped up and voiced concern. Groupthink has led to many high-cost oil and gas mistakes, several of which I outlined in the Oilfield Survival Guide.

When going along with the group, there is comfort in the feeling that the optimal decision has been made. Without debate and critical evaluation, this feeling is often an illusion of comfort. Although "yes men" make group decisions easier and instill a sense of harmony within the organization that the right decisions have been made, they lead to low-quality decisions that are not fully vetted and often prevent innovation. Additionally, "yes men" are more prone to ignore red flags that the decision is incorrect because independent thinking has been suppressed. This is especially critical in the oil and gas industry because of the universal law of real-world operations:

> **No operation goes exactly as planned.**

Since we are working to maximize cash flow with cost reduction, we want to encourage everyone to voice opinions, controversial matters, alternative solutions, and novel ideas before taking action to make fully vetted high-quality decisions. We also want to encourage people to take action to help make things work when challenges arise. Cost reduction efforts are not focused on harmony. The primary objective is safely reducing costs and delivering quantifiable results.

We are looking for critical evaluation, which includes vigorous debate, constructive arguments, and independent thinking. We seek breakthrough ideas that do not come from consensus or groupthink. We want to encourage paradigm shifts and valuable change in legacy thinking, methods, designs, approaches, and actions that lead to efficiency improvements and cost reduction to maximize cash flow.

## *Texas Sharpshooter*

When a man randomly shoots at a barn and draws a target around the tightest cluster of bullet holes, that man could be a Texas Sharpshooter.[10] The term refers to the human tendency to jump to conclusions based on data clusters. By focusing on similarities, ignoring differences and randomness, we make an inaccurate or incomplete conclusion.[11]

The term's origin is credited to a "legendary Texan" who randomly fired his rifle into a barn. Then, he painted a target over each of the bullet holes. Afterward, he invited the townspeople to show them his skills. They were so impressed that they declared him the "greatest sharpshooter in the state."[12]

This often plays out with geologic analysis, drilling analysis, production analysis, and completion design variables perceived to drive production performance. With a large data set of wells, it is easy to high-grade the best wells and look for similarities. Then conclude, certain controllable parameters are responsible for production performance, while ignoring known and unknown subsurface variables.

In the oil and gas business, we are all guilty of being Texas Sharpshooters to a certain extent because it is difficult and expensive to collect all the data to paint a complete picture of what is happening downhole. There are too many variables. Therefore, we use the data we have, which is incomplete. There is an old saying in the oil business, "We will know the truth when the last well is plugged." This means everything will be known about a specific play once there are no more wells to introduce new data to prove anyone wrong.

## *Not Invented Here*

One of the most significant impediments to maximizing cash flow is the human proclivity to diminish the ideas and work of others. Learning from others is a logical approach. However, it is common among technical staff to have a strong bias towards their own ideas and refuse to learn from the successes and mistakes of others. In every cost reduction initiative I have been involved with, not invented here (NIH) syndrome has been one of the primary human prejudices negatively impacting progress.[13]

A lot of money can be made while avoiding financial pain by studying the successes and failures of others. Why learn the hard way? Especially in the oil and gas industry, where problems are extremely expensive.

Unfortunately, pride, ego, jealousy, envy, and anger all contribute to NIH syndrome, which, if not addressed, will financially destroy an oil and gas company from within.

The NIH mindset can result in reinventing everything regardless of cost. If the goal is to create value and make money, it should not matter where a good money-making opportunity originates from or who discovered a tactic to avoid expensive problems.

However, to the people suffering from NIH, it does matter. When value comes from colleagues, subordinates, adversaries, competitors, or any place that is disliked or perceived to be a threat, it is diminished, disparaged, and dismissed. This reactive devaluation bias, in which other people's ideas are diminished, is standard NIH syndrome.

## *Backfire Effect*

During our cost reduction journey, I will present information to challenge current beliefs about shale oil and gas operations. When presenting information that questions current beliefs, it is not uncommon for people to more strongly support their original beliefs. This is the backfire effect. People initially reject new information that contradicts their current beliefs.[14] If you identify strongly with a current shale operational practice, and a different approach is presented, it is human nature to immediately reject or ignore it, no matter how much data is presented.

## *Reactance*

During cost reduction efforts, reactance is a typical response. Reactance is the tendency to do the opposite of what is asked.[15] People do this because nobody likes to lose freedom. When humans feel that they are forced to do something, especially if they don't want to do it, they become hostile and resist. The more we push, the more they push back.

In many cases, this hostility is enacted in a Machiavellian way so that it is not openly apparent. Remember, nearly 20% of the workforce is taking action against the interests of their company.[16] During cost reduction efforts, this percentage is likely to be much higher.

It is a natural human reaction to push back when a sense of freedom of choice is limited, or we are told what to do. A reflex action occurs. After big meetings where new rules, regulations, procedures, or initiatives are rolled out, it's common to hear banter among the team, which can be summarized into one sentence: "Don't tell me what to do!"

We need to be highly self-aware so that when we feel potential reactance, we address it individually and internally. As detailed in Chapter Two: Shale Economics, a review of the "27 Reasons for Shale Cost Reduction" was partially motivated by a desire to address reactance.

If just one of those reasons resonates with each member, the probability that the team will experience reactance is significantly reduced. When people have a vested interest in cost reduction, cash flow, or any big-picture objective, individual bias can be mitigated to help move our company towards its strategic goals.

## *Tribalism*

Early in my career, over 20 years ago, I worked overseas, often on shale exploration projects in which we utilized U.S. shale technology to test new unconventional targets. In one remote outpost, the country manager told me that if I wanted to have a good experience, I should not discuss three things, "Don't engage in the discussion of religion, politics, or money, and you will not have any problems working here."

He was right, and I avoided discussing these topics with locals or other expatriates. Religion, politics, and money are highly sensitive subjects where people have strong beliefs. I witnessed other Americans engage in heated discussions on these topics, and it never ended well. Calm discussions on religion or politics often escalated into screaming matches, sometimes resulting in individuals being relieved of duty.

Deep loyalty to one's group is considered tribalism, and it can have positive and negative results within an organization.[17] Humanity is tribal and evolved over thousands of years with a preference for forming tribes. History has taught us that human survival was and is very much based on forming tribes. Group loyalty and cooperation in achieving the tribe's goals are rewarded. Disloyalty is often punished. Humanity evolved with this framework, and the modern human mind has a tribal bias.

All humans have a tribal mindset which can be an asset. However, sometimes tribalism can result in selective exposure to information or biased processing of new information. This can impact cost reduction in multiple ways. For example, if an anti-cost reduction tribe forms within the company, they may avoid seeking information to reduce costs or process cost information with a tribal bias. Objective facts may be dismissed or not viewed with a rational mindset. Inconvenient facts might be ignored if they do not support tribal beliefs.

Years ago, I managed a cost reduction initiative across multiple basins in the United States, where tribalism was a major problem. Within one of the basins, engineers, geologists, and field operations staff were strongly opposed to cost reduction. They felt costs did not need to be lower and could not be any lower than current levels. As a result, they resisted all efforts of the cost reduction initiative.

A lot of time and energy was invested within that basin to help them reduce costs. The anti-cost reduction tribe was not interested. Working with these folks individually was productive, but when they bound together, it was nearly impossible to get anything done. The tribe thought they could fight the cost reduction initiative until it ran its course and the company moved on, but they were mistaken.

The decision was to release all anti-cost reduction tribe members in the office and in the field. Then move forward with the cost reduction effort. This anti-cost reduction tribe was shocked that they were all fired. They could not believe it because they thought their experience was irreplaceable.

If our company must generate positive free cash flow, remove individuals who do not support cost reduction. Shale operations do not work without ongoing cost reduction. If we do nothing, costs will increase to the point of making our project uneconomic.

Cost creep is always occurring. If someone has a tribal belief against cost reduction, it will be almost impossible to convince them otherwise. It's not worth the time or effort. Do yourself a favor and cut anti-cost reduction tribes out. Just make sure you get them completely out, including and especially all anti-cost reduction field personnel.

## *Bystander Effect*

In an emergency, the probability that someone will take action decreases as the number of people present increases. This is the bystander effect, and the most well-known example occurred in Queens, New York in 1964, when Kitty Genovese was murdered while many neighbors witnessed it unfold. Not one person took action.[18]

The same situation takes place in corporations with layers of management and executives. If the corporation is struggling financially and no one takes significant action to save the entity or project before it is too late, the bystander effect may have contributed. Individuals are less likely to take timely action when other people are present, especially in workplace settings when everyone thinks someone else will take responsibility and solve the problem. This is also one of the reasons why most people do not enact Stop Work Authority.

This diffusion of responsibility may occur because people feel someone else is more qualified to act, their contribution is unwanted, or it is not worth the risk and effort. In terms of cost reduction efforts, the bystander effect is common. Often, it is unintentionally encouraged by the corporate culture or by specific members of management.

## *Delusion*

Fixed beliefs that conflict with reality are considered delusions.[19] It is easy to become delusional in the oil business because much of what we deal with is in the subsurface. It can take time to determine what is real and what is not. This uncertainty can permeate through everything, including costs.

> **See things the way they are, not the way we think they are or want them to be.**[20]

It is not uncommon for undesirable situations to damage cash flow by increasing actual costs or negatively impacting production performance. Delusion can cause us to convince ourselves that these situations are one-off "freak events" that should not have occurred and will not occur going forward. Therefore, actual costs are adjusted, and production performance is filtered to remove these undesirable realities, further adding to a self-reinforcing delusion.

## *Hindsight Bias*

Many things that are hard to predict occur in the oil and gas business. However, these events are often perceived as predictable after they occur. This phenomenon is known as hindsight bias.[21] When people convince themselves they can accurately predict the future, it usually leads to making low-quality decisions and ignoring risk.

The subsurface contains significant variability, and every well is different. Hindsight bias is pervasive in the industry because the mind tries to make sense of the variability with incomplete information.

Maximizing cash flow with cost reduction is a solid strategy to address this uncertainty, mitigate risk, and deal with incomplete information. Although it is impossible to predict the future consistently, documenting, in writing, how decisions are made can help mitigate hindsight bias and should be incorporated into a thorough cost reduction campaign.

## *Sunk Cost Dynamics*

Sunk cost dynamics play a significant role in shale operations and related decision-making. A cost that has occurred that cannot be recovered is considered a sunk cost. When people continue with an operation solely due to previously invested resources, it may be a sunk cost fallacy.[22]

Sunk cost dynamics can manifest within the shale business in various ways. For example, acquiring high-cost acreage for development that later turned out to be less desirable than initially thought has resulted in operating companies drilling to hold acreage that should have been allowed to expire.

During drilling, sunk cost dynamics can manifest when deciding to TD a well early due to drilling difficulties. It is not uncommon for the reservoir department to view all costs incurred as sunk costs and recommend to drill ahead no matter how much money has been spent or level of drilling difficulty encountered. Asking an engineer if you should keep drilling is akin to asking a barber if you need a haircut.

A similar situation occurs during fracturing when dealing with a difficult stage or interval within the lateral. It is almost sacrilegious to skip a frac stage. The mindset is that all costs are sunk costs, and we should make decisions based on forward economics. The problem with this approach is that, eventually, the company will run out of money. This mindset has contributed to multiple companies' financial demise.

---

**If the goal is to maximize FCF, then in specific cases, TD'ing early or skipping stages is the best option.**

---

Remember, our goal is to reduce risk and make the maximum amount of money. The goal is not to reach TD on every well no matter the risk, reward, and associated cost. The same goes for placing 100% of the job on fracturing treatments.

The goal is not to place 100% of the sand regardless of the risk/reward ratio. The goal is to make the maximum amount of money, not take the maximum amount of risk. High-risk, high reward is not a prudent strategy. We want low-risk value-creation opportunities. This business is risky enough.

## *Law of the Hammer*

If all we have is a hammer, everything looks like a nail. This is the essence of the law of the hammer: an overreliance on a familiar tool to solve all problems.[23] Law of the hammer is pervasive across the industry, and each discipline has its own favorite tool to solve problems, no matter what they are.

## Oil and Gas Problem Solving

| Discipline | Favorite Hammer |
|---|---|
| **Drilling** | Longer Laterals |
| **Completion** | Bigger Fracs |
| **Production** | Choke Size, ESPs |
| **Geology** | More Logs and Pilot Holes |
| **Land** | Expanded Area of Interest |
| **Geophysics** | Advanced Seismic |
| **Petrophysics** | Additional Core |
| **Reservoir** | Science Experiments |
| **Management** | Better Talent |
| **Executives** | Positive Free Cash Flow |

We all have our favorite hammers. Maximizing cash flow with cost reduction requires a combination of all hammers.

## *Ostrich Effect*

Avoiding information that is perceived to be unpleasant is considered the ostrich effect.[24] This cognitive bias is present in the shale industry when people tout high-performing wells or big IP's but stick their head in the sand when it comes to cost overruns or poor well results. This bias is also relevant regarding near-miss events.

High-cost situations are an opportunity and should not be swept under the rug or ignored. Information avoidance is not going to help reduce costs; it is going to contribute to further problems. When running multiple rigs and bringing many wells online, it is common to ignore the undesirable wells and focus on the great performers. Staying positive is essential for morale. However, the issue with this approach is that if problems are not addressed, they often expand, especially when dealing with near-miss occurrences.

## *Status Quo Bias*

A preference to maintain current methods because any change is perceived as negative is considered status quo bias.[25] Once operators and engineers find a method that works, they are highly reluctant to change. All operating teams have status quo bias because it is very hard to get to a point where the shale machine is working with few operational problems.

When we find something that is working, we are very hesitant to change it. The problem is that although it may work operationally, it might not work financially. There is often an economic disconnect.

With money, anyone can hire contractors or employees to drill and complete wells in the United States. Money solves most problems in terms of delivering wells. Teams do not need to be exceptional to accomplish this.

However, consistently delivering highly attractive economic wells separates exceptional operators from the herd. The problem with status quo bias is that even if current results are economical, over time margins get eroded. Think of the "27 Reasons for Cost Reduction." Sticking with the status quo virtually guarantees our assets will become uneconomic. Evolve or get eliminated.

### *Authority Bias*

In 1961, Stanley Milgram, a professor at Yale University, conducted an experiment in which participants were instructed by an authority figure (scientist in a lab coat) to administer shocks to a person if they gave incorrect answers to questions. The perceived voltage ranged from 15 to 450 volts. The participants administered harmful shocks to the point that they thought the person receiving the shocks was dead. They did this because an authority figure told them to do so. This is an extreme example of authority bias.[26]

We are influenced by people in positions of authority and tend to attribute more weight to their views, regardless of content. Throughout my career, I have worked with many authority figures in the industry, from technical subject matter specialists to benchmarking and turnaround experts. Some added value, others destroyed it.

Many years ago, a technical subject matter expert was guiding the company I worked for toward expensive manmade proppant. At the time, we were running over 100 rigs, and the decision would have increased CAPEX hundreds of millions. Of course, if it worked, it would have been worth the investment. However, we had recently tried this method and not only was it much more expensive, but production-performance was the same or worse than pumping low-cost natural sand.

A meeting with management and technical staff was called and the expert presented his case to transition to manmade proppants. He had done this successfully for other operators, and the data he presented was convincing. The problem I had with it was primarily the cost. I was surprised that no one across the company was challenging him. Maybe they were suffering from the bystander effect.

Therefore, I decided to challenge his proposal in front of the entire team at the meeting. The expert became rattled, then angry, and did not do a good job defending the technical or economic merits.

Challenge authorities to determine depth of knowledge and biases. When working with subject matter experts on problem-solving, asking detailed questions on the problem is a useful method to determine depth. If someone has truly struggled with a problem, they will know the answer to extremely detailed questions. If they were managing the process but not actually grinding out the solution or involved in the intricacies of the problem, they will not know the answer to extremely detailed questions.

## *Outcome Bias and Normalized Deviation*

Outcome bias and normalized deviation should be discussed together due to the interlinked role they play in near-miss events. Outcome bias is assessing a decision based on its results or outcome rather than the quality of the decision.[27] Normalized deviation leads us to accept deviations from correct behavior as they are repeated without incident.[28] Over time, these deviations become normal operating practices.

Texting while driving is a perfect example of outcome bias and normalized deviation. Everyone is doing it and not having an accident, so the behavior is normalized, even though it is common knowledge that driving while texting is six times more dangerous than drunk driving.[29] People drive and text successfully, with minor near-misses, which has become normalized; that is until a catastrophic accident.

In all industries, near-misses occur. Unless addressed thoroughly, near-misses will cause people to take more risks because of outcome bias and normalized deviation. Near misses must be captured, discussed, and permanently woven into the company's intellectual capital to prevent dangerous thought processes from manifesting into expensive and painful events.

Artificial intelligence is a valuable tool we can use to address near-misses and the associated cognitive biases. To capture the value, we must document their occurrence. As previously discussed, consider documenting near-misses in our Operational Issues List. We can then program the details into our Virtual Company Man to alert our team.

---

# ACTION
## Identify Cognitive Biases

---

Decisions determine our success or failure. Information and its interpretation when making decisions contain cognitive biases. The oil industry is highly susceptible to these errors in thinking due to the uncertainty in the subsurface environment.

The human brain has limits, revealed as cognitive biases when attempting to simplify complex subsurface data. Once we learn to spot cognitive biases, we will see them everywhere, from the wellhead to the boardroom.

Bias can determine success or failure depending on how it controls our perception of reality. Enhancing self-awareness by reflecting on one's thoughts can add significant value. Consider running through a mental checklist of the cognitive biases we reviewed to identify the level of impact on your thinking. Often, in the oilfield, we feel the need to make quick judgments and decisions. Unidentified biases can make it seem like we are making the right decision at the time, only to be disappointed by the results. After the situation unfolds, things look obvious, and our decision looks flawed.

If we think deeply about the dynamics during the analysis and decision-making process, was bias a factor in our decision? Most likely, it played a part. When we are in these situations, it pays to consult others to get different opinions and viewpoints. This process can help identify bias by talking through various scenarios to make the best decision with the information available.

--------------- **ACTION** ---------------
## Identify Intelligence Traps

It is common for highly educated people to be more exposed to risk and not manage it properly. Intelligence, university degrees, credentials, expertise, and experience can work against us by increasing hidden bias and flawed logic. When discussing matters, highly intelligent people can better articulate and defend their position, convincing others of a course of action.

When things do not go as planned, intelligence can destroy additional value when people are unable to learn from mistakes because intelligence is used to convince others and themselves that it was not flawed thinking but some other factor. This is commonly referred to in the oilfield as outsmarting oneself or, academically, as the "intelligence trap."[30]

The term was first used by Edward de Bono during talks and in his books on thinking as a skill.[31] The oil business is suspectable to the intelligence trap because we cannot see what is happening in the subsurface, and many aspects are uncertain. Therefore, the person with the strongest opinion or the most believable, gets their way, regardless of whether it is the optimal course of action.

Any position can be defended convincingly. The higher the level of intelligence, the better the ability to do this. However, this is not good decision-making or risk management. If we are looking for truth, consider building systems to explore all potential scenarios regardless of bias, interest, or believability.

# ——————— ACTION ———————
## Expand Transparency

Consider increasing transparency to reduce risk and costs across the company, from the field to the office. Seems easy, but the nature of oil and gas operations makes it challenging. Remote work environments, with a rotational workforce consisting mostly of contractors and vendors, make it structurally difficult but not impossible.

If the shale industry had a fixed traditional manufacturing facility, it would be easier. For example, we could place management and engineering offices directly on the factory floor, which many top manufacturing companies do. Thereby maximizing transparency, communication, and understanding.

Seeing what is happening in the oilfield is not easy. The problems are complex and not captured in reports or morning meetings. Daily operations reports are not what is happening on location. They are a biased, redacted version of the truth. The more time we spend in the field, the better we understand this dynamic.

When engineering and management hear about a problem from the field, the issue does not look the same compared to seeing the problem directly. It is very different. From the office, it often seems that the people in the field do not know what they are doing, which is untrue.

Because of this and other dynamics, the field often works to keep problems from the office and hide things. The mentality is, "The office does not need to know what is happening out here." I have heard this statement or a similar derivation at every company I have worked for when I spent a significant amount of time in the field to the point that people were comfortable speaking to me openly.

In the past, I have been mistaken for a roustabout because I would sometimes perform manual labor alongside my field team. As a result, contractors, vendors, and others spoke openly in front of me. This removal of the human tendency to filter information is hard to achieve in a large corporation, so we must do everything to remove dynamics that prevent the truth from reaching executive leadership.

> **Making decisions without the whole truth**
> **results in undesirable outcomes.**

Doing everything possible to maximize transparency will help mitigate this dynamic. The more transparency exists within an organization, the more respect and trust can develop, enabling communication regarding what is happening on both sides of the operation.

If the field is not transparent about the problems they are dealing with, we cannot address them and manage the risk. This is important because we will be making many changes to reduce costs. Leveraging modern technology can greatly improve transparency. Regardless, there is no substitute for directly seeing the issues in the field or having experience dealing with them firsthand.

## Actions to Maximize Transparency

- Automate and systemize transparency with technology.
- Expand the use of the "Operational Issues List."
- Build the Knowledge Bank for our Artificial Intelligence.
- Construct the "Solutions Book."
- Increase, improve, and encourage open communication.
- Share the maximum amount of information.
- Eliminate secrets.
- Provide opportunities for vigorous debate across the team.
- Install cameras on location.
- Implement the "Daily Visual Report."
- Have field employees attend management meetings and vice versa. Mix it up between leadership and field hands.
- Supplement standard daily reports with operations-specific detailed data capture reports.
- Engage the workforce for solutions and implement them.
- Eliminate layers of management.
- Identify and eliminate fiefdoms.
- Utilize communication tools, including mobile apps.
- Capture the maximum amount of operational data with enhanced telemetry combined with drilling, completions, and production applications, including data analytics.
- Visit field operations regularly and randomly. Establish a presence of managers, engineers, geologists, and other office personnel on location.
- Probe deep into operations and identify problems.
- Establish transparency key performance indicators (KPIs).

We do not want to be in a situation in which we think everything is going great, the numbers look amazing, but the operation is teetering on the brink. The NASA Space Shuttle Challenger accident is an example of this type of situation.

Viton rubber O-ring failures on the solid rocket boosters occurred multiple times before but the launches appeared successful. Success blinded NASA executives and contractor, Morton Thiokol, from seeing the correlation between ambient temperature and O-ring failures.[32]

The decision-making process, communication structure, safety factors, and culture were flawed. Looking at the launches and successes before the accident would lead us to believe everything was going great with the Space Shuttle program. However, the results do not tell the whole story. It is a very common mindset on oilfield locations to think and say:

*"You can't argue with success!"*

I hear this flawed statement all the time during discussions and disagree with this type of thinking and mindset. There may be hidden problems, near-misses, and other undesirable actions or circumstances in the field that are obscure, unknown, or not addressed. The information may be hidden or suppressed from reaching leadership. A successful outcome does not mean our operation is a success.

What if we had a massive near-miss that only a few know about? A successful outcome may further bury inefficiencies and risky actions from the light of day, contributing to even higher levels of risk and associated costs.

───────────── ACTION ─────────────

## Capture Intellectual Capital

Institutional memory is the unrecorded collective wisdom of our company.[33] This includes all the tools, tactics, techniques, methods, and knowledge acquired over many years, often through spending billions of dollars. This knowledge outlasts the people who first developed it as long as there is enough overlap between the old and new people circulating through the company.

The problem with institutional memory in the oil and gas industry is that due to heavy contractor and subcontractor dependence, the rotational nature of positions (days-on to days-off), the 24-hour operating cycle, and heavy turnover (extremely high on the services side), a lot of institutional memory is lost.

Additionally, a good portion of institutional memory is tacit knowledge, which is intelligence that is difficult to put into words and includes instincts, intuition, experiences, habits, and know-how.

It is common for people to be unaware that they have this knowledge. Therefore, it is challenging to capture and share. Successful shale operations are full of tacit knowledge, stuck inside the heads of the 350+ people involved, often the critical factor between economic success and failure. It is the experience in the minds of the team that provides a competitive advantage and reduces risk.

Another dynamic I would like to address is knowledge that

is intentionally held back. Many contractors do not want to share all their knowledge because they feel that if they do, it is easier for the operator to replace them. Much of the competitive advantage between vendors is tacit knowledge, so there is an incentive to hold it back.

The thinking is, "If I tell you everything, then you won't need me. This knowledge is what I bring to the table." I hear some version of this every week in the field and in the office with contractors.

Therefore, regarding some of this hidden high-value knowledge, whether tacit or not, there needs to be an incentive to share. For example, years ago, I was at a shale artificial lift conference and in one of the sessions on lift selection and engineering, the speaker handed out hard copies of the presentation. When he got to the most critical slide, I tried to follow along in my handout, but the page was blurry. It was unreadable. I leaned over to several colleagues and asked to see their handouts. It was the same.

When it was time for questions, I mentioned to the presenter that the slide with the most critical information came out blurry; maybe there was a problem with the photocopy machine when the presentation was printed? What he said next was a shock and revelation to everyone in attendance. He said, "No, there was no problem with the copies. I made that slide blurry intentionally. Next question." Wait a minute, I said. "Why did you do that?" He said, "If I give you everything, how am I supposed to make a living?"

Bam! Suddenly, it became clear. He wanted to be hired for consulting work. That was the incentive. In this case, the solution was easy. It's harder when a good amount of the knowledge we need to capture is from people already working for us, and we are already paying. Money is the easiest way to incentivize people, but it's not always possible.

The goal is to turn institutional memory into intellectual capital, intangible assets that create value for the organization. Operators spend billions of dollars annually to drill, complete, and produce wells. This generates tremendous amounts of highly valuable knowledge, often lost because minimal effort is taken to capture and leverage it to maximize FCF. We need this data to help reduce costs and facilitate innovation.

## Actions to Capture Intellectual Capital

- Leverage the "Knowledge Bank" and "Solutions Book."
- Use mobile apps and digital technology to increase written communication and social interaction between people.
- Audit every operation and vendor operation comparing written processes versus actual practice.
- Shadow people. Hold "Show-Me in the Field" sessions.
- Implement reverse-mentoring in which a contractor is assigned as a mentor to a Sr. level member of the operator.
- Conduct video recordings of operations and review them.
- Encourage storytelling (war stories) about problems, solutions, successes, and mistakes during meetings / calls.
- Start internal knowledge videos, blogs, wikis, and podcasts.
- Measure intellectual capital with financial data to track the progress of capturing and growing knowledge.

# ACTION
## Build Checklists

One tactical method to capture valuable information and incorporate it into the enterprise quickly is to expand the use of checklists. This can be a simple one-pager with bullet points and a title for a given task or operation that we want to address. For example, if we want to strengthen frac stack swap-overs on simul-frac or super zippers, which can have accidental shut-ins costing millions, consider adding checklists to our process. After every operation, tweak the checklist with additional knowledge, rules, actions, and reminders that address communication, equipment, and inspection.

Procedures and checklists should be living documents that grow with our company. Over time, these systems, brought into existence in the form of a checklist, become highly valuable assets. To acquire this knowledge costs millions and a lot of risk exposure over many years. Don't lose it.

When enhancing written tools, we want to be careful not to create bureaucracy with additional paperwork or other aspects that get ignored. Therefore, try not to make checklists robotic. They must connect with our teams' value system and be human-friendly. Do not hesitate to reference past situations in detail that we want to avoid. Additionally, we could digitize the structure, incorporate an app, or use other modern tech to bring our checklist to the next level. Our Virtual Company Man artificial intelligence system is designed to integrate checklists to help assist the team.

---

# ACTION

## Address Risk Transfers

---

Over the past two decades, trends in our industry have resulted in transferring as much risk as possible to the operator. This has occurred for a variety of reasons. Part of the dynamic involves taking direct control over operations previously handled by vendors or third parties.

For example, historically, frac companies provided all services related to the frac including equipment, chemicals, sand, and fuel. Due to fire risk and compressed margins, frac pumpers wanted to get out of diesel logistics and fueling operations. This risk was transferred to operators who bore the risk through new vendors but also looked to capture margin. As the industry evolved, operators began to unbundle all aspects to reduce costs, providing additional options, such as eliminating diesel for other forms of fuel and power.

However, a lot of risk was transferred in the process. As we look to reduce costs further, we must pay attention to the risk that is being transferred and put proper systems in place to manage it. A lot of analysis needs to be done to make the right decisions. For example, it is not always beneficial to unbundle everything all the time. Risk must be calculated and assessed.

Additionally, we need to address internal aspects of risk transfer between divisions. The following chart identifies a few critical items to consider strengthening that are handed off between groups within an operating company, as well as the associated risks transferred.

## Internal Risk Transfers between Operator Divisions

| Division | Items | Risk Transferred |
|---|---|---|
| **Field Construction** | **Pad and Road** | Location structure, stability, quality, stormwater drainage. Risk transferred to drilling, frac, production. |
| **Drilling** | **Casing Quality** | Production casing quality or casing that was drilled through which will see frac pressure. Risk transferred to frac operations. |
| **Drilling** | **Casing Make-up** | Correct casing running / install. Risk transferred to completion operations. |
| **Drilling** | **Wellhead** | Correct wellhead installation which will see pressure and connect to frac stack. Risk transferred to frac ops. |
| **Drilling** | **Cement Placement** | Cement quality & placement to enable isolation and plug-and-perf operations transferred to frac. |
| **Drilling** | **Lateral Placement** | Lateral placed in the optimal geologic target. Risk transferred to frac and production operations. |
| **Drilling** | **Wellbore Trajectory** | Trajectory enables completion operations and artificial lift plans. Risk transferred to production. |

## Internal Risk Transfers Continued...

| Division | Items | Risk Transferred |
|---|---|---|
| Fracturing | Perforation Placement | Correct placement of perforations for long-term production. Risk transferred to production operations. |
| Fracturing | Frac Plugs | Optimal plug selection for risk reduction during millout. Risk transferred to post-frac millout and flowback ops. |
| Fracturing | Treatment Chemicals | Chemicals administered for $H_2S$, scale, gummy bears, and other long-term production issue prevention. Risk transferred to production. |
| Millout | Plug Removal | Proper plug removal. Risk transferred to flowback and production operations. |
| Millout | Sand and Debris Removal | Removal of sand and debris to safely isolate pressure, install completion packer and/or flow well successfully. |
| Facility | Equipment | Correct equipment installed properly with necessary specs. Risk transferred to flowback and production. |
| Facility | Build Quality | Facility construction risk transferred to flowback and production operations. |

———————— **ACTION** ————————
## Establish Zero Tolerance Rules

I give people multiple chances when mistakes are made, even when I explain exactly what needs to be performed and the person does the opposite. However, at some point, if running a safe, low-cost operation is critical to the ongoing success of our company, critical aspects of the operation must have strict rules. Rules that, if broken, result in disciplinary action, including personnel release, loss of future work for vendors, or being permanently banned.

In the Oilfield Survival Guide, I review 50 principles for working in our industry based on over 1,000 situations. Many should be considered zero-tolerance. However, the details regarding specific rules should depend on company concerns, past problems, and current operations.

For example, if we have a problem with consultants exceeding max pressure during frac jobs, which could result in destroying $8.0 million wells, consider establishing max pressure breaches as zero tolerance. Therefore, the next consultant to exceed max pressure on a frac will be released. Word travels fast; based on experience, this problem will be less of an issue going forward.

Another common high-cost problem for zero-tolerance rules involves personnel lying or withholding information. Lack of timely information often results in high-cost events. It is common for undesirable events in the field to be left out of the daily reports, not conveyed to leadership, or explained in a less-than-truthful manner. It is also not uncommon for people to flat-out lie about stuff that happens in the field. Many people who work in the field believe that what goes on in the field stays in the field. It is not anyone's business, especially the office personnel. Addressing this mentality is critical to consistently run a low-cost operation. We cannot afford for people to conduct business like this because these situations will turn our economics upside down, and we won't even know what happened.

Twisting the truth, withholding information, and lying should be zero-tolerance. For example, recently, during a frac job for an offset operator, one of the consultants accidentally closed a master valve on a well that was being fractured. Pressure did not exceed max iron ratings according to the gauges, so the field team thought everything was okay. The lead consultant did not report it to management in the office. When office personnel found out, all the consultants on location were released. It was a difficult decision because several of them were excellent and had been with the operator for over a decade. They were also good friends with company management. It was a hard decision all around, but it was the right decision, and it sent a message to everyone.

If you hide information, you will be terminated. It does not matter who you are or how close you are to management. In this case, none of the consultants reported that one of the them made a mistake and shut-in the active well. I guess they were trying to protect their friend in the field. However, this action suggests that they will hide information, and that cannot be tolerated.

# Zero-Tolerance Rules (example list)

- Safety, environmental, and ethical violations
  - o *"If you do not use Stop Work Authority, then you'll be released."*
- Theft, kickbacks, bribes, favoritism, inappropriate use of funds
  - o *"If you use rig diesel for your truck or steal diesel, then..."*
- Negligent valve, stack, or BOP operation resulting in damaged pipe, cut wireline, overpressure events.
  - o *"If you cut wire, then..."*
- Stuck and Lost-in-hole (LIH) situations
  - o *"If you lose equipment downhole, then..."*
- Casing mistakes, including damage, CVD, handling, make-up (torque, turn, speed), tally, running, landing.
  - o *"If you incorrectly make-up the casing, then..."*
- Cement miscalculations, including design, additives, density, amounts pumped, displacement, field execution.
  - o *"If you miscalculate the cement job volumes, then..."*
- Exceeding Max Pressure
  - o *"If you exceed max pressure on the frac, then..."*
- Wireline mistakes (incorrect depths, pre-set plugs)
  - o *"If you perforate off depth, then..."*
- Supervision violations during critical operations, including not being on the floor, at the wellhead, or at the frac stack.
  - o *"If you are not at the wellhead during installation, confirming the technicians are installing components correctly, then ..."*
- Automated settings during production including safety valves, electronic shut-in set points, high-low controller.
  - o *"If the safety shut-in system is not set correctly, then ..."*
- Production facility violations, including patrolling truck during loading, using H-Braces, going on the tanks, overloading, draining liquids to the ground or containment liner.
  - o *"If you drain liquids to the ground or to the liner, you're fired."*

----------------------- ACTION -----------------------
## Form a Risk Committee

---

Due to recent events and the increased risk operators are exposed to in the world, consider forming a risk management committee within the company. The risk committee could include executives, managers, engineers, and field employees. Consider meeting on a regular basis to assess potential risk exposures and address change in our industry.

Recently, I have heard of a few shale operators creating new risk manager positions within their organization to help address and capitalize on this dynamic. This new position becomes especially critical if an undesirable situation occurs on a multi-well pad. The role prevents putting the full burden on current executives. It also protects against situations similar to those listed in the "CEO Departures due to Risk Management Dynamics" table we reviewed earlier.

> **Risk management is one of the most underappreciated processes across all organizations and industries.**

When initiating a cost reduction campaign, it is a good idea to involve the risk committee, chief risk officer, or risk managers in all aspects of the plan because we will make many changes to reduce costs. When change occurs, risk is introduced, sometimes unintentionally.

If we are unsure whether we need someone focused on risk

as their top priority, consider the collapse of Silicon Valley Bank (SVB). In April 2022, SVB's chief risk officer (CRO), Laura Izurieta, stepped down.[34] At the same time, the Federal Open Market Committee began raising interest rates. The CRO position was not filled until January 2023.[35] The bank collapsed in March 2023. Therefore, during one of the riskiest periods for this company, they had no CRO. Of course, managing risk in soft assets is not the same as hard assets, but there are many similarities, particularly with our capital-intensive industry. Therefore, the lessons are transferable.

*The Wall Street Journal* has called risk management, or the CRO, the "most thankless job" or "least appreciated role in American business."[36] The reason is that "when success means averting danger, it's hard to notice when someone's not on the job until it's too late."[37] I agree that risk management is not given the credit it deserves across all industries.

In many cases, it is forgotten about, especially when everything is going smoothly. Unfortunately, it only takes one mistake to eliminate an entire company, sometimes an entire industry.

The situation at SVB was avoidable with proper hedging (standard practice) during the months of interest rate increases before the collapse. For example, SVB could have purchased fixed-for-floating interest rate swaps or taken several other actions, but they did not.[38]

Throughout 2022, SVB decreased its hedges and had zero swaps.[39] As mentioned previously, the foundation of a maximize free cash flow strategy is strong risk management. Before executing cost reduction actions, perform thorough risk assessments. During this process, it is expected to find things that we are not comfortable with and, therefore, must address.

During cost reduction campaigns, occasionally, in the beginning, costs might increase because certain operational aspects regarding risk need to be addressed before we press on the costs and drive them down. This is part of the process of reducing risk. During our initial risk assessment, it is a good practice to dig deep until we find many things that we do not like. We do not want to go into a risk assessment hoping to find nothing—that mindset is a mistake. Dig for unaddressed risk until finding something—then dig for more. Finding and addressing risk is a core part of our cost reduction system.

The culture at SVB had the opposite mentality. It was suggested to be "one of aggressive executives ignoring what the overwhelmed, understaffed risk department had to say and forgetting the most elementary lesson in finance:"[40]

> **"There is no such thing as fabulous returns without potentially fatal risk."** [41]

In the shale cash flow game, can we take an $8.0MM well and reduce it to $4.0MM? Yes, absolutely, but the risk must be addressed. The goal is to reduce the risk and the cost. I have no interest in operating at the upper right-hand corner of the risk matrix—neither should you.

Oh, and one more thing: while the U.S. government might bail out a bank for improperly managing risk, it won't bail out an oil company. Address risk or get eliminated.

———————— **ACTION** ————————
## Bring In Subject Matter Experts

During the Manhattan Project, the World War II secret operation, with origins at Columbia University (my alma mater), to build the first atomic bomb, there were many challenges that almost resulted in the project's failure.[42]

The operation was a U.S.-based multi-site, multi-state operation, in some respects similar to multi-basin, multi-state shale operations. The Manhattan Project's three primary sites were Oak Ridge, Tennessee; Hanford, Washington; and Los Alamos, New Mexico.[43]

The Oak Ridge site produced uranium-235, the Hanford site produced plutonium-239, and the Los Alamos site built the bomb. The project employed 130,000 people at its peak. Major General Leslie Groves, who led the project, employed the smartest people he could find, including Robert Oppenheimer, a theoretical physicist selected as the director of Los Alamos.[44]

The project developed two types of bombs: a simple gun-type and a complicated implosion-type. Manufacturing both uranium-235 (Little Boy gun-type) and pluotonium-239 (Fat Man implosion-type) required uranium ore, the primary raw material for the project.[45]

Most of the uranium ore for the bombs came from the Republic of the Congo and was available because of the genius of engineer Edgar Sengier, CEO of High Katanga Mining Union, a natural resource company that operated in the Congo. Edgar proactively started stockpiling uranium ore in Staten

Island New York City, years in advance, to prevent it from falling into enemy hands.[46] About 99.3% of natural uranium's mass is uranium-238, and 0.7% is uranium-235.[47]

For the Manhattan Project to be successful, sufficient quantities of uranium-235 and plutonium-239 had to be produced quickly and cost effectively for multiple bombs. However, the Oak Ridge site was having a lot of operational problems producing uranium-235 for the gun-type Little Boy. The refining process of uranium-235 was incredibly slow, inefficient, and costly.[48]

Additionally, the gun-type bomb needed too much uranium-235: 85 pounds for the target and 55 pounds for the projectile. There would only be enough material for one bomb. It would take years to manufacture additional material for a second uranium-235 bomb. This would not meet the timely WWII need of the U.S. army to deter the enemy with the threat of multiple bombs.[49]

However, Pluotonium-239 was easier and quicker to produce in sufficient quantities for multiple bombs and only needed 13.6 pounds per bomb. Unfortunately, the simple gun-type design would not work with plutonium because of pre-detonation. The weapon would melt before coming together.[50]

This was the reason why the plutonium implosion bomb had to work for the project to be considered a success.

The problem Los Alamos had was that they could not get implosion to work. The failure of the gun-type plutonium bomb and difficulty with implosion caused Oppenheimer to consider resigning from the project.[51]

However, his colleagues changed his mind, arguing that

without him, there would be no bomb.[52]

Oppenheimer decided to do the exact opposite of resigning by taking massive action. He created two new divisions, one to address the physics of implosion and one to address the explosive lenses.[53]

He brought in many subject matter experts, going around Captain Parsons, the engineering head of the Ordnance Division, which infuriated him. Captain Parsons did not think implosion could work.[54]

If Oppenheimer had not taken action and brought in multiple subject matter experts, bypassing the head of engineering, it is possible the Manhattan Project would have failed. Oppenheimer knew he had to take massive action because we were in a race with the Nazis. There was a sense of urgency to deliver results or get eliminated, literally.

Few people would have gone around military engineering leadership to get the job done. I am sure he made a lot of enemies doing that, which may have worked against him later, but it was the only way to deliver results imminently—there was no time for anything else.

In my over two decades working on highly technical projects across the United States and the world, it is rare to see technical people ask for help or suggest bringing in new people to assist in solving problems—I have never seen it.

Engineers and operations personnel are too proud to ask for help. They often feel it is not their place to ask, as they were hired to get the job done, not ask for more people. This type of Oppenheimer-level action must come from the executives.

---
## ACTION
## Be On the Front Lines
---

Digital technology and telemetry improve every day; that's human progress, but there is no substitute for physically being on the front lines. Merging quantitative analysis from the office with real-world intelligence from the field helps us make better big-picture strategic decisions and systematically address operational risks, preventing expensive mistakes.

Traveling to multiple wellsites is time-consuming and difficult to do when there are so many things going on in the office. However, there is no way around doing this if we value firsthand information. From a risk awareness perspective, it is essential to interact with the field team and observe things directly. Secondhand information will always have bias.

A company with a frequent management presence on location conveys the importance of operations. Our team might not be fully engaged or understand the bigger picture, especially when efforts are being made to maximize cash flow with cost reduction. People do not like change, particularly when it is coming from above and is difficult to implement. I have seen this in every basin across the United States.

When the field team regularly sees leadership on the front lines, in the middle of the action, fully engaged in the battle of daily oilfield struggles, it motivates them and removes any belief that we are not serious. The benefits of our management team spending additional time in the field outweigh the inconvenience.

# Benefits from Increased Field Presence

1) Protects shareholders.
2) Firsthand knowledge leads to better decisions.
3) Reexamine assumptions in business and financial models.
4) Eliminate field-to-office information filters and bias data.
5) Shape 360° opinion on contractors and processes.
6) Identify value creation opportunities.
7) Connect to the daily struggles of our team.
8) Get viewpoints and talk directly with the workforce.
9) Solidify field culture with a regular presence on location.
10) Manage risk with a complete picture from the front lines.
11) Get a good feel for what's going on.
12) See problems in real-time and solve them in real-time.
13) Keep leadership practical and grounded.
14) Demonstrate hands-on management vs. hands-off culture.
15) Observe things for yourself.
16) Get different perspectives from multiple settings.
17) Provide detailed oversight on core business and personnel.
18) Understand intricacies and hidden obstacles.
19) Assess employee abilities and intelligence on location.
20) Identify exactly who needs to be released or promoted.
21) Learn precisely how things are working.
22) Absorb tacit knowledge.
23) Get into the details of the business.
24) Show importance with actions, i.e., our engaged presence.
25) Establish a competitive advantage.
26) Answer questions and address concerns.
27) Personally thank individuals for their hard work.

Spending significant time on location may be difficult if there is a lot going on in the office. If having a regular presence is not possible, find people you work with in the office whom you trust to do this work and report back. The dynamic in the field, especially for large entities, is that valuable information is often suppressed from reaching the executive level. Much of this information has to do with risk.

To address this, many companies have office engineers go to the field on Fridays. Unfortunately, this process has been abused. In fact, it can destroy value and credibility. The reason is that it's common for engineers to work in a game of golf, sporting clays, a nice lunch, or other events with a vendor while they take a tour of field ops before heading home early for a nice long weekend.

The field visit is "fake news," and everyone in the field knows it. The Friday "show-my-face-tour" is often taken like a safari vacation to an exotic location to observe the oilfield workers in their natural habitat. Some tourists don't even get out of their truck.

Rarely will the office tourist get deeply engaged in a complex operation, help make the field team jobs easier, or provide solutions, nor does the tourist want to find anything that could jeopardize weekend plans.

Friday is the worst day to visit the field for this reason. It's better to show up randomly and frequently. Go to critical operations where there is risk and engage deeply in the details. When making operational changes, there must be oversight to ensure the process is going as planned and there are not high-risk situations or individuals exposing the company.

As a former independent oil and gas operator, I could not afford problems. But who can? Based on experience, it is more effective to show someone how to do something rather than tell them. Increased front line presence enables us to enhance our systems and incorporate what is actually happening into our written procedures and artificial intelligence.

People used to always ask me why I spent so much time in the field on location, in shops, pipe yards, and manufacturing plants. I usually answered with something like this:

---

▶ **Why does Elon Musk sleep on his factory floor?** [55]

▶ **Why was Robert Oppenheimer hands-on in assembling the first nuclear weapons?** [56]

▶ **Why did Rockefeller build his own barrels?** [57]

**Because they could not afford to fail, neither can I, and neither can you.**

---

As we expand operations with more rigs and wells, we can leverage economies of scale to achieve cost advantages and efficiencies. We can expand our workforce to help oversee operations and manage contractors. However, in our business, the risk expands too, sometimes significantly.

John D. Rockefeller, one of the wealthiest people in modern history, as CEO of Standard Oil with 20,000 domestic wells and over 100,000 employees, would join his employees in their work.[58] Even at Standard Oil's peak, Rockefeller would personally diligence vendor invoices and dispute charges.[59]

---------------- **ACTION** ----------------
## Establish Clear Responsibility

---

Our friend, Admiral Rickover, one of the world's greatest engineers, twice awarded the highest civilian honor in the United States, the Congressional Gold Medal, believed in detailed oversight and personal responsibility on all issues across his operations involving thousands of workers, contractors, and subcontractors.[60]

Unfortunately, over the past several years, there has been a societal mindset transition away from taking individual responsibility. This mentality has made many inroads across the world. The mindset shift away from personal responsibility in society and popular culture has intensified this dynamic in the industrial workplace.

Within business and industrial settings, the transition is not 100% positive regarding risk management and performance. When things go well, people continue to demand individual recognition and financial compensation. However, when undesirable situations occur, people look to diffuse blame and place it on the group, upper management, the company, the weather, or the subsurface environment.

When looking at the big picture, many companies may unintentionally encourage this type of diffused responsibility behavior by not establishing clear responsibilities for each individual. In the oil and gas industry, these dynamics increase risk because no one is identified as responsible for specific aspects of the operation. In the field, where individual responsibility is critical and every second counts, this mindset will contribute to delayed action or no action.

In 1961, during congressional testimony on safety and personal responsibility, Rickover said:

*"Responsibility is a unique concept: it can only reside and inhere in a single individual.*

*You may share it with others, but your portion is not diminished. You may delegate it, but it is still with you.*

*You may disclaim it, but you cannot divest yourself of it. Even if you do not recognize it or admit its presence, you cannot escape it.*

*If responsibility is rightfully yours, no evasion, or ignorance or passing the blame can shift the burden to someone else.*

*Unless you can point your finger at the man who is responsible when something goes wrong, then you have never had anyone really responsible."* [61]

In today's world, it is unpopular to point the finger at one person. Establishing individual accountability is often frowned upon by today's expert consultants and risk managers as not promoting a team environment.

While both sides have valid concerns, I lean towards Rickover's view on personal responsibility and suggest establishing clear responsibilities for each individual, especially those responsible for significant aspects of the operation including consultants, foreman, superintendents, geologists, and engineers.

In 1979, during congressional testimony on the Three Mile Island nuclear accident, Rickover said:

> *"The practice of having shared responsibility really means that no one is responsible."* [62]

I agree with this statement. It is prevalent in the oilfield in that shared responsibility diffuses responsibility across multiple people, resulting in situations in which no one person really cares or loses sleep if something goes wrong.

However, multiple people can be 100% responsible for the same thing, like the common oil and gas practice of having two tested barriers for pressure control. Sharing responsibility between two people is not the same thing as two people being 100% responsible. As Rickover said:

> *"You may share [responsibility] with others, but your portion is not diminished. You may delegate it, but it is still with you."* [63]

This is critical to understand in the shale business because there are 350+ people involved in daily operations per wellsite, and a lot of responsibility is delegated but that does not remove the engineer or wellsite consultant's responsibility to make sure the work is performed correctly and safely. You cannot delegate responsibility and then wash your hands of it. You are still responsible.

Rickover went on to say in his testimony to U.S. Congress:

*"If responsibility is rightfully yours, no evasion, or ignorance or passing the blame can shift the burden to someone else."* [64]

This is critical to understand because in the shale industry people claim ignorance or pass the blame when things do not go as planned. During large scale operations, it is not uncommon to hear, "I have never seen that before" or "This is the first time that has happened" or "Man, I can't believe it, this never happens," particularly from contractors and vendors when something does not go as planned.

I hear statements like this almost every week. It's cringe-worthy to me but commonly said by people looking to claim ignorance or diffuse responsibility to the unknown and unforeseeable.

During cost reduction efforts, we will look at innovative technologies and methods. It is prudent to establish clear responsibility during the planning and discussion phases. Demanding accountability commands respect, helping us reduce costs beyond current benchmark levels.

Establishing exactly who is responsible for every aspect of our operation will help command everyone's full attention and incentivize service providers to bring the absolute best they have to offer in terms of personnel, equipment, and time, which is always in short supply.

Structuring clear responsibilities and expectations is a strong nonmonetary incentive that should be utilized to create value, reduce risk, and reduce costs.

---
## ACTION
### Enhanced Human Selection
---

Having the right people, as discussed in Chapter One, is the most critical aspect of our strategy to increase shareholder value by maximizing FCF and reinvesting the proceeds into high-return assets.

In this action item, "Enhanced Human Selection," let's dig deeper into several positions critical to success, including foreman, superintendents, engineers, geologists, and most importantly, wellsite leaders.

The wellsite leader, often called the Company Man or Wellsite Consultant, is typically a day-rate contractor working on a rotational basis and is often the highest-paid person on location. This person is continuously on the front lines and is the face of our company on location. The wellsite consultant can operate with a significant amount of autonomy due to the nature of the business.

Bad decisions, mistakes, inattentiveness, or lack of detailed and timely communication by wellsite consultants can impact our company severely, to the point of elimination as a going concern, which has been the case for many operators.

From the corporate executive position, decisions by the wellsite consultants can result in an executive getting terminated, as was the case for BP CEO Tony Hayward.[65] Think about that for a moment. The CEO of one of the world's largest companies was forced to resign in shame because of the decisions of the wellsite leaders on just one of the many wells

BP operates. This is not a unique situation regarding the impact key personnel can have on corporate executive leadership and the financial future of an enterprise.

I know of several relatively recent situations where wellsite consultant decisions and actions resulted in the financial destruction of multiple shale oil and gas companies. These situations are more common than generally thought.

In speaking with colleagues across the industry during times of operational difficulty, I often ask the question, "What was your process for selecting the wellsite consultants?"

The response is usually of confusion, then anger, "What do you mean? I didn't choose the guy. I don't know him! I don't know the guy, damn it!"

Herein lies the problem. It is common and accepted in the industry that there is not a structured, robust process by the operator in vetting and selecting wellsite consultants, arguably the most critical position on location. Sadly, more diligence and vetting go into selecting a cup of coffee than into selecting a wellsite consultant.

It is common practice for the wellsite consultant to be recommended or placed by a firm, referred by a friend or former colleague, suggested by another vendor, or handled by the field office. Often, they just show up at the last minute to fill a void. From where exactly? No one quite remembers, especially when things go wrong.

There are many paths a wellsite consultant can take to get on location. Unfortunately, almost all of them entail little, if any, structured diligence and detailed vetting by the operator. The current industry practice and business model for sourcing and placing wellsite consultants is full of risk.

If this dynamic is not properly addressed, managing risk and maximizing cash flow becomes painfully difficult. Weak wellsite leaders can introduce significant volatility into the process of cost control and reduction.

Since this position is crucial for the safety and financial success of every shale operator, the selection process must be rigorous, meticulous, vetted by the entire team, and controlled by the operator. Let's address several aspects of what I refer to as "Enhanced Human Selection."

## *Operator Controlled*

The process for selecting the right candidates must be controlled by the operator. Since the operator must live with the consequences of the wellsite consultant's actions or lack thereof, the operator must control the process from start to finish. Under no circumstances should this process be outsourced. I am strongly against allowing an engineering firm or any other consultant supply company to select the consultants.

There is significant risk and conflict of interest if you allow the firm that sources consultants to control the selection process. There are many dynamics at play in terms of selecting the right person, vendors, and tactics that must be navigated by the operator. Additionally, if we allow the firm to place the consultants, that person's loyalty is to the firm, not to the operator. We do not want to be in this situation. Some firms will force you to only use their selection of candidates for this very reason.

In conversations with firm owners, they openly admit that they want to use their people because that is who they trust and who is loyal to them. This makes perfect sense for them but not for the operator. This is a risk management tactic for the firm supplying consultants since the consultant operates under the firm's insurance. However, it is the operator who is exposed, especially if there is an undesirable situation.

To address this dynamic, some operators have switched back to an employee structure for the wellsite leader position, eliminating the firm. Others identify the consultants they like and then allow them to work through a list of firms that have a signed MSA, for insurance.

The operator is in a much stronger position if candidates are initially identified and vetted by the operator through an organic process.

## *Selection Committee*

When selecting a wellsite leader, several people should be involved in the decision, including foreman, superintendents, engineers, geologists, and upper management. It is common in large companies for the selection process to only include one person. This is a missed opportunity to minimize risk by leveraging multiple perspectives on the candidate and the expertise of the team.

Consider forming a selection committee that votes on the decision for each candidate. Many must interact with this person in high-stress situations and live with the decisions made. Therefore, the selection process should be inclusive, encompassing both field and office personnel across operations, geology, and management. With modern communications technology this process can be accomplished quickly at no additional cost.

## Minimum Technical Requirements

Wellsite leaders with limited experience in the basin or formation of interest make me nervous. I am not opposed depending on the situation and the person. New people to a basin, formation, or inexperienced folks can be integrated successfully with a complete understanding of strengths and weaknesses. However, it is prudent to have a list of minimum requirements that a person must meet before being considered for the role of wellsite leader.

Specific minimum requirements are a matter of company preference, but they should be formalized and incorporated into the structured selection process.

Having a copy of the candidate's current resume is crucial; you would think every operator has this. However, in the past during challenging situations in which leadership is interacting with a wellsite consultant, I have asked for a resume before the meeting and have been met with blank stares and scurrying to slap something together.

In these situations, I always wondered what diligence was done to hire this person if we didn't even have a resume. Often, the answer is—none.

## Procedure Testing

During the vetting process, consider reviewing a written procedure with the candidate. Ask questions and propose hypothetical situations. This will also help assess the person's communication abilities.

Another option is to remove several key steps in the procedure and see if the candidate can identify the omission. Skipping a step can be fatal in this business. If a candidate cannot identify a critical step that has been removed, maybe it's best to move to the next candidate.

## *Typing Test*

One of the areas I see wellsite leaders struggle with is typing daily reports, responding to emails and texts, or handling daily team app communications in a timely manner. Since this is a critical aspect of the position, consider administering a typing test to see if the candidate can type in an acceptable amount of time.

Part of the reason daily reports have become so generic is that many people cannot type. Some consultants struggle so much with this issue that they spend the majority of their day dealing with the reports.

Although this seems trivial, it can expose the operator to risk if the consultant is so overwhelmed with the reporting requirements that they cannot spend time on the more critical aspects of the job.

## *Reference Check*

Calling references is a must. Most people list three. Once we have a list of three from the candidate, ask for a few more, including the past several operators the candidate worked for.

It is common to list references from clients that are not

recent, which is fine. I just think it's best to speak with the last several operators that this person worked for to see if there are any red flags or issues.

## Vendor Check

Vendors are a great source of information when evaluating a wellsite consultant. There are many people in this world who deal with their management differently compared to people who report to them.

If a consultant has a reputation for mistreating vendors, it's a major problem. Our goal is to reduce risk and maximize returns. If the vendors, who encompass the majority of the team on location, are not treated with respect, the environment will become toxic. As a result, risk and cost go up.

## Negative 360s

A typical 360-degree feedback process is a review in which an individual is evaluated by people who work around them, including coworkers, direct reports, and managers. The process is designed to obtain multiple points of view.

A negative 360 is when we search for the most critical feedback from all angles (vendors, peers, clients) on the entity under review. When evaluating a wellsite consultant, it is helpful to find both good and bad feedback.

Wellsite consultant provided references most likely deliver the best possible feedback, which is why they are listed. That's one side of the spectrum. To get to know someone, it's a good practice to get the other side of the spectrum from the negative 360 before placing the individual on location, ultimately responsible for millions of dollars and more importantly, people's lives. Negative 360s are also a good tool to incorporate into "Vendor Analysis Deep Dives" which we will discuss in the next chapter.

## *Drug Testing*

As part of the selection process, consider robust drug testing. In addition to initial testing, hold random drug tests to identify elements of unnecessary risk. Some consultant supply firms require this, but simple drug tests are easy to beat. Consider testing using hair, which is proven to identify an array of users exposing our company to unwanted liability.

Expand random testing to employees involved with the operation, including geologists and engineers. Surprisingly, I know of several situations in which a key geologist or engineer was using heavy drugs, which significantly impacted the operation and financial future of the company.

The situation was discovered through police involvement outside of work. If random drug testing were robust within the company, the situation would have been caught before the damage was done.

## *Background Check*

People in leadership positions who directly deal with millions of dollars on a daily basis require significant trust from the operator. Trust is built over time. A thorough background check helps build that trust. I want to know everything I can about the people I work with. There is too much risk in this business not to have people who we can trust, especially when no one is looking. As an owner or executive, an unscrupulous manager, geologist, engineer, or consultant can ruin our lives if we put ourselves in that position.

## *Character Assessment*

Finding individuals with a personality that matches our company culture and values is key. The wellsite consultant sets the culture on location. A company could be the greatest place to work in the office, with the best culture in the shale industry.

However, none of that matters if the wrong wellsite leader is on location. All the hard work in building a culture and brand within the industry is irrelevant on the front lines if the company man is toxic.

If open communication, respect, safety, and efficiency are essential, those same values must be a priority for the wellsite consultant, or we will have problems. In many cases, if we do not pay close attention, these problems are hidden until after we hire the guy and send him out there.

Conducting a 360° evaluation by speaking with vendors, regulators, the hands, and other consultants that this person worked with can help determine who they are from a character perspective.

Often, just asking the question, "How are they to work with on location?" will get the information we need. I would rather have a person with the right character and temperament than the right experience. It's far easier to address experience than it is to change someone's personality.

## *Physical Abilities*

Many jobs in the field require physical abilities. The wellsite consultant is no different. Some operators overlook this issue. I require my wellsite consultants to be at the wellhead, at the frac stack, on the accumulator, on the rig floor, at the pits, on the blender, and a host of other places that require the person to get out of the trailer and walk up difficult stairs on the equipment and around location.

As crazy as this sounds, there are wellsite consultants who never leave the comfort of the trailer house, sometimes due to physical inabilities. I do not require my consultants to run a timed 5K or obstacle course, although that's not a bad idea. I do have a conversation about what is physically expected during the evaluation process.

## *Predictive Assessments*

Each person is unique, and it is hard to predict how they will work out once we bring them on board. Due to the risk and remote aspect of the industry, it can be difficult to get to know someone without spending a significant amount of time with them on location. This is often not possible, especially when scaling up operations with rotating crews and people coming and going.

One option to address this is to have candidates take predictive assessments. There are several companies that provide web-based tools to help assess candidates on a variety of topics. These assessments can be administered efficiently online. There are so many options that we can tailor a package to fits our needs. Assessments can be administered in as little as five minutes.

## Predictive Human Assessments

| Predictive Test | Measures |
|---|---|
| **Cognitive Intelligence** | Problem-solving, attention to detail, critical thinking, ability to learn, apply new info, and make good decisions |
| **Risk Traits** | Ability to follow procedures, honesty, safety minded, work habits, trustworthiness. Ability to stop unsafe personnel, actions, and operations |
| **Character** | Personality, anger, attitude, patience, ethics, greed, entitlement |
| **Work Values** | Communication dynamics, aggression, cultural fit, emotional control under stress, emotional stability |
| **Safety Temperament** | Focus, resistance, anxiousness, impatience, cautiousness, impulsiveness, distractibility |
| **Motivation** | Motivation dynamics. Determine needs, drives, interests, values, inspirations |

There is value in having our potential and current workforce take these assessments. For example, if someone is quantified as a potential high-risk individual with anger issues, impulsiveness, and lacks attention to detail, do we really want them on our location? I don't. Do we want to find this out after a costly event? Absolutely not. A resume and interview will not reveal these undesirable attributes.

Furthermore, self-reflection is a powerful tool, which these evaluations help facilitate. Predictive human testing, analysis of the results, and discussion with our team strengthens our culture on what matters most. Finally, if we acquire assets and people with those assets, consider having our new team members take these tests as part of the process of getting to know each other.

## Building the Best in the World

Our goal to increase shareholder value 100X is not impossible, but it is obviously very hard. Many things must go right. To have the best chance, the finest people must be on location, on the front lines, working hard day after day.

If we have weak people on location, achieving high returns over long periods will be very difficult. Finding the best people takes work. I know people are reading these recommendations and thinking it's impossible because it takes too long.

Not true. The process can be accomplished in a few days. Some might think, if I need a guy, I need him immediately. This is common, which is why a bank of people vetted and selected should be assembled. We need that anyway, regardless of how rigorous our selection process is.

Another option that a few operators have embraced is rotating in engineers. If a consultant cannot make it or someone calls in sick and we don't have a replacement, consider slotting in one of the office engineers. Usually, this person is a known entity we can trust. Getting these leaders on the front lines will strengthen our team and keep our finger on the pulse of what is happening.

Additionally, when the office team knows that at any moment they may have to go to the front lines and execute, they will be fully engaged during the regular course of business. Furthermore, placing engineers as company men and flowback hands can act as a cost reduction tactic.

Finally, when people know we are looking for the absolute best and our selection process is rigorous, questionable candidates will self-select out of the process, further reducing risk.

There are many consultants who have been involved in unethical practices and have been fired as a result. However, to everyone's amazement, they continue to get hired by other large publicly traded shale operators. I can only assume those entities do not have a rigorous selection process.

The best consultants take their work seriously and appreciate a rigorous selection process because it keeps risky consultants from working alongside them.

---

**The best want to work with the best.**

---

——————— **ACTION** ———————

# Monitor and Document Vendor Relationships

Establishing strong working relationships with service providers is essential as long as the relationship remains professional and ethical.

When someone becomes too close to a vendor or contractors get too close to each other, it makes me a little nervous. However, I am not necessarily against it. I want to know, in any case. As long as we know about the relationship, it can be documented and monitored.

Unnecessary risk is introduced when certain people in decision-making roles (consultants, pumpers, foreman, superintendents, engineers, managers) are close to a vendor or each other, and leadership is unaware of it. Therefore, it becomes hard to weigh the relationship dynamic when critical decisions are being made.

Business development professionals are trained to get close to operator personnel to influence decision-making when it comes time to award work. This is not a bad thing as long as no backsheesh or kickbacks are involved. There is value in establishing good working relationships and trust between contractors and salespeople for many reasons.

Service providers can destroy well economics if they are not on a friendly basis and willing to work with us during difficult times. Some vendors use every persuasion method available to win work, and they often win, but many are rarely willing to work with us when, for whatever reason, the actual ticket exceeds the estimate.

When this happens, I never forget and document the experience not only with the service company itself, but also with the individual service company personnel.

Constructing a "Vendor Relationships" database that tracks vendor dynamics, including personnel history, previous work relationships, after-work vendor interactions, days-off relationships, gifts of any kind, troubled job negotiation results, disputed ticket resolution, and bid / actual spreads, can add value when it comes to cost reduction.

When dealing with hundreds of vendors and contractors, monitoring everyone is challenging. It is not uncommon for people to take advantage of this and use their position to award work purely based on friendships, regardless of cost or performance.

This is often justified based on historic working relationships and trust, which can go a long way but rarely results in low-cost operations. The oilfield is a small world, and word travels fast once it is known that a particular operator is awarding work to all their "friends."

---

**If you don't think business relationships need oversight, then you don't get out much.**

---

Some might think documenting the details of vendor relationships is overboard, but it definingly adds value when everyone knows that all relationships are being watched. Million-dollar decisions are made every day by key people at multiple levels of the operation in positions of power.

For example, if a person managing drilling or completions operations is spending their days-off hunting with the primary salespeople of company vendors who are being paid millions of dollars, they need to disclose that, and the relationship needs to be documented and monitored.

If we don't know these relationships exist, it will be hard to detect any potential bias when discussing aspects of the operation that involve the vendor and any personnel in question.

Additionally, when there is a close relationship between a member of our team and a service provider, we need to take that into consideration during our internal vendor audits. It is common for field tickets to receive less due diligence and scrutiny when an engineer or company man who is approving those tickets is spending "extra" time with the vendor.

For example, if our engineers and company men just went on a 3-day fishing trip with a vendor and now they will diligence $10 million worth of work from that same vendor, it is hard not to let a few things slide. It's human nature to reciprocate when someone does something nice for you.

The farther we are from the front lines of our operation, the harder it will be to see these relationships and get the truth. Unknown personnel relationships add significant human complexity to an already technically complex operation, contributing to higher costs across our entire operation.

## ——————— ACTION ———————
## Identify Conflicts of Interest

Running a low-cost operation that continuously looks for opportunities to maximize cash flow requires a workforce with a clear mind. Conflicts of interest cloud the mind, leading directly or indirectly to higher costs.

The foundation of our plan is to ensure we have the right people in place. Those who are engaged in conflicts drain cash from the company in a variety of ways.

## Capital Draining Conflicts of Interest

- Making business decisions at the company's expense for personal gain or other self-dealing.

- Management, employees, or consultants with undisclosed financial interests in businesses employed by the company.

- Using company resources for personal benefit.

- Inappropriate interactions or romantic relationships with colleagues, vendors, regulators, or subordinates.

- Hiring friends and family, and other forms of nepotism.

Clear policies are essential. Collective team decisions also help reduce conflict of interest risks.

Additionally, testing for conflicts, regular audits, and disclosing interests to all stakeholders, especially non-op working interest owners and investors regarding affiliates, help prevent issues, concerns, and questions.

# ACTION

## Analyze Personnel Expense Reports and Discretionary Spending

How someone spends money can tell us a lot about how they think, what type of person they are, what they value, and what they respect.

Giving the workforce the ability to incur expenses that the company will reimburse is an excellent way to see who has the company's interests at heart.

What people say and what they do is often very different. When management and employees make decisions on how to spend the company's money, captured in detail with an itemized expense report with receipts, the document can act as a powerful tool—a microcosm of how they make larger decisions with the company's resources.

If someone cannot manage small personal expenses when spending the company's money, how can they be expected to make prudent decisions when more significant sums are at stake?

Establishing business expenses and discretionary spending guidelines can help guide the workforce on what is acceptable and reimbursable. Another option is to continuously discuss the importance of maximizing cash flow for the company and making expense decisions that align with our goals. Then, let people expense whatever they think is appropriate. After a few months, we can analyze the expense reports and release people who are taking advantage of the privilege.

It is not uncommon for people to expense as much as possible to see what they can get away with. From a management perspective, expense reports can be utilized to make personnel decisions as long as itemized receipts are required for miscellaneous incidentals and discretionary spending.

When looking at detailed expense reports to form an opinion about an individual, it is not only the amount expensed that should be analyzed but also what is expensed.

Years ago, I worked in a role that required significant air travel. The company I worked for had a policy that you could expense the cost of any meal; it was at each employee's discretion as to the price of the meal and it could be any amount. One of the engineering managers I worked with would always order a +$300 bottle of French wine with each dinner meal and expense it to the company. I thought it was ridiculous. The executives also thought so and mentioned it to me multiple times.

Before I finalized my expenses, I would ask myself if what I am expensing is directly contributing to profitability and helping the company create value. If the answer is no, I would not expense it. You can justify any expense to yourself, but will it stand up to stakeholder scrutiny?

Needless to say, the high-roller wine connoisseur manager was one of the first people to be released during a workforce restructuring. It was definitely not a surprise to me, but it was to that manager, who couldn't believe he was being let go. It blindsided him. How he spent the company's money was a factor in his release.

It is always interesting and entertaining to read stories in the news about employees who go wild with their expense accounts. Providing our team with corporate expense accounts and monitoring what they expense is a window into their soul. It can help us really get to know who they are and how they think about the company and its success.

Many people view employment at a company from the perspective of what they can extract from it, not what they can contribute to it. It is very difficult to know how each person views their relationship with the company. It is always interesting to read stories about highly paid executives at publicly traded companies who abuse their expense accounts for years before being caught.

At this point in my career, I spend each dollar prudently as though every move I make will be published on the front page of *The Wall Street Journal*—which is something I recommend to others. Think about every expense and ask the question before making a purchase:

*"Do I absolutely need to buy this to get the job done safely?"*

If the answer is yes, the next question to ask is:

*"Are we getting the best price?"*

Sometimes, the answer is no, but I might buy it because we need it immediately. Therefore, we must factor in the time aspect of needing something urgently in the middle of an ongoing operation or daily production problem that needs to be addressed quickly, or there will be a cash flow impact.

---
## ACTION
### Legal Data Mining
---

Over 40 million lawsuits are filed each year across the United States.[66] 95% of cases are filed in state courts.[67] This data provides an opportunity for enhanced risk management and cost reduction.

Court filings have a wealth of information, including pleadings, motions, hearing transcripts, and trial records. With a focus on oil and gas cases, ask the legal department to compile a monthly summary of cases with as much technical detail as possible.

For example, oil and gas equipment failures are a significant issue that has contributed to the bankruptcy of multiple companies. If a lawsuit is filed against an OCTG (Oil Country Tubular Goods) distributor, manufacturer, or related company anywhere in the United States, this is critical information that could help us make a prudent decision on tubular bidding and selection.

Envision a scenario in which multiple casing failures result in an operator switching suppliers and filing a lawsuit. A large amount of this casing then enters the market and becomes available at an attractive price.

Knowing the big picture because our team monitors legal filing puts us in a position of power versus competitors to not select this casing or at least to ask prudent questions. Additionally, what if the casing failure lawsuit involves casing that we just ordered for the next 25 wells but have not run? It might be time for an enhanced third-party inspection or to reject it and find another manufacturer.

A company can do a lot by data mining court filings and distilling the associated information. These documents contain highly valuable knowledge on undesirable operational situations to learn from, build tactical strength for our team, and enter into our artificial intelligence (AI) systems, namely our Virtual Company Man (VCM).

Our AI needs information to be able to provide value for our company. Therefore, we need to gather data from every possible source. Legal filings are a great source of knowledge that we can feed our Virtual Company Man, manually entering enough oilfield knowledge into our systems so that our AI can operate at the edge of what is known.

It would be time-consuming and impractical for our operations team to sort through thousands of legal documents each month, pull out useful information, and then remember all of it every day, when executing operations. This is impossible for a human to do, but not for artificial intelligence.

With the VCM acting as a digital advisor packed with relevant information, we can set it up to send us notifications throughout the day, and as a repository, we can access on-demand, with legal filings as one of our AI's many sources.

Very few people monitor this information. Even fewer distill the data, look for trends, and aggregate the findings into timely, usable information for the operations team. Leveraging our AI systems, we can get this information into an actionable format and push it out to our team across our assets as they execute operations.

---
## ACTION
---

## Use MSA Redlines to Identify Risk

---

During Master Service Agreement (MSA) processing, it is common to have redlines regarding aspects of the contract. This process is an opportunity to learn more about potential risks in a specific operation, service, or piece of equipment.

Based on the redlines, there are often reasons for the desired changes due to past undesirable situations experienced by the vendor during competitor operations. It pays to dig deeper during negotiations and ask questions to find out why the changes are requested and what happened in the field.

Many vendors convey that they do not have problems. Therefore, getting a salesperson to share all their dirty laundry is unlikely. I have had luck getting detailed operational information during MSA negotiations since the legal department is compelled to explain why they want changes so that we can get a contract signed. We can use this information to strengthen our operation, address issues, discuss past events, and be proactive.

If possible, do not outsource the MSA negotiation, do not use a middleman's MSA, and do not sign blanket Terms and Conditions (T's and C's) agreements unless we have no option and are in a forced-to-sign situation—which is a separate problem we must address to reduce our risk exposure. Let's do our own MSA, negotiate it in-house with our people doing the negotiation, and share the revelations with the operations team.

## ——————— ACTION ———————
### Insurance

Speaking with the CEO of an offset operator to a former area of interest in Permian Basin, the topic of risk arose. I noticed his company does not run safety valves, so I asked him about it. His response was classic:

> *"I don't run safety valves because*
> *that's what insurance is for."*

I laughed, but he was serious. Insurance is part of risk management, and how it dictates operating strategy is an individual decision. On production facilities, we install multiple safety shutdown systems, discussed in further detail in Chapter 9: Production Actions.

I sleep better knowing that our algorithms and self-contained backups will shut in the wells if there is a problem. Nobody wants to wake up to a spill or other surprise event.

However, things can still happen, so insurance is necessary, but we don't operate relying on it. During risk mitigation discussions, I have heard people say on location and in the office, "That's why we have insurance." An attempt to end the debate and move forward with little or no action. I don't like these comments because they encourage a culture opposite to what I prefer.

People who make those comments don't understand how insurance works in the real-world. If there is a situation in which we need to use insurance, there is a good chance they will deny coverage for a number of reasons. Then we will have to sue them. That could drag on for years and work out unfavorably while incurring millions in legal fees. Ultimately, dealing with insurance could cost more, which is why some operators go self-insured.

The other thing about insurance is that it takes work. It is not a set-it-and-forget-it exercise. Someone in the company should be speaking with insurance account executives regularly to ensure our insurance is aware of all our activities, policy dynamics, renewals, changes in coverage, updates, regulatory impacts, and provider requirements. All this stuff is changing and evolving as things happen across our operation.

Set a goal to speak with insurance at least once a month. Over the years, I have learned a lot doing this. Surprisingly, large operators with robust coverage do not take the time. For example, if we have a spill and don't notify our insurance within a specific period, they will deny coverage, a common occurrence. There are all kinds of nuances like this that will leave us exposed.

Over the years, the insurance assessment questionnaires have become much more detailed. For example, to get coverage it is required by some providers to answer intricate questions about the casing specs, frac designs, and cycling stresses due to multi-stage operations—just to name a few. Insurance asks detailed questions about casing, wellheads, and other specific items and actions because of the problems and failures occurring across the industry.

On the production side, they are asking questions to

identify spill riskiness of an operator and a host of other related production risks. For example, they ask how often are producing wells checked and exactly how they are checked. This is asked multiple times in multiple places on the applications.

Many people have gone to "pumping-by-exception," and pumpers might not visit every well every day. The method has become popular, but it can put more exposure on us and our insurance, depending on how it is executed. We will review how to pump-by-exception and simultaneously reduce risk later in the book.

> **Seek multiple perspectives to reduce risk and make money not mistakes.**

Dealing with insurance can add insights into what insurance companies are concerned about based on their experience. A different perspective can shed light on something that might be overlooked. Information from our insurance provider on issues and risks other operators are dealing with will help us avoid similar situations, reduce mistakes, and reduce costs.

Part of the value of having insurance is leveraging the knowledge of the people providing coverage. Building solid relationships with the agents and account executives is highly beneficial. Adding or increasing coverage based on what we are doing or plan to do is a viable risk management tactic; however, I would not run an operation depending on it.

---
## ACTION
### Claim Events Analysis
---

Insurance brokers and insurance companies have a wealth of information regarding historic and recent claim events. Consider requesting a claims analysis on claim events in our area of interest (AOI) across the competitor landscape. If we can obtain a significant amount of data, a statistical analysis could be performed to identify areas of overlooked risk, areas of high risk, or timely issues affecting service providers, vendors, contractors, other operators, and the industry.

For example, is there an uptick in pressure control claim events involving the failure of a specific piece of equipment, like frac stacks, wellheads, or vessels?

If so, more diligence might be warranted on this aspect of the operation. In general, if something were to fail during operations, it is often not an isolated event. This means that it does not just happen to one operator or service provider.

> **Undesirable events are often dismissed as anomalies.**
> **We cannot afford to make that assumption.**

Whenever speaking with insurance brokers, I always ask what has happened recently in the industry. I ask many questions to get a feel for current claim events in all aspects of the business. Then, I leverage that information to enhance safety and reduce costs across our operations.

—————————— **ACTION** ——————————

## Subsurface Protection

There are three common strategies regarding vendor-supplied insurance protection: always reject it, always elect it, and sometimes elect it.

Surprisingly, several large public operators have a policy of always rejecting vendor-supplied insurance, regardless of the situation. Rejecting the insurance will save money because we do not have to pay the premium, but from a risk management perspective, I don't like it.

Always rejecting the insurance regardless of the situation has two major impacts commonly overlooked by financial analysts or engineers performing the cost-benefit calculations.

First, always rejecting the insurance can convey a risk culture that every well is the same and every situation is the same, which is not the case. For example, vendor-supplied protection options, which are often discussed on location when a subsurface tool is delivered, are moments when risk dynamics are discussed on the front lines and a transfer of risk action takes place.

If we have a policy that our company never elects the insurance, the conversation on location, often in front of team members, will go something like this:

Tool Vendor: "Hey, boss, I just dropped off the agitator."

Company Man: "Thank you."

Tool Vendor: "Do you want the Lost in Hole protection?"

Company Man: "They never get insurance. It's policy."

Tool Vendor: "Crazy. I heard this is a well from hell."

Company Man: "Yup. Doesn't matter. Some bean counter in Houston figured that it's better for the company to never get insurance. The wellbore could be in the process of collapsing, and they would not elect the insurance."

Tool Vendor: "Stupid, I guess that's why they get paid the big bucks. It's better for my company, really. You will have to pay the full cost of the tools if DBR'ed or lost in hole. I will let our people know that you never get the insurance."

Company Man: "Exactly. Not what I would do given the situation with this well, but they don't look at things like that. It's all the same. That's the mentality. Where do I sign?"

Tool Vendor: "Right there. Just check the box saying you refuse protection. Good luck. Probably not going to see that tool ever again. Haha."

Do you see the problem? Even if we do not want the coverage, I do not think it adds value for our culture or respect on the front lines if we have a blanket policy to never elect coverage. Plus, on a high-risk well do we really not want the coverage, even if the coverage cost is minimal?

Second, once all the vendors know that we never elect the coverage, they will use that information to manage risk on their side of the equation. They offered to share the risk and partner with us, but we rejected it. Therefore, they transferred 100% of the risk and washed their hands of it. If the tools are lost or DBR (Damaged Beyond Repair), they will get full value. The vendor has no skin in the game, a critical aspect when working with partners of any kind.

When an operator elects tool coverage, which is usually self-insured by the vendor, the vendor is more engaged in our success. If we fail, there is a direct immediate monetary impact on both us and the vendor.

Often, lost in hole (LIH) insurance is 50/50, so we split the risk. Vendor insurance is a significant incentive for the vendor to do everything possible for our success. For example, if there are two tools in the shop going to two different clients, one that rejected LIH and one that accepted LIH, the better tools and vendor personnel might be incentivized to go to the client who elected LIH because they have a lot of money in the hole and they want it back. "Who cares (wink-wink) what goes to the client who rejected the LIH. If they lose it, we get the big money." I have heard this statement in the shop.

Unfortunately, sometimes vendors send questionable equipment to location that has a high probability of failing because they are incentivized for the client to lose it or DBR it since it has been fully depreciated, and they need a new tool or hefty invoice to meet monthly revenue projections.

Losing a very expensive tool on its last leg or at the end of its life is a monetary windfall for the service company. I know this sounds shocking to some and Machiavellian to others, but people do what they are incentivized to do. Is it really a surprise?

Years ago, a highly respected oil and gas executive asked me for help in raising capital for a new company that he wanted to start. I introduced him to several private equity investors and attended many pitch meetings with him.

Several investors were interested. However, when it came time to discuss how much money he was willing to invest in his new company alongside the prospective investors, he was not interested in putting up much. He thought his reputation, intelligence, and time were more than enough. That was a problem. No sophisticated investor will back a guy asking for money if that guy does not have money on the line. Investors want management to have significant skin in the game so that the project commands the team's full attention.

This same dynamic occurs when vendor-supplied protection insurance is used. Some vendors, particularly directional drilling companies and other tool companies, include 50/50 LIH insurance in their standard rates. There is no extra cost. This says a lot to me that they are willing to put skin in the game and go 50/50 with me as a true partner. This is what I like to see, but it's not common. Usually, there is an extra cost, and sometimes they offer no LIH or only cover 25%. That could be a red flag, depending on the situation.

If I am going to partner with a directional drilling company and plan to stick with them for a long time, it certainly makes me feel good that they are true partners in the BHA at 50/50.

Usually, in these situations, if my well is having problems, the directional company management is fully engaged because if something goes wrong, we are both going to lose money. As opposed to a zero-sum game situation if we get stuck where the vendor gets rich off the operator losing. There is already enough of that in this business.

So, what do I do? Typically, I get insurance. I always ask for protection and get it if offered. I value predictable costs, as do most investors, so if there is an issue, I am minimizing the ugly surprise costs that bust up the AFE.

One way to look at the dynamic of what to do is to perform a quantitative cost-benefit analysis on an area-by-area basis. Calculate how often we have a LIH/DBR event and the cost of that event, and then compare it to the cost of the insurance.

Apply the analysis to all the drilling and completion operations to see what that looks like. If we have a massive operation and never have issues, then it makes sense to rarely get LIH/DBR protection or other supplementary LIH insurance from a 3rd party.

However, this analysis is quant-focused. As previously discussed, there are qualitative aspects to consider. Also, as the industry moves out of core areas and drills longer laterals in more challenging areas with more expensive and complex BHAs, the risk goes up.

From a risk intelligence perspective, consider getting quotes on DH tool coverage from independent 3rd parties. Insurance agents are interesting and enlightening to speak with. They often have a unique story to share about recent operational events. The future is certainly not the past.

———————————— **ACTION** ————————————

## Stress Test

After the 2008 financial crisis, bank stress tests were put in place by the Federal Reserve to determine if large banks ($50B) can withstand negative economic stress. The analysis is conducted under hypothetical situations, and the scenarios are specific, such as a war in Saudi Arabia or a hurricane hitting Texas. Additionally, the scenarios include multiple things happening simultaneously, like an earthquake with a tsunami hitting California, combined with a 30% stock market crash.[68]

These are black swan events, rare but not impossible, especially when considering the recent past with the dot com bubble burst, negative oil prices, ESG movement, global pandemic, and Russia invading Ukraine.

Detailed hypothetical situation stress testing can also be applied to shale operations. Oilfield stress testing should be tailored to the area and type of operations engaged in. Banks are required to perform quarterly stress test calculations and semi-annual to annual reporting of stress test results. For oil industry stress testing, consider performing full-team hypothetical testing on a scheduled basis or as conditions warrant. When walking through hypothetical stress testing situations with the team, present the event as it could unfold. Then, have the team discuss actions to address the issues. During this process, we will find weaknesses in thinking, strategy, planning, procedures, equipment, and actions that must be addressed.

# Stress Testing Scenario Ideas

| Operation | Hypothetical Situation |
|---|---|
| **Planning** | Commodity price collapse. |
| | Development area results underperform production forecasts. |
| | Unfriendly competitor is fracturing multi-well pad sites offset the next several drilling locations and is unwilling to work with us. |
| **Supply Chain** | Casing and thread inspection reveals defects across significant amounts of pipe planned for delivery to multiple wellsites. |
| | Key piece of equipment or vendor is no longer available. |
| | Manufacturing flaw is identified in newly set vessels, treaters, tanks, valves, or trees. |
| **Pre-spud** | It is discovered multiple legacy wells have pressure on surface casing offset upcoming multi-well drilling location. |
| | Several pads become unstable after recent weather event. |
| **Drilling** | Significant metal shavings are found on ditch magnets. |
| | Hung up when picking up off bottom. |
| | Gas starts bubbling up from cellar. |
| | Anomalous negative test after production casing cement job. |

# Stress Testing Scenario Ideas continued...

| Operation | Hypothetical Situation |
|---|---|
| Toe prep | Unable to establish injection. |
| | Wellhead appears to be sinking. |
| Frac | Unexpected pressure on backside during frac. |
| | High pressure leak between lower master and wellhead during frac while pumping. |
| | Wireline stuck in stack. Then leak develops. |
| | Offset abandoned or inactive vertical well starts spraying gas and fluid from below ground, stuffing box, or wellhead. |
| Millout | Leak on wellhead during millout. |
| | Stuck during millout. |
| | Coil tubing or stick pipe parts at surface. |
| | Metal found on return magnets. |
| Flowback | $H_2S$ detected in tanks, gas, or water. |
| | Multi-well pad high pressure leak is blowing gas directly onto closely spaced adjacent wellheads all of which are online. |
| Production | Leak outside secondary containment, and fluid has made its way into a creek. |
| | Loss of control on multi-well pad w/high $H_2S$ offset to neighborhood. |
| Workover | Obstruction while running in hole. |
| | High $H_2S$ detected on floor of workover rig. |
| | Well starts flowing during intervention. |

———————— **ACTION** ————————
## Strengthen Stop Work Authority

When properly implemented, incentivizing Stop Work Authority (SWA) could mean the difference between FCF positive and negative. SWA is a policy companies can utilize to enable any person to stop unsafe operations immediately. SWA is widely known in the U.S. oil industry and globally to some extent, as it is part of oilfield culture. However, people often hesitate to use SWA in real-life situations.

Stopping individual work that is risky or incorrect is not unusual. Helping an individual you know or a coworker from getting hurt and stopping them on a man-to-man basis is common and occurs regularly. Where SWA often fails is in shutting down an entire operation for safety or stopping a task involving multiple workers. These larger-scale situations carry more risk and is when most workers hesitate to use SWA.

It is easy to talk about SWA and say we have a program that supports it, but it is difficult to get people to use it during serious situations. I have thought deeply about this dynamic due to my research and being involved in situations in which SWA worked, was used but not effectively, and was not used on location. In year one of my career, I witnessed an event where a blender tender shut down an entire operation for safety but was overruled by the company man and an oil company manager. It was a significant mistake by those who overruled SWA and left an impression on me regarding how things transpire in the real world compared to the ideal world.

## Why Do Workers Not Exercise Stop Work Authority?

1. People want to be a hand. Get the job done. Not stop work.
2. This is a massive multimillion-dollar operation; how can I shut it down?
3. It is expensive to shut the job down. Don't want to be responsible for cost or consequences of stopping work.
4. Don't want to make trouble.
5. The operator is very safe, so they wouldn't do something wrong. Therefore, whatever they're doing, it must be okay.
6. False sense of safety.
7. Don't want to be wrong in front of everyone.
8. Don't want to be ridiculed.
9. Not sure if the operation needs to be stopped.
10. Bystander Effect: Someone else will stop it if necessary.
11. Never exercised stop work before. Not sure what to do.
12. Not sure they are authorized to act.
13. Want to go home. It's not my business. Not my problem.
14. Don't want to get involved in something.
15. It's not my responsibility.
16. Worried about repercussions. Nervous about getting fired.
17. Don't think their boss will back them up.
18. Discouraged by certain people on location or management.
19. Toxic culture or company man / boss.
20. Environment on location does not welcome Stop Work.
21. Other people have exercised Stop Work Authority before and they were not respected. They got heat, push back.
22. Not worth it to me. No benefit. All I see is negatives to me.
23. No incentive.

SWA is especially important during cost reduction efforts for many reasons, and it is critical that the workforce is comfortable utilizing it regularly, if necessary, and without hesitation. Since we will make changes to be more efficient, the operation will move quicker in some respects. Therefore, everyone needs to be fully engaged.

For example, if we have a $10.0MM well cost and target a 40% reduction to $6.0MM, the operation could be occurring in half the time. Therefore, specific tasks will demand the team's full attention, and our SWA program must be powerful.

To address the reasons why someone might hesitate to use SWA, the plan is to incentivize SWA. Any stigma or negative beliefs about SWA must be eliminated.

To build a culture in which SWA is celebrated, regularly visit the team on location during drilling, completions, and production to explain how vital SWA is to you and your family—make it very personal.

In the past, I have explained that our company pays people to shut the job down. Most folks are surprised when they hear this. When someone uses SWA, I would write a detailed note to the person and include gift cards. Then, publicly thank the person and get into the details of what happened.

Finally, we would address the situation going forward to prevent it from happening again. People do what they are incentivized to do. Consider doing everything possible to incentivize people to be safe, work with proactive behaviors, and make good decisions.

# Methods to Encourage Stop Work Authority

1. Construct a high-quality, powerful SWA card with a strong personal message, company principles, and safety rules, signed by the CEO, and distributed to all personnel including employees, consultants, contractors, vendors, managers, and executives.

2. Add one full page addressing SWA in every written operational procedure and checklist.

3. Incorporate SWA into our artificial intelligence systems.

4. Have executives, managers, and engineers speak at safety meetings on SWA and encourage its usage.

5. Incentivize the use of SWA.

6. Incorporate the use of SWA into measurable safety goals.

7. Remove any direct or indirect goals, dynamics, or programs which disincentivize reporting near-misses, incidents, undesirable situations, or any SWA type event.

8. Instruct wellsite leaders to mention SWA at least once per day to a group outside of the typical safety meeting.

9. Have every person on location sign a SWA pledge.

10. Run SWA exercise drills to allow people to practice the actions of shutting down the entire operation for safety.

11. Video interview people who used SWA for training.

12. Recognize people who use SWA.

13. Write a detailed personal letter to individuals who have used SWA and thank them for taking action.

14. Feature people who used SWA in the company newsletter.

15. Create a SWA award or recognition program.

16. Start safety meetings recognizing people who used SWA.

After researching over 1,000 undesirable oilfield situations, I believe the majority could have been prevented if a strong-willed person stepped up and used SWA.[69] However, this would only work if the company and contractors had a powerful SWA system and culture that respected SWA. It is worth the effort to put this system in place. It will go a long way to protect everyone and keep top executives off the previously reviewed "CEO Departures due to Risk Management Dynamics" list.

Unfortunately, sometimes, people are incentivized to not stop undesirable situations. I know several service company owners who have become wealthy at the expense of operators' mistakes. Field operators are savvy enough to understand this. I hear remarks on it all the time. It is done jokingly, but there is always a little truth in those comments. People say what they think, even under the guise of humor.

In general, I do not think people do undesirable things intentionally, but what happens is that they stand back and let things happen. Sometimes, just to see something go wrong. It's a form of amusement or retribution. It's the same reason pranks and accident videos are so popular on YouTube.

Additionally, interpersonal dynamics on location can prevent someone from stepping up and using SWA. Consider the Deepwater Horizon accident. Many people could have shut the job down, preventing the accident. The CEO at the time, Tony Hayward, almost certainly would not have lost his job.[70] I am sure he will think about that accident every day for the rest of his life. No amount of money is worth that.

---
# ACTION
## Start Small
---

If massive cost reduction results are needed immediately, it is tempting to make radical operational or design changes within the first week of a cost reduction effort. I caution against this approach. Significant changes to the drilling program or frac design can yield substantial savings, but risk is involved.

Before making big changes, it is best to perform thorough due diligence, not just in the office from a technical perspective, but also in the field with the team on location.

It pays to give everyone time to discuss how to implement the changes successfully. However, even with thorough diligence, the actions may not work as planned, resulting in increased costs or other unintended consequences.

Starting a cost reduction campaign with big actions that fail does not invigorate the team or set the stage for long-term success. Consider initiating a cost reduction campaign with small changes and small wins.

It's not flashy, but it builds confidence and momentum. To get sustainable multi-million-dollar savings per well, we must deliver results that encourage and engage the broader team, especially the vendors.

Generating momentum is crucial to get everyone on board. Once the collective wisdom decides to engage 100%, I am convinced that nothing is impossible.

However, getting everyone engaged is challenging. The initial changes and cost reductions will get much attention.

Many team members will use the initial cost reduction actions as a frame of reference or comparison for their ideas and potential contributions.

If we start with a million-dollar cost savings on the frac operation, it could discourage many team members from contributing. They may think that their ideas will not make a hill-of-beans difference because they are not at that level of impact. As a result, they may withdraw from our effort.

Starting small will help engage everyone. Every dollar counts, as shown in Chapter One, with how saving $50 can grow to $100,000,000.

---

**Hitting consistent cost reduction singles and doubles helps set the stage before swinging for the fences.**

---

For example, I started one cost reduction campaign with a sanitation cost reduction idea from two field personnel that we implemented successfully. The idea saved $40 to $60 per day during drilling and completion operations. Then, we publicized it in the corporate newsletter kicking off the cost reduction campaign across the broader company.

Once this came out, ideas came in from everyone. We needed every dollar of savings, and this small action helped engage the team with all kinds of ideas and actions to reduce costs. You never know if a seemingly small idea could end up being a cost reduction and cash flow game-changer.

# The Language of Cost Reduction

The theory of linguistic relativity, also known as the Sapir-Whorf hypothesis, suggests that the language we speak influences how we think. Therefore, learning a new language can change how we view the world, how we process information, and what actions we take.[71] As someone who has worked across the United States and the world on complicated business matters, I agree with this principle.

Culture includes the attitudes, behavior, values, goals, and beliefs of a group. Language influences culture, and culture influences language.[72] The relationship is complicated and intertwined. This dynamic is present not only across different countries but also across different companies.

In a sense, maximizing cash flow with cost reduction is a language that can have a powerful impact on how our team thinks and acts. Strengthening our skills in this matter will facilitate a culture that looks for cash flow. Risk management is the foundation. For the remainder of the book, we will look at specific tactics to achieve our goals.

## Risk Management and Cost Reduction Hierarchy

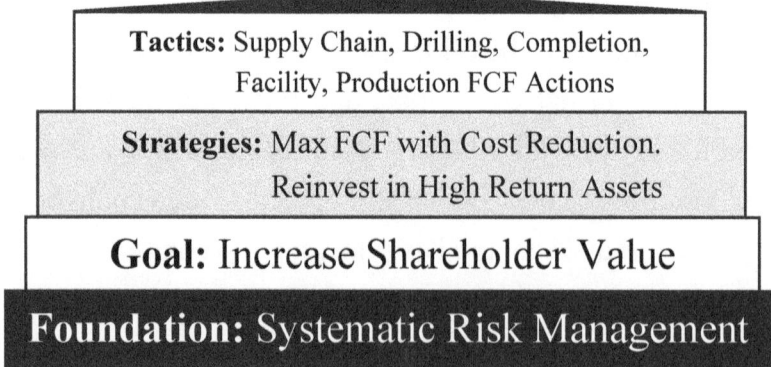

**Tactics:** Supply Chain, Drilling, Completion, Facility, Production FCF Actions

**Strategies:** Max FCF with Cost Reduction. Reinvest in High Return Assets

**Goal:** Increase Shareholder Value

**Foundation:** Systematic Risk Management

# 5

# SUPPLY CHAIN ACTIONS

—

The oil and natural gas industry is a small world and I feel fortunate to have worked with many great people throughout my career, including two oil and gas legends and billionaires, John Hess of Hess Corporation and Aubrey McClendon of Chesapeake Energy Corporation and American Energy Partners.

These two industry icons have very different personalities. However, during my time working for them, I noticed several fundamental traits they had in common. Both worked incredibly hard, highly valued safety, the environment, and maximizing free cash flow with cost reduction. These similarities are noteworthy, but it is not what I found most interesting.

The dynamic that stood out the most with John Hess and Aubrey McClendon was the value they both placed on information. High-quality, insightful information distilled in a specific way to make decisions that create shareholder value and avoid mistakes. The ability to use information to reduce risk, accurately predict the future (a challenging thing to do), and take prudent action feeds directly into the shale oil and gas supply chain and related tactics in this chapter.

Supply chain roles and responsibilities vary considerably between companies, depending on scale, risk tolerance, vertical integration, and level of involvement within each division. Therefore, a focused supply chain management (SCM) group can play an extensive role at one company while non-existent at another.

Operators can function without dedicated supply chain teams because various supply chain functions are handled by default within drilling, completions, and production groups or not handled at all. The latter is frequently the case, to the financial detriment of investors, especially during periods of growth.

When ignored or not handled by operators, SCM functions are irregularly addressed by contractors and vendors, unbeknownst to operators, due to ignorance of these activities, usually diminished or dismissed as trivial.

We must address this dynamic. Operators need to take direct control of critical SCM functions and not depend on service providers, especially when it comes to quality control of casing strings, wellheads, BOPs, and frac stack equipment, which I refer to as "life and death components."

As a result, many of the suggested actions in this chapter may seem overwhelming to the operator who does not have a dedicated SCM division.

Before jumping into specific tactical actions, let's review a few potential high-level functions of an oil and gas supply chain group. The standard definition of SCM is the process of controlling the flow of equipment, raw materials, contractors, vendors, and service providers across the business. In its broadest sense, it's the entire process of turning raw materials and services into a producing well.

In a stricter sense, SCM is focused on specific aspects of the process, primarily procurement, inventory management, and logistics.

In the context of maximizing cash flow with cost reduction, I feel SCM delivers the most value by providing a service to the division teams in an expanded capacity. Please see the table titled "Cash Flow Critical SCM Functions" on the next two pages, which lists functions that an SCM division could be responsible for in an expanded role.

Legacy operator SCM groups focus solely on managing tangible items. Modern SCM teams should add the supply of information to their responsibilities. Aggregated, streamlined, and flowing information can be one of the most highly valued assets an oil and gas company holds.

Incorporating the information and knowledge our SCM group gathers into our artificial intelligence will enable our operations team to execute at a level beyond our current ability. Competitor benchmarking alone can be a game-changer once incorporated into our systems.

# Cash Flow Critical SCM Functions

| SCM Function | Value Creation Impact |
|---|---|
| **Benchmarking** | Monitor and measure industry and competitor strategies, processes, trends, actions, and performance over time relative to our company. Develop KPIs, identify areas to address and improve. Incorporate benchmarking knowledge into artificial intelligence systems. |
| **Competitor Analysis** | Put competitors under a microscope. Pick apart their entire operation. Dissect every little step and detail to gain insights and intelligence. Learn from and avoid competitor mistakes. Organize and label data for artificial intelligence databases. |
| **Market Forecasting** | Predict business cycles and commodity prices to make strategic decisions on growth, procurement, and contracts. |
| **Procurement Strategy, Purchasing, Logistics** | Determine internal needs, run bid process, rank vendors, select vendors, address risks. |
| **Negotiate Contracts** | Structure favorable terms and prices with service providers. Process MSAs and distill valuable safety and technical information with operations teams. |

# Cash Flow Critical SCM Functions

| SCM Function | Value Creation Impact |
|---|---|
| **Vendor Relationships and Information** | Expand approved vendor list. Increase intelligence, leveraging vendor knowledge. Collect information and supply it to the team. Label information for artificial intelligence databases. |
| **Quality Assurance and Control (QA/QC)** | Prevent failures, defects, and ensure performance verifying quality of critical equipment and materials including casing, wellhead, BOPs, frac stack, valves, vessels, sand, chemicals. Incorporate into AI. |
| **Standards and Procedures** | Develop and evolve standard operating procedures, quality standards, checklists, processes. Incorporate into AI. |
| **Safety and Ethics Programs** | Construct safety procedures, policies, guidelines, and manuals to prevent undesirable situations, conflicts, fraud. Integrate into AI. |
| **Enforcement** | Confirm compliance externally and internally for all relevant programs, procedures, and protocols. |
| **Vertical Integration** | Identify areas for vertical integration. |
| **New Technologies** | Uncover new technologies and operational practices. |

---
# ACTION
## Extreme Benchmarking
---

Our primary objective is to build a 100X value-creation shale machine. Generating maximum cash flow and reinvesting it in high-return projects over long periods requires incredible diligence. There is little room for mistakes. We cannot make mistakes of any significance because the setbacks inhibit the compounding effect.

Therefore, we must learn from the mistakes and successes of others, including the collective wisdom in our industry and peer group. This way, we avoid mistakes while leveraging successes and advancing the compounding process. This philosophy can be summarized with the metaphor "standing on the shoulders of giants."[1]

The metaphor means we will use the knowledge gained by others to make further progress. It dates back to the 12th century to Bernard of Chartres, a French philosopher.[2] However, it is often credited to Isaac Newton, an English polymath, from a letter he wrote in 1675. "If I have seen further, it is by standing on the shoulders of giants."[3] This is considered Newton's most famous statement, demonstrating his modesty.[4]

Benchmarking is a great tool to help accomplish this by comparing our performance to the performance of others in our industry over time. There are many levels of benchmarking. The most common is corporate benchmarking, which looks at profitability, unit costs, FD&A costs, growth, sustainability, and a number of other financial performance metrics. This type of benchmarking is often found in investor reports, analyst reports, and other financial publications. While this level of benchmarking is important and must be done to ensure one does not fall behind one's peer group, it is not enough.

To attempt 100X returns, we need to perform what I call "Extreme Benchmarking." Extreme benchmarking is aggregating the most detailed data we can obtain on competitors and vendors, then comparing that information to what we are doing or thinking of doing while extracting valuable intelligence (insights, experiences, learnings) to avoid mistakes, navigate around the low-value traps, and leverage successes to push further, beyond what is currently possible.

The level of benchmarking detail I am referring to is extreme. For example, we are not scratching the surface of CAPEX by looking at competitors well costs per unit and comparing them with a selected peer group. That is straightforward; everyone does it to see if they are in line with the group. There are not many actionable insights to be gained other than answering the question on whether we are high or low compared to the competitors under examination.

Instead, we start with the same high-level comparison and then dig deep to get into competitor cost strategies, designs, geology, operational processes, vendors used, materials, equipment, anecdotes, performance, and so on. We drill down as far as we can go.

For example, if we perform extreme benchmarking on competitor wellsite pads, we would compare all competitor pad designs, shapes, materials, equipment used, stabilization methods, tons of Portland or CKD, tons of rock, inches of rock spread, fence type, gate type, cattleguard designs, construction time, wellhead spacing, wellhead wing valve angle (safety-critical), whether power or pipeline is set before rig mob, microgrid dynamics, field gas fuel source details, where do they put the facility, when are they building it, how does the pad handle stormwater, issues with the pad during drilling or completions, which vendors are doing the work, where are the materials coming from, and so on. This extremely detailed technical information is compared to our baseline.

Aggregating this detailed information and putting it into a usable format takes work. It is not easy or quick, but the value is tremendous. Our "Competitor Analysis Deep Dives" and "Vendor Analysis Deep Dives" which we will discuss in the next two action items, feed into our extreme benchmarking efforts, and help make this analysis possible.

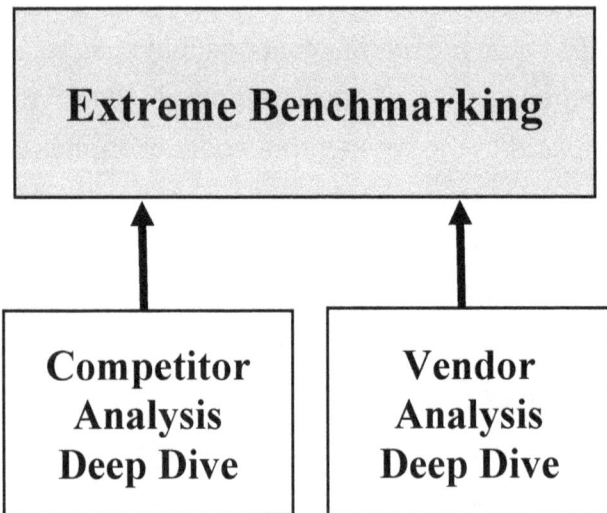

# ACTION

## Competitor Analysis Deep Dive

A competitor analysis deep dive is the process of zeroing in on one specific oil and gas company, aggregating as much detailed information on that competitor as possible to understand every aspect of their business. We want to put the competitor under a microscope and pick apart their entire operation. Dissect every little step and detail to gain insights and intelligence. We are looking at not only the operation itself but also the experiences that they have had, both positive and negative. The competitor's experiences provide a tremendous amount of value. We want to live vicariously through them.

In almost every area of the United States, there are at least five to ten public and private competitors within proximity from which to start. A direct offset competitor is a good choice for the initial deep dive. Once we have multiple deep dives under our belt, we can perform competitor deep dives in other areas, geologic targets, and basins. There is a wealth of knowledge across the industry from which to leverage if we are willing to put in the effort to collect the data and build it into a format our team can utilize.

An idealized system would be one in which our team produces competitor analysis deep dives regularly (perhaps monthly), with the analysis presented for discussion and debate. The meeting would be followed up with a detailed booklet for additional study by the teams. The goal is to create a pipeline of actionable information from competitor experiences that flows through our company.

This pipeline of competitor deep dives will not only elevate our operation but also provide the dual benefit of identifying potential acquisition candidates that aid in our 100X returns goal.

The opportunity to perform detailed competitor analysis is exciting, but at this point, one may think, "Extreme benchmarking and competitor deep dives sound amazing, but where do I get all of this detailed competitor information?" Let's review the multiple sources of competitor intelligence.

## Well Data Sharing Agreements

Consider reaching out to an offset operator and negotiating an information swap. For example, trade five wells worth of data from our wells for the equivalent amount of data from their wells. We may be pleasantly surprised when discussing this option with other operators. Usually, there is at least one multi-well pad or well they are interested in getting the data on and they are willing to trade for it.

## Minerals and Non-Operated Working Interests

Buy minerals, take a small lease in an upcoming drilling unit, or participate with a fractional non-operated working interest. With any of these options or others, negotiate for information rights.

Although these options require a small investment, it can be minor compared to the value of the information obtained, as long as we use the information and it doesn't just sit on the shelf, which unfortunately is more often the case. One strategy that has worked in certain basins is to take a small lease in an upcoming competitor unit or any offset section and then protest the competitor's hearings.

Location exceptions are a good one to protest for this reason. Agree to lift the protest for information rights. Most people will agree, especially if they are in a time-sensitive situation, which is often the case. Just ensure our letter agreement is very detailed on the data that we want. It is also good to ensure the deal includes providing data in a timely manner. Otherwise, they will drag it out for a year.

## Field Reconnaissance

One of the benefits of our industry, in terms of gathering competitor information, is that our factory floor is not hidden behind a wall. It's in the open air and easily visible from public roads. If there is a competitor deep dive that we are preparing for, go to their locations and spend some time watching operations from the road.

Competitors and iced-out vendors did that to my operations all the time and there was nothing I could do about it, although sometimes I wanted to throw tomatoes at them. Regardless, make a list of all the contractors that our competitors are using and then start making phone calls.

## Oil and Gas Software as a Service (SaaS) Platforms

Several oil and gas SaaS platforms have datasets built from public filings that are a great starting point for competitor deep dives and benchmarking. Different SaaS platforms have different options and abilities, so it adds value to have subscriptions from multiple vendors.

If we can afford it, subscribe to all of them. After several months to a year, determine which ones add the most value, then cancel the services that don't work and keep the services our team enjoys. If there is an interesting SaaS option, I would not hesitate to subscribe on a trial basis.

## Regulatory Filings

Based on the location of the well, operators are required to file multiple documents that become public information. These filings are the source for most of the SaaS databases. However, the databases do not capture all the documents, nor do they capture all of the information in those documents. When performing competitor deep dives it's best to get our hands on the raw data. In this case, the actual documents with all the intricate details.

Digging into public documents can help put the puzzle together on competitor operations. If the wells of interest are under federal jurisdiction, we might be able to find a considerable amount of information on file with the Bureau of Land Management (BLM). Applications for permits to drill (APDs) and the associated sundry notices have significant detailed information that can be useful.

## Hearings and other Legal Situations

Oil and gas development-related issues that cannot be resolved between operators or the public usually end up in a formal proceeding and are heard before a judge.

In these situations, interesting information can come to

light and become publicly available. For example, if two operators feel that they are both best suited to develop a drilling spacing unit (DSU), they will have to make the case at a hearing before a judge or state authority.

During the proceedings, engineers, geologists, and other experts will have to explain why their company is the most efficient operator, will add the most value for the mineral owners, or has the best track record and provide details as to why that is the case. Then, they will have to get through cross examination by aggressive attorneys. This is under oath, so competing operators will ask each other everything and anything that they want to know.

This is a great way to get information. Our competitor's technical experts are on the stand under oath, testifying as to why they think they are the best operator. There is no better time to press for information to see what they think differentiates them and what competitive advantage they believe they have.

The most valuable cases are those that are transcribed and have many exhibits with thrilling cross-examination of engineers, geologists, and managers. Transcriptions are usually available to the public. Depending on the county, state, and circumstances, this information can be found online or obtained physically at the courthouse. Sometimes, this information can be challenging to locate, but it's there.

## Investor Relations Materials

If our competitors of interest are public companies, there is a wealth of data available on their website, usually under the investor section. I enjoy reviewing investor presentations and reports. Some companies remove historic presentations and only provide the most recent data.

However, historic presentations are usually found on an investment-related website or from an online service provider that maintains a database of these documents.

Most of the information in investor presentations does not provide the level of detail we seek, but it does provide directional guidance. Usually, the smaller the company, the better the quality of detailed information provided. Listening to the quarterly calls or reading the transcripts also adds value.

## SEC Filings

In addition to investor presentations, I recommend digging into all Securities and Exchange Commission (SEC) filings. Sometimes, we can find useful information in the management discussion and analysis section, notes section, or under footnotes.

On the financial side of the business, I worked with a group of analysts who, no matter what document was given to them, whether from an investment bank or consulting firm, would immediately look at the footnotes before reading anything else. It's a good approach because the devil is in the details.

## Conferences and Events

If there is a conference occurring anywhere in the world that has relevance to our business, a company representative should be in attendance, gathering information. I have traveled across the United States and the world, collecting and synthesizing information from multiple venues for both John Hess and Aubrey McClendon because both highly valued timely, detailed intelligence.

The key is to send someone who can get the information needed from the conferences, events, presentations, talks, meetings, side conversations, dinners, or wherever the event takes place and process the raw data into valuable material so that the broader team and executives can utilize it.

The most interesting information is often conveyed face-to-face in one-on-one conversations with people we are meeting for the first time, so we don't want to send someone who is not personable and interested in this type of work.

After events, put together a report, memo, or written summary that includes the raw information and intelligence synthesis with conclusions, including suggested actions for decision-making.

I understand that some people reading this may think that this level of intelligence gathering is excessive. However, I argue the opposite. For example, during the normal course of business, the president of the United States and the federal government send members of the executive branch, the Central Intelligence Agency (CIA), the Federal Bureau of Investigations (FBI), and others all over the world to gather information to manage risk and help make good decisions.[5]

The United States cannot afford to be in the dark or blindsided by anything. The CEO of an energy company should look at business the same way due to the risk and capital intensity of our operation.

Furthermore, there is a substantial additional benefit that

the U.S. government is not really focused on, and that is cost reduction and free cash flow. These events more than pay for themselves in terms of the information gained to assist in maximizing cash flow with cost reduction and making good decisions.

## Local Meetings and Roundtables

In terms of industry events, there is a focus on the large annual conferences for new ideas and information exchange, and as a result, the interest in attending these events gets much attention from staff and management.

However, smaller basin-level meetings and roundtable events happen more often and can provide an attractive information-rich environment. People are more willing to talk openly at smaller events. Therefore, when looking for interesting discussions to attend, do not forget about the smaller regional gatherings.

## Social Media and Blogs

The internet provides a venue for social information sharing about any topic of interest. Fortunately, or unfortunately, the oil and gas industry is not excluded.

Therefore, it is common to find significant data on social media sites and blogs when looking for competitor information. Although most field personnel are discouraged from posting this information, they still do it. Everyone loves taking pictures and posting them online for likes and comments. Significant data can be gathered from a single photograph if you know what you are looking at.

For example, it is believed that the developers of Stuxnet, the highly advanced computer worm that targets SCADA systems, used pictures from a news publicity tour of an Iranian nuclear facility to design a computer worm to attack that facility.[6]

Iran is very proud of their nuclear program. They have a "National Nuclear Day" and showcase videos of the facilities on their TV news channels. A press tour was conducted in 2008, and during that tour, pictures were taken of the proud engineers and technicians that were published by the Iranian government. In one of the pictures, part of a SCADA screen could be seen showing six groups of centrifuges, each group having 164 elements.[7] This configuration was a perfect match to the Stuxnet computer worm that had a line of code with an array attacking this specific centrifuge assembly.[8]

Stuxnet is designed to look for Siemens SCADA systems that control gas centrifuges, targeting programmable logic controllers (PLCs) that interact with variable-frequency drives (VFDs).[9]

Stuxnet changes the rotational speed of the centrifuges to spin outside of their safe operating range to physically destroy the rotor tubes while displaying that everything is working fine on the screens.[10]

This would be similar to running oil and gas submersible pumps (ESPs) above max recommended frequency (Hz) or rapidly turning them on and off until they break while holding the SCADA human-machine interface (HMI) display constant and not sending any text alerts that there is a problem.

Stuxnet technology could destroy oil and gas infrastructure

if programmed accordingly. This was the first time malware physically destroyed industrial equipment in the real world and potentially hurt people.[11]

It is believed that all of this was done with the aid of a picture. As you can see, images can be powerful for competitor analysis if we know what we are looking at.

## Oil and Gas Aggregation Websites

There are several oil and gas industry websites that are worth following for relevant data. Although the operational information might not be detailed enough for what we are doing, the articles can be a good starting point or help put the puzzle together on competitor intelligence.

## Trade Journals and Magazines

Our industry has many excellent magazines and trade publications that are highly valuable to the collective wisdom. The data from these resources is timely and insightful for risk management and value creation.

However, it is common for most team members to have little time to read all the material. Therefore, to capture the data and integrate it into the organization, it is a good idea to task someone with summarizing these publications and sending the summary to the team.

Then, discuss the most valuable insights during morning meetings or throughout the month when meeting with team members. This is also a good barometer to see who is engaged and interested in our industry and who is not.

## Technical Papers and Publications

Many of the greatest scientific and technical discoveries in human history were disclosed in technical papers and publications. These discoveries include the Copernican system, electricity, penicillin, the diesel engine, nuclear fission, and shale gas, just to name a few.[12]

In the oil industry, technical papers and materials contain practical insights, engineering breakthroughs, and actionable intelligence published regularly by multiple societies, entities, and individuals. These materials are a great source to help manage risk and maximize cash flow, and they must be utilized by our team.

There are too many papers and publications for each team member to sort through and read each month. Therefore, a system should be set up within our company to leverage this data and get it into the hands of the right people in our organization.

I suggest assigning the task to a group of individuals on the supply chain team or other technical group that can process and integrate this supply of information into the human capital of our organization, including our AI.

Someone needs to identify the relevant papers and summarize them into actionable intelligence for the company to digest quickly and easily. These actionable summaries could be made available on the company intranet or sent out regularly to appropriate team members.

In many cases, technical papers have information relevant to efficiency and cost reduction efforts without the paper authors realizing the significance or intention. As a result, the paper's title and abstract mention nothing about cost reduction when, in fact, the material has cost reduction implications that add value to our free cash flow goals.

In 1939, an article titled "Disintegration of Uranium by Neutrons" was published in Nature based on its discovery in Nazi Germany.[13] The article explained nuclear fission. If American scientists and engineers were not engaged, and reading technical papers and publications, while adversaries were, it's possible the United States would not exist today.

## Acquisition Candidates

On our journey to 100X, asset acquisitions will be part of the process. Therefore, we must continuously look for potential candidates on and off market. Private equity and private investor-backed companies want to sell in a timely manner and are motivated to make a deal. They want to educate potential buyers on their company and the value they bring to the table. These companies want buyers to get comfortable with their assets and operations.

Therefore, approaching private oil and gas companies and asking for information outside of a formal sales process is often desired by them. These companies want us to know what they are doing so that we will consider buying them.

As long as we act in good faith with a potential acquisition interest, competitor analysis of these companies should be part of our process. Hopefully, we do not need to sign anything to acquire the information, but if requested, confirm that the agreement does not prohibit data usage.

# —— ACTION ——
## Vendor Analysis Deep Dive

Vendors have a wealth of information and intelligence on oilfield operations, far beyond what most clients think. Since operating companies are in control and feel they know best, vendor knowledge is rarely utilized, especially by oil company engineers and field personnel.

Vendors see a broad landscape of the successes and failures of their service and their competitors across the industry beyond the scope and scale of the operator. We must leverage this to manage risk, reduce costs, and maximize free cash flow.

For example, a few years ago, I was getting ready to complete several wells in a new area when an offset operator got stuck milling out frac plugs. I had information rights and was following the situation, living through the ordeal vicariously. It took over a month and $1.5 million to rectify the problem.

To get to the bottom of what happened, I was talking to everyone, trying to confirm whether it was a reservoir situation (crossflow, lost circulation), a frac plug issue, a procedural mistake, or a combination of issues. During my diligence, I gained a tremendous amount of intelligence on post-frac plug millouts in this particular formation, all of which I incorporated into written procedures.

Additionally, I found a vendor technician who had performed detailed analysis on thousands of plug millouts.

This person had the ability to perform statistical analysis on every aspect of the millout. He had all the variables, including situations in which operators got stuck. Only someone with a significant amount of time and data (scope and scale) could put this analysis together.

The 10,000-Hour Rule, coined by Malcolm Gladwell in his book "Outliers: The Story of Success" suggests that it takes around 10,000 hours of intensive practice to achieve world-class expertise in any complex skill.[14]

This vendor checked the 10,000-hour box regarding frac plug milling expertise and analysis. Based on his study, it was easy to see what the offset competitor did wrong. They ran a composite frac plug that was a "Top-5 Most Likely to get Hung Up or Stuck" during millouts and missed several key operational steps when pulling out of hole (POOH).

The issues were related to coil trip speed relative to debris pile movement and annular velocity. There were a few more mistakes made, but the point I am highlighting with this story is that a vendor had the data, analysis, and ability to generate actionable insights.

With the average cost of getting stuck during a millout at $1.5 million on the low side and the total loss of the well on the high side, can we afford not to access this level of vendor intelligence? Absolutely not.

Therefore, we must ask our vendors to provide a detailed analysis of all their related operations. Consider making a few requests to our vendors, as listed on the next page.

# Initial Requests to Leverage Vendor Intelligence

✔ Provide a detailed analysis of the successes and failures of the service or equipment that we are using or plan to use.

✔ Do you have a competitor benchmarking study to share or discuss on this service or equipment?

✔ Can you provide a quantitative analysis on this service?

✔ Please send case studies.

✔ What operators are you working for?

✔ Please provide references that I can call and discuss their experience with your service and competitor services.

✔ Please send me best practices on this service or operation.

✔ Tell me about recent problems you are seeing in the field regarding your service and others.

✔ How can I improve my procedures relative to your service or equipment to address areas of concern?

✔ Are there any issues with any other vendors I am using?

✔ What technical analysis is your company currently doing? Can you share the analysis?

✔ Who is the technical expert on this operation? I would like to meet him or her in person.

✔ What group within your company performs statistical analysis on this service? Please arrange a meeting.

✔ I would like you to present ideas on how we can reduce costs using your technology and processes.

---
## ACTION
## Reverse Engineering
---

There are many techniques to reverse-engineer a product, process, system, or operation. Competitor deep dives, combined with vendor deep dives, put us in a position to not only execute on our benchmarking efforts, but also to initiate a reverse engineering program.

Once we understand what our competitors are doing, we can leverage their experiences, designs, processes, both good and bad, and incorporate valuable elements, while avoiding costly mistakes. The knowledge gained during this process will help us advance.

Reverse engineering competitor designs and operations helps us conserve capital and forces us to think more deeply about what we are currently doing. Many top companies across all industries utilize reverse engineering. There are independent firms that specialize in providing reverse engineering services for every detailed aspect of competitor products and operations.

The goal of our reverse engineering program is not to copy our competitors, but to build a stronger operation, leveraging the deconstruction process and intelligence gained from multiple approaches by various competitors focused on the same target formation. Once this is accomplished, we can expand our program to include competitors working on additional target formations and in regions outside of our direct area of interest.

From a geophysical perspective, the foundation of our industry is based on reverse engineering the construction of the Earth. We have the finished product (Earth), and we must work backward. We look beyond what is obvious on the surface to deconstruct the internal workings of the planet and our target interval to understand how it was built over millions of years.

Although I am not a geologist by degree, I enjoy building my own geologic models and performing detailed geologic analyses when time permits. Thinking about what transpired on this planet over millions of years to the present day is one of the more enjoyable aspects of working within the energy and natural resources industries.

Reverse engineering the formation of target intervals, and the detailed geologic analysis required reminds me of one of my favorite quotes from the Book of Job:

---

"WHERE WERE YOU WHEN

I LAID THE FOUNDATIONS

OF THE EARTH?

TELL ME,

IF YOU KNOW SO MUCH." [15]

---

*The next time you drill across a fault, ponder this passage.*

--- **ACTION** ---

## Cash Flow Speaker Series

"Talks at Google" is an interview speaker series hosted by Google for employees.[16] The program invites authors, scientists, influential thinkers, and others to discuss their work. Talks are held virtually and in person across Google offices worldwide.[17] Most are recorded and shared with the public. A similar speaker series geared towards cost reduction and cash flow can add significant value for our team if it is managed and executed correctly.

Many entities hold speaker series, including top business schools, Goldman Sachs, Toyota, Salesforce, IBM, Adobe, American Express, BlackRock, and Standard Industries, just to name a few.[18] Consider constructing an internal speaker series to further maximize cash flow.

### Potential Speaker Series Guest Lineup

- Oil and gas business leaders.
- PE investors with portfolio companies of interest.
- Technical staff from potential acquisition entities.
- Wall Street energy analysts.
- Institutional investors.
- Thought leaders and innovators.
- Consultants and subject matter experts.
- Wellsite leaders that have something to share.
- Technical paper authors who interest the team.
- Vendors and service providers.

The series can be a mix of virtual and in-person events held at corporate headquarters and in field offices, streamed internally. The talks could be recorded for future viewing within the company. The series should include speakers on a variety of topics relevant to our company's interests and goals. Consider lining up one speaker per week or about 30 to 40 speakers per year when accounting for holidays and other corporate events.

It will not be hard to find people who want to speak at our company. Vendors are always looking for a platform to showcase their latest work. However, the key to keeping it interesting is having a broad lineup of the right speakers, including all aspects of the business. Consider what we want our speaker series to achieve and then design a speaker lineup with that goal in mind.

Technical and operations team members are commonly underappreciated by their employer. This dynamic can be leveraged by inviting competitor innovators and idea generators to our speaker series, identified from technical journals, articles, or interactions at industry events.

To keep things interesting, change the venue. With modern technology, we can hold the talks anywhere and post the recordings on our company intranet. If an interesting speaker cannot make it to our office, go to them. A good speaker series will not only enhance the intelligence of our most valuable asset, our people, but also add value to our company brand, attracting new investors. The more people we speak with, the more value will flow our way.

───────────── **ACTION** ─────────────

## 10 Bids then Buy

It is common practice in the oil industry to bid work to a minimal number of vendors, preferably the lowest number that is optically acceptable by upper management. The standard bidding practice is called "Three bids and a buy."

This means we bid three vendors and select one. Increasing the number of vendors increases the complexity and associated workload. Managing people, personalities, and information takes time and effort. Plus, most operators already have a preference on whom they want to use.

Managing three vendors is an easy way to engineer the bidding process to ensure we work with whom we want and optically satisfy anyone watching with the appearance of a competitive bidding process.

It is convenient to put the preferred vendor up against a high-dollar vendor and a vendor nobody likes. This way, the choice is simple, and we check the box for market price verification. We feel good that we are not paying top dollar and move forward.

Executives reading this might think, "My team bids more than three; we bid everyone. I should know because we just finished the frac bid process."

The executive would be correct for two items: frac crews and drilling rigs. In general, frac crew and drilling rig bidding, due to the CAPEX impact and larger strategic obligations, involves a much more elaborate, in-depth bidding process. For the remaining services, which collectively can cost just as much or more than frac horsepower and a rig, it's typically "Three bids and a buy."

To address this issue, consider setting a goal to try and bid at least ten vendors for each service. This is considerable work. Dealing with at least ten vendors for each service will become a full-time job for a short period; then, it will end. Therefore, this can be accomplished without adding staff. However, it is much easier to achieve with a supply chain team in place, but it is not required.

The value in bidding out everything to the maximum number of vendors is not only to see the full spectrum of pricing and pricing structures but also to maximize information flow and intelligence into our shale machine.

Talking to everyone on the vendor side is highly valuable. We will learn far more bidding ten vendors compared to three. Think about it: we will deal with over 3X the number of people, 3X the technical information, and 3X the amount of pricing data. Our phone will not stop ringing with vendors telling us how bad of a job our current vendors are doing on someone else's location. That's good; we want to hear the gossip and what happened on other locations so we can address it and make sure it does not happen to us or make a vendor change if needed.

Additionally, with the goal of bidding ten vendors for each service, we will have to actively look for more vendors to speak with. This forces us out of our comfort zone.

Many think, "In my area, we do not have ten vendors available for each service." That's good if it's the case because we have reached the limit of whom we know. Now we must expand our contacts.

Our current vendors do not want us talking to everyone because they cannot contain information flow, therefore, they lose control. They want to keep us in the dark. The more people we speak with, the more pressure it puts on our current preferred vendors to reduce pricing.

If vendors know an engineer or company man is a:

*"Three bids and a buy guy,"*

the pricing they offer will be higher, or the structure will be unfavorable compared to what is presented to the operator who finds everyone and bids out the work aggressively and relentlessly to the maximum number of service providers.

The other dynamic is that if we only speak to a limited number of vendors, our current service providers will increase pricing on us more aggressively. Vendors in this business know which operators are lazy when dealing with service providers, who do not return vendor phone calls, who are impossible to get in front of, who only use their vendor friends, and who are corrupt. Those types of operators will pay for it with higher pricing and unfavorable pricing structures.

It's okay to keep using whom you want, but we must talk with everyone. We must bid everyone.

Dealing with all these vendors can provide tremendous value if we can capture that value and provide it to the team. The next action item offers several options to help make this a reality.

# —————— ACTION ——————
## CRM and VRM SaaS Systems

The most innovative vendors manage client relationships with the help of customer relationship management (CRM) software. These cloud-based systems help manage every aspect of the client-vendor relationship, from the initial meeting to mobilizing services to location.

Vendors leverage the full value of client interactions by documenting discussions, insights, intelligence, bidding, feedback, and operational performance with CRM platforms.

Operators, however, rarely capture any of these insights other than price data. In terms of risk, cash flow, and cost reduction, significant value is lost. The larger the organization, the more of a problem this is. For example, let's compare two oil companies, one called Family Oil and one called Big Oil.

Family Oil is a small operator who talks to all the vendors and drills a few shale wells per year. Family Oil does not have a supply chain group. One person deals directly with the vendors and is also on location with the consultants executing operations. Big Oil, on the other hand, has a large supply chain group who talks to all the same vendors and runs the bid process working with the operations group.

All else equal, if we were to bet on who has a smoother, lower risk, lower cost operation, I would put money on Family Oil. The reason is that the intelligence gained from vendor interactions during the planning and preparation phases to drill, complete, and produce is not lost at Family Oil. It's maintained with one central person, start to finish, and that person is on location executing alongside the field team.

Therefore, all those critical insights, data, anecdotes, recommendations, and intelligence that often mean the difference between success and failure, are not lost within the annals of the organization.

---

**Establishing systems to centralize intelligence gathered from multiple divisions will reduce risk and cost.**

---

At Big Oil, when the supply chain group or operations group meet with vendors throughout the business, no one else has access to the intelligence, especially on the front lines.

Since Big Oil has supply chain groups, operations groups, management, executives, plus the entire field team, including foreman, superintendents, and rotating wellsite consultants with new people every other week, Big Oil is running operations blind to vendor level intelligence gained by various people throughout the value chain. Since vendors provide services for every aspect of shale operations, this is a problem.

Family Oil has the intelligence centralized with one person, and everyone is interacting with that source. Additionally, that source is on the front line with the field team. Therefore, the intelligence is continuously front and center.

Big Oil does not have that benefit because they are operating at scale. Big Oil must take action to address this dynamic with a centralized vendor relationship management system (VRM) that everyone can access 24/7.

Software as a Service (SaaS) entities that provide CRM platforms can configure them for buyers, or there are SaaS platforms that provide buyer-focused VRM services.

If we do not want to spend the resources on an outside system, a database or internally developed cloud-based system could be implemented. The benefit of the more advanced systems is that additional apps can be layered into the systems, including analytics and artificial intelligence.

## VRM System Expansion

VRM cloud-based systems not only take operations to the next level but also help modernize deal flow. This will further help us achieve our 100X goal.

Furthermore, VRM systems can help manage and document interactions with midstream entities, mineral owners, surface landowners, and regulatory groups. Carefully managing relationships with all stakeholders, particularly surface landowners, can help reduce operational costs and avoid expensive, undesirable situations.

It is a good practice to document interactions with these people in the field and office with a cloud-based system. If an oil company is not organized in this respect, it is easy to get taken advantage of by shrewd outside parties that prey on disorganized oil and gas operators.

These individuals leverage the size and scale of a large operator, with poor information flow between people and departments, against that operator for monetary benefit.

---

# ACTION

## Passive to Active Sourcing

---

As an operator, it is easy to get unintentionally lulled into a passive vendor-sourcing model without realizing it. Passive sourcing is when we primarily select vendors that come to us.

Aggressive, hungry vendors with savvy sales professionals are constantly bombarding operators because it works. The best vendors might not be assertive because they already have a full plate. As with everything in life, it is hard to get the best because the best are already booked.

Daily tasks, requests, and constant fire drills can consume most of a typical workday, so there is no time to actively look for vendors. Therefore, we select vendors that put themselves in front of us. That's okay; we all do it, but let's not structure our entire operation this way.

Switch from a passive to active vendor sourcing model. Active sourcing is more work compared to passive sourcing. Active sourcing requires identifying vendor contacts, calling lots of people, checking references, performing due diligence, watching vendors in action, getting to know the specific people from each vendor that will be on our location executing the work and so on.

If we want to put the best possible team together, it takes effort and will not happen if we spend our time sitting behind a comfortable desk in a climate-controlled office, selecting vendors based on salespersons that take us to lavish lunches and apply other manipulative tactics.

# ACTION

## Employ Vendor Minds

It is common to have a mental model that vendors are like tools in a toolbox. We select a vendor and then insert them into our operation to get the job done. There is an old saying in the field, "They pay me from the neck down." In other words, the mindset is, "I am not paid to think or provide feedback. I just do what they tell me, even if I don't think it's the best way."

To put this in context, it bears reminding that vendors receive so much aggression from hostile operator reps and their own leadership that they shut down from providing valuable suggestions to maintain sanity. If a person is trying to help but the feedback is hostile, there is no incentive to continue; in fact, there is a disincentive.

To address this and leverage vendor intelligence, it will require modifying the intellectual relationship with each vendor and the culture on location. When a vendor comes to the office or on location, they must believe they are valued as an individual, and we want to know what they think.

When I am paying for a person to be on location, I want the value from the whole person, especially the brain. I am paying for it. I want the vendor to use all their brain power, experience, and intelligence to ensure success.

Task the team to engage every vendor to provide cost reduction solutions. Explain what we are trying to do and ask for help. It is humbling to ask for help but empowering to the person you ask.

---

# ACTION

## Negotiate

---

If we max out the number of vendors bid, bring in additional vendors, bid aggressively, and engage vendor intelligence to assist in cost reduction, there is a good chance this will not be enough to get costs where they need to be. The next step is to take the top bids and start negotiating.

Shortlist the bids to the top two to five, depending on how many vendors were bid, then work with each vendor to try and get costs down further. There is a good chance, depending on market conditions, that concessions will be made. However, we may need to get a little creative on the adjustments.

For each service, there will be sticking points where a vendor cannot move on price. This does not mean that there is nothing more to do. Most vendor invoices have multiple components which impact the final actual price. Look at each of these secondary components for opportunities.

## Alternative Secondary Vendor Negotiation Points

- Downtime, stand-by rates
- Mob, trucking, delivery charges, mileage, travel time
- Personnel charges, technician fees
- Backup equipment charges
- Repair costs
- Damaged beyond repair (DBR) charges
- Third-party markups
- Special circumstances charges (pressure, rate, weather etc.)

# ── ACTION ──
## Map Vendors

Many vendors do not have a clear address where they are based and where equipment, materials, and personnel travel from to service our work. Most vendors closed local yards during the downturn and now travel great distances.

Clarify where each vendor is based and create a map. Include all current vendors and then add potential vendors to the map to see what our alternatives are, especially those closer to our operation. Many good vendors do not have a sales team or digital presence, so it pays to drive around the development area to identify smaller, locally based options.

This map will help reduce our costs in several ways. First, the map will help with vendor selection by taking proximity into account. Second, when negotiating on price, many vendors will reduce time and travel charges to match those of the closest competitor if we ask for it.

Third, it will help confirm invoiced travel miles, time, and related expenses since this line item has been abused as a profit center and rarely verified when approving field tickets. Fourth, the map will help assess downtime potential if delays might be possible. Fifth, the map will help make real-time decisions if something is needed quickly and an alternative vendor is closer to the wellsite. Finally, with all the vendors mapped, we can assess our collective risk for getting things to location. Since managing time is one of the best weapons to reduce cost, it must be addressed from all strategic and operational angles.

---
## ACTION
## Source Local
---

All else equal, I prefer to source equipment, materials, and services as close as possible to the location of each wellsite. Currently, I work on wells in the United States; therefore, looking at my situation from the fifty-thousand-foot view, I prefer to source everything from within the United States. This is especially true for pressure control equipment and any components that contain pressure, including OCTG, wellheads, frac stacks, vessels, and flowline components.

To get more granular, two critical items on my checklist that every operator must purchase are casing and wellheads. These overlooked, high-risk components are frequently selected based on availability, cost, or salesmanship. The standard mental model is that OCTG (drill pipe, casing, tubing) and wellhead components are all the same, commodity items, which are interchangeable with minimal thought. I strongly disagree and spend considerable time on the due diligence of these components.

For example, due to the risk and severity of a casing or wellhead failure, I prefer to source these components from vendors where I can easily visit the factory to personally view the items under construction, ask questions, obtain information, and address concerns.

Casing is commonly purchased through distributors who maintain an inventory, with whom I want to have a strong relationship. To a greater extent, I want a stronger relationship with the casing manufacturers and the manufacturer's technical teams. I must know precisely when and where my casing was manufactured and where it has physically been since it left the factory. This is why I prefer 100% Made in the USA; I can track critical dynamics more easily.

If I can get my casing and other crucial items on location without worrying about them getting picked up with sketchy cranes in high winds at foreign ports, loaded heavy-handedly onto jalopy cargo ships of unknown origin, traversing the planet's rough oceans for months, transferred ship-to-ship at the Panama Canal, dropped off for additional processing at another facility, threaded at a third place, with couplings from fourth place, I will choose the lower risk option.

I do not have a problem purchasing non-domestic pipe if I can track where it's been and visit the overseas factories. The problem is that I get uncomfortable when I see foreign language on casing, wellheads, and other components. I do not know what it says or means.

When visiting overseas factories, I have a harder time getting a feel for the quality and workmanship. I can't just start asking people questions and perform my typical due diligence. And if something does not look right, it's hard to take a picture and text it to a foreign manufacturer to ask questions in the middle of the night, as I would do with a domestic mill before running it into the subsurface.

Unfortunately, I have been involved in multiple casing and wellhead failure situations and have become hyper-aware of the many dynamics regarding this part of the business, which is out of sight and certainly out of mind when we are on location installing these life and death components. To simplify this aspect and many others, I prefer to keep things close, with massive oversight.

## Geographical proximity reduces costs and risks.

The closer the vendors are to the wellsite, the fewer misunderstandings there are and the more economical things tend to be. When I say economical, I am referring to value in terms of quality, effort, speed, and time dealing with each aspect. A large percentage of CAPEX is a function of time-based service charges from vendors in the form of minutes, hours, and days that end up on a field ticket and final invoice, extracted relentlessly and unforgivingly.

Keeping things as close as possible to the wellsite helps reduce these critical time-based numbers, especially when things do not go as planned, something breaks, or someone needs to run back to the shop "real quick." Every extra minute adds up against our efforts to get costs as low as possible.

Additionally, geographical proximity improves quality. When our operations are close to the vendor's central place of business or factory, there is often increased oversight from vendor management. It is also less of a hassle to get the proper tools or components to ensure success. When the vendor shop is far from the wellsite, it is common to make do with what is on location because of the inconvenience, time, and cost to go back to the shop to replace or fix something—even though that would increase the probability of success.

—————————— **ACTION** ——————————

# Audit Vendors

Every active vendor must be audited on at least an annual or semi-annual basis. Audits should be conducted separately from the operations team to get an unbiased independent analysis and to ensure there are no shenanigans occurring. Consider splitting the audit into three domains: invoice examination, performance review, and facility inspection.

## Invoice Examination

During the annual audit, the first step is to pull all field tickets and final invoices from the last twelve months for the vendor under examination.

Confirm the field tickets match the final invoices. I have caught vendors changing final invoices after approved field tickets were signed, so we want to do a quick check and double-check with internal accounting transactions to confirm there is nothing questionable going on internally.

Second, compare the original bid prices and bid structure to the invoices to confirm they match. It is common for vendors to quietly change pricing and pricing structure throughout the year or for vendor reps to make pricing mistakes that get carried over, so we want to perform thorough due diligence.

Third, chart the invoices and line items broken down in as much detail as possible to see what each cost is doing over time. Are we building knowledge and experience as we do more work with this vendor, which is realized with efficiency and lower invoiced costs?

Or is the vendor taking advantage of us? This will be clear looking at cost trends on a line-item and total cost basis.

Fourth, add non-cost items to the analysis, including vendor personnel names who are on location executing operations. Are certain vendor personnel more efficient than others? Is one person responsible for more mistakes?

Fifth, consider developing company-specific, service performance metrics, for additional vendor insights. Much analysis can be done here.

Finally, it is good practice for every service to have more than one vendor performing the work so we can audit both vendors simultaneously and perform all sorts of interesting quantitative analysis between vendor competitors.

Artificial intelligence-based software can help reduce the data gathering, processing, and analysis workload. Or old-school excel modeling can get the job done. A combination of both methods will likely be most revealing.

## Performance Review

Our quantitative analysis from the invoice audit provides the foundation for our performance review. The more detailed the field tickets and final invoices, the more we can dig into the performance of our vendors.

If vendor invoices lack transparency, it should be addressed by having the vendors modify the field tickets with additional detail to perform deeper diligence before they are approved and signed.

However, even with this additional detail, it is not

uncommon to have less than optimal information included on the field tickets and final invoices. Therefore, we need to look at daily operations reports and recorded real-time well data. Additionally, we must speak with the wellsite consultants and engineers who oversee the work.

Launch an online survey to assess the vendor under audit and send it to everyone who interacts with the vendor, particularly the company reps who sign the field tickets.

This can be accomplished with several cloud-based systems for free or minimal cost. Then, construct another survey and send it to the vendor to assess their experience working with our company.

Gear the questions to address safety, ethics, performance, and opportunities to reduce risk, improve quality, and reduce costs. Include all the vendor technicians that are on location based on the tickets.

Next, speak with the company reps who signed the tickets to get in-person feedback and confirm they signed and approved the tickets under audit. If there are any shenanigans, we might catch them here.

The primary goal is to find opportunities to reduce risk and maximize FCF. Most operators do not realize that the field tickets and final invoices are a treasure trove when it comes to data mining for cash flow because the invoices are detailed operational facts of cash leaving our company, which can be utilized to find opportunities to reduce costs, improve performance, or make vendor changes.

If the field team favors a vendor based on performance, we should see it on the invoices. If we are making good vendor selection decisions, we should see it on the invoices. If a vendor is building experience, we should see it on the invoices. However, the opposite is also true.

> **In an industry that is highly subjective, invoices tell the truth.**

## Vendor Facility Inspection

With a detailed understanding of vendor-specific invoicing metrics, combined with a performance analysis of that vendor, travelling to the vendor's place of business is the next step.

It is an industry tradition for vendor representatives to visit operators' offices. However, I prefer to go to the vendor's place of business to get a better understanding of the value they bring to the operation. For example, there have been several instances in which visiting the vendor's shop revealed a dilapidated enterprise, a glorified middleman, minimal standard operating procedures (SOPs), a ghost company, and high-risk practices that transferred onto location.

Based on the services under audit, they should have been performing QA/QC, repairing items, and inspecting items, but there was not much of anything to be able to do that. In another example, I found that the vendor facility was storing my OCTG precariously, potentially causing damage.

## Final Report

After the audit, construct a written report on the overall findings and discuss it with the team. This is an opportunity to identify and plan cost reduction actions for our company and the vendor. Additionally, meeting with vendors to discuss their successes and areas for improvement can add value if they are open to constructive feedback, which most top vendors are.

When the vendors realize the depth of the analysis occurring behind the scenes and the desire to continue working with them to build a stronger, more productive relationship, they will be motivated to provide higher-quality service. This may result in better people and equipment sent to our wells.

Hawthorne Works, a telephone and consumer products factory outside of Chicago employing 45,000 workers, commissioned a study to see if workers would be more productive under different levels of light.[19]

Productivity improved when changes were made by observers and then fell when the study ended, and the observers left. It is believed that the performance improvement occurred due to someone showing an interest in the workers.[20]

This has been demonstrated in multiple additional studies and has become known as the Hawthorne effect, also called the "observer effect," in which individuals under observation modify their behavior due to an awareness of being observed.[21]

People are affected and act differently when they are being watched. In my opinion, people do a better job when they get more attention, know there is oversight, have management interest, and receive feedback. Auditing vendors regularly helps accomplish this, which further reduces our costs and maximizes our free cash flow.

--------- ACTION ---------
## Supply Chain Shenanigans

95% of U.S. businesses are affected by theft.[22] Approximately 33% of corporate bankruptcies in the United States are linked to theft.[23] Globally, the top two industries for theft by median loss are mining and energy.[24]

Theft is a serious issue impacting cash flow in the oil and gas industry due to the amount of money flowing through the system, the remote aspect of oilfield operations, and the trust we place in managers, employees, vendors, and contractors.

Theft in all its forms can push a shale company into bankruptcy or, at the very least, materially erode cash flow performance. Therefore, we need to understand and address the many ways theft can occur in the shale industry to prevent it, discourage anyone that is considering it, and catch anyone who has engaged in or is currently doing it.

"Supply Chain Shenanigans" is my lighthearted way to refer to the rather serious topic of theft, fraud, corruption, and many other unscrupulous tactics (illegal and legal) utilized to quietly extract value out of the enterprise by managers, employees, and contractors. Shenanigans are occurring at every company. No organization is impervious. If you somehow think that your company is the exception, you don't get out much.

With everything that needs attention in this business and the amount of risk involved in oilfield operations, the last thing we want to think about is theft, but that's why it's pervasive in our industry. It's not a focal point and many do not think it is happening.

As a result, people take advantage and do not think they are going to get caught. If they do get caught, the consequences are often not that bad. They just get fired and have another job in a week because hiring diligence in the oil and gas industry by vendors and operators is minimal.

Due to the amount of money involved in oilfield operations, it is simply too tempting for many people not to take advantage. Remember, 75% of employees have stolen from their employer, and 90% of significant theft losses are from employees. [25]

If the average worker is only productive for 60% of the day and the average office worker is only productive for three hours per day, they have a lot of free time to engage in shenanigans. [26]

When people feel angry, upset, unappreciated, or underpaid, a good employee or vendor may decide to give themselves a pay increase and engage in shenanigans to facilitate it. It is also common for people to justify their actions as good or necessary for the company even though they are highly questionable but not necessarily illegal.

To address some of the gray areas and make it clear what is not acceptable, consider renewing or enhancing the current company code of conduct with a modern look and feel to help resonate with today's workforce.

With a strong foundation in place, represented by a code of conduct, a solid company culture, and a reputation for catching shenanigans, let's review some of the most common concerns, red flags, and occurrences.

## Invoicing Shenanigans

- Inflated invoices submitted by vendors.
- Incorrect price book, with higher prices utilized for tickets.
- Wellsite consultant signs field tickets without confirming accuracy or is intentionally approving incorrect tickets.
- Random mistakes on tickets that always favor the vendor.
- Double billing with minor differences on invoice header.
- 3$^{rd}$ party costs included in service as per agreement, but 3$^{rd}$ party submits invoices to operator separately.
- Invoices are broken down into multiple smaller invoices.
- Fake invoices submitted below approval threshold.
- Signatures, stamps, stickers forged on field tickets.
- Approved field ticket amount is increased on final invoice.
- False invoices generated to cover up embezzlement.
- Employee creates sham company and submits invoices.
- Vendor employee creates sham company and submits invoices to vendor, which are then passed to the operator.
- Vendors take advantage of consultants approving invoices digitally on mobile devices, which are harder to scrutinize.
- Employees or vendors use modern invoicing tech systems and mobile apps to their advantage, against operator.
- Vendor interacts only with specific operator personnel.
- Unclear reasons for a service or invoice.
- Employee or consultant becomes unsettled if someone else from the company interacts with the vendor.
- Employee, consultant, or vendor cannot produce support docs or confirm accuracy during audit process.

- Vendor claims missing or stolen records during audit.
- Vendor uses tactics to circumnavigate the approval process.
- Vendor submits invoices late or waits until after crew change to minimize or avoid diligence.
- Vendor trickles in additional invoices after the original field ticket was already approved.
- Employee opens credit cards in the company name for personal use and pays with company funds disguised with fake invoices.
- Contractor fails to pay subcontractors for work that was already paid by operator, so subcontractors submit additional invoices directly to operator.
- Change request on vendor payment details.

## Kickback Shenanigans

- Vendor invoices are above market price.
- Invoices do not match bid prices.
- Consultants or employees go on trips with vendors.
- There is pressure to use a specific vendor.
- Continued use of vendor with poor track record.
- Employees or consultants collude with vendor to submit and approve fake invoices.
- Employee or consultant hints, implies, suggests, or requests kickbacks, favors, trips, meals, trinkets, or other gratuities.
- Complaints from vendors that kickbacks are occurring.
- Minimal competitive bidding.
- Consultants or employees are related to vendors.
- An unnecessary middleman is involved in the operation.

- Employee provides max price point to vendor in exchange for kickback on spread between cost and invoice.
- Consultants or employees are too close to vendors.

## Bidding Shenanigans

- Undisclosed relationship with vendor.
- Winning bid is high compared to expectations.
- Bid information is leaked to favored vendor.
- Bid is not representative of actual costs.
- Actual costs are higher than bid, vendor claims operational changes increased costs, bid assumptions were incorrect, or any one of a number of excuses to explain higher invoices.
- Pattern of rotating winning bidders.
- Collusion between bidders.
- Vendors no-bid and then become subcontractors.
- Winning bidder is subcontracting the majority of the work.

## Conflict Shenanigans

- Invoice approval process is outsourced to the same entity that supplies wellsite consultants to the operator.
- Engineering firm overseeing operations also has an undisclosed ownership or relationship with vendors.
- Wellsite consultant also owns or has an undisclosed interest in a vendor on location.
- Employee or vendor has an undisclosed side business.
- Employee, consultant, or contractor acquires a current vendor or interest in a provider without disclosing it.

- Employee, consultant, or contractor receives compensation directly or indirectly from a vendor.
- Employee or contractor has an interest in a tool, process, material, or anything utilized by the operator.

## Operational Shenanigans

- Vendor brings unnecessary people to location.
- Vendor trains people on our location without notification.
- Vendor brings equipment to location that was not requested and charges for it.
- Vendor uses our location as a staging pad for other clients.
- Vendor subcontracts the work without notification.
- Vendor masquerades as direct provider but is a middleman.
- Consultant, vendor employee, or employee is running a side business while working for us.
- Vendor is using our well to conduct an experiment, run an unproven subcontractor, test a new procedure, try a new material or chemical, run a new tool or piece of equipment, run a slightly modified tool, or do something that was not agreed to or disclosed and approved.
- Vendors create problems, intentionally or unintentionally, which are blamed on "the oilfield."
- Vendor creates problems which he then fixes; gets rewarded with bigger tickets and respect from operator as a "problem solver," even though he created the problems.
- Vendor claims to not know about well conditions, pressures, temps, depths, or any variable impacting their ability to perform.

This list is by no means an all-encompassing list, as shenanigans and red flags are just about endless, especially as technology improves, and new methods are used to address weaknesses in systems. However, no matter how advanced control systems are, people determined to take advantage for personal benefit will find a way. Therefore, we must make it very difficult and do everything possible to catch them.

If analysis suggests that over 30% of bankruptcies are linked to theft[27], we do not have a choice but to take massive action against all forms of shenanigans. I wonder how many companies at the precipice of bankruptcy realize it was theft all along. Unfortunately, once a company gets to that point, it is often already too late.

Globally, crude oil is the largest stolen resource and is very common in the United States.[28] Recently, a North Dakota man caught stealing $2.4MM worth of oil summed it up this way:

> *"There are a lot of trucking companies out here*
> *that skim oil, everybody does it, it's very easy,*
> *you can manipulate the numbers."* [29]

Unfortunately, this is true. It is easy to skim oil. Therefore, we must be more clever than the criminals. On the next page is a chart titled "Notable Supply Chain Shenanigans," detailing a few relevant real-world shenanigans from the energy industry. As the data illustrates, people, internally and externally, will construct innovative ways to take advantage of the operator. Another reason why having good people is critical.

# Notable Supply Chain Shenanigans

| Category | Fraud Details |
|---|---|
| **Invoicing** | Oilfield vendor submits +1,100 fraudulent invoices over several years, each under $5K; a level they knew received minimal review before approval. **($4.5MM fraud)** |
| **Invoicing** | Oilfield vendor worker forms LLC, creates fake invoices for purchases of equipment, parts, tools, reimbursed by vendor, then added to operator invoice. **($450K fraud)** |
| **Invoicing** | Oil executives submit false invoices for fake operations which they self-approve to receive company funds for personal benefit including travel, hotels, country clubs, fishing equipment, and hunting. **($1.5MM fraud)** |
| **Invoicing** | Oil executives created false invoices to support inflating revenue before & after going public. **($70.0MM fraud)** |
| **Invoicing** | Oilfield office manager created false invoices to conceal electronic funds transfers from company accounts to third parties for her own benefit. **($1.2MM fraud)** |
| **Invoicing** | Oilfield dispatcher created false waybills and invoices showing fake deliveries that were hidden by attaching them to legitimate waybills and invoices. **($800K fraud)** |
| **Invoicing** *and* **Stolen Oil** | Field supervisor submits 116 fake invoices for services never rendered. Then he stole 50 loads of oil from his employer and sold it to a reclaimer. **($400K fraud)** |
| **Kickbacks** | Oilfield managers hired salesmen to sell products and received kickbacks from commissions. **($150K fraud)** |
| **Bidding** *and* **Kickbacks** | U.S. Strategic Petroleum Reserve contractor conspired with sub providing confidential bid info for competitive advantage during 50 bid processes. Contractor received kickbacks from sub. **($15.0MM fraud)** |
| **Conflict** *and* **Invoices** | Engineer acquired control of a service company which was a vendor for the Oil Company he worked for. Then he invoiced the Oil Company for maintenance, which little work was performed, billed for equipment never delivered, and double billed. **($1.3MM fraud)** |

*This table was constructed referencing multiple sources. See Sources, Chapter 5, note 30, infra*

—————————— ACTION ——————————

## Ethics Hotline

In a 2018 interview on CNBC, Warren Buffet, the world's most successful investor and CEO of Berkshire Hathaway, discussed the topic of identifying corporate bad behavior in relation to his experience with problems at Wells Fargo, Salomon Brothers, and GEICO. In his opinion an ethics hotline is the single best tool for identifying unethical practices. Below is a transcript from the interview.[31]

**Buffet:** *"We have 377,000 people working for Berkshire and right now I don't know whether five of them or ten of them or 20 of them – but I'll guarantee it isn't zero – are doing something wrong. My job is to act when I hear about anything, and to make sure we've got some systems, so we do hear about things. We get about 2,000 contacts through what we call 'the hotline' a year. I get anonymous letters and those are the two best sources for finding out where something is wrong. I mean, it's better than having 100 people crawling all over the books. Anonymous [letters] sometimes they sign them, I received one last week, you know.*

**Interviewer:** *"You received one last week about something happening at Berkshire?"*

**Buffet:** *"Yeah, sure, but I mean I receive them all the time. They're going to come in. Sometimes people just* *don't like the person working next to them. They come in for a lot of frivolous reasons. But you have to look into what it is. If you look at 100, and 99 are, you know the guy next to me has bad breath or something like that but that's the way you do find them, overwhelmingly is tips basically."*

**Interviewer:** *"How do you track down every one of those tips?"*

**Buffet:** *"Well we have an audit department that sorts out the ones that they think that I ought to see. You know, so Becki Amick is in charge of the audit department. And incidentally the larger companies, they have their own groups too but those people cannot only write their own company but they can write us. And some of them just come in, you know, a letter comes into Warren Buffet Chairman. It's usually not signed but that's okay. I mean, obviously when they get very specific and say, this is going on or that's going on, that's what happened at American Express in 1964. They had a field warehousing subsidiary. They were getting calls from a guy at a bar in Bayonne to the head of the - and he was saying, 'The tanks are phony. Go to this tank and go to that tank and you will find that it's not filled with solid oil.' And the guy didn't want to hear it. And he didn't want to tell his boss. And then it just gets worse and worse and worse."*

Think about this for a moment. The top investor in the world, and one of the richest people by any metric, takes time out of his day to deal with letters and calls coming in from an ethics hotline.

Berkshire Hathaway has three systems in which people can report issues, a toll-free number, letters to the Chairman, and a dedicated website. A third-party organization provides the website and toll-free number in which issues and concerns associated with unethical or illegal activity can be reported anonymously and confidentially.[32]

This system also allows people to follow up on their report anonymously. Berkshire Hathaway has woven these systems into their "Code of Business Conduct and Ethics." [33]

The fact that Buffet personally reviews the calls, web submissions, and letters suggests he values this method of finding problems—tips. He knows this based on experience and studying issues at other companies.

The example he refers to at the end of his comment is the famous Salad Oil scandal of 1963 that caused over $1.82 billion (2023 dollars) in losses to American Express, Bank of America, and many international trading companies.[34]

Allied Crude Vegetable Oil company of New Jersey was able to obtain loans based on its inventory of salad oil as collateral. To do this, American Express would validate how much oil Allied was holding and write a warehouse receipt.

That document allowed Allied to go to a bank and offer it as collateral. Once Anthony De Angelis, the owner of Allied, noticed how weak the inspection process was from American Express, he started to supplement the oil with water to fraudulently expand his inventory. He built a system to transfer oil from tank to tank to further fool auditors, and he forged receipts.[35]

With the money from the loans, Anthony bought as many soybean oil futures as he could in an attempt to corner the market, owning the physical commodity and the futures. As Warren Buffet mentioned, a guy called in to American Express to report that the tanks were not full of oil, but it was ignored.[36]

When the fraud was finally discovered it was so big it nearly took down American Express. Warren Buffet took advantage of the situation and purchased 5% of the company.[37]

There are many lessons from studying this case. First, every oil and gas company should have an ethics reporting system that is widely advertised to employees and vendors. Second, each report needs to be investigated. Third, if audit systems are weak or non-existent, people will take advantage of it, as Allied Crude did.

If Warren Buffet takes time out of his day to review ethics reports from individual employees and vendors, then no matter what we are doing or how busy our day is, we can make time to address potential ethical issues. Even though 99% of them may be frivolous, it is still worth it because it only takes one to destroy our company, reputation, or end up on the "CEO Departures due to Risk Management Dynamics" list.

We must do everything possible to prevent this from happening to us. This is why Warren Buffet exercises such extreme diligence when it comes to matters like this. With a large operation, he knows unethical actions are happening, as he said, "*I'll guarantee it isn't zero.*"[38]

---
# ACTION
## Strengthen Compliance Systems
---

Due to the capital intensity of our industry and the ability for weaknesses in compliance systems to destroy the company, consider installing robust systems with multiple checks and balances, especially on all payments, before any cash leaves the company. For example, regarding invoicing, systemize the process with strong rules. Below are three to consider:

## 1. At least two people, one from the field and one from the corporate office must approve each invoice before it is paid.

When a field ticket is generated, have the company rep on location be the initial approver with at least one more approver, familiar with the work, located in the corporate office.

If there are only two approvers, then the second approver must not be in the field. For example, if a consultant approves a ticket, and then the superintendent approves the final invoice, do not pay that invoice until someone in the corporate office looks at the ticket and confirms it is correct. The field operation needs oversight, and if the field or office is the only approver, it opens us up to shenanigans.

Furthermore, if an invoice is only approved in the corporate office without someone from the field also approving the invoice, it opens the company up to corporate office fraud.

Based on recent fraud cases, many fake invoice scandals and other invoicing shenanigans occurred exclusively in corporate offices by people working within and around the invoicing software systems, far removed from actual field operations. It seems digital technology may reduce field-based theft but increase office-based theft.

## 2. Never sign a field ticket unless all items are confirmed correct.

One would think this rule is a given. Don't sign digitally, don't sign physically, don't stamp, or put the well sticker until every single price and service variable is confirmed correct, right? Unfortunately, this is not what is happening.

The reality in the field is much different. It is common for the field representatives on location overseeing the work to not know what the exact costs should be. They don't have anything from the oil company office clearly outlining the bid prices and negotiated cost structure.

Significant work goes into getting our costs down with bidding and negotiating, but when it comes time to approve the tickets, there is very little diligence.

The vendors know this, and it is easy to take advantage of, which many do. The first line of defense is the company representative on location who oversees the work and confirms that it has occurred as planned. That person in the field, often the wellsite leader or company man, is one of the only people in the entire operation who can confirm that all the variables are matching with what actually occurred.

A vendor could easily pencil-whip the hours, amounts, equipment, service, personnel, and many other items on the field tickets. It's too tempting not to. We would not know for sure regarding many of these elements unless we were on location watching the work.

Therefore, the field representative must know what all the costs should be so they can double-check the costs and the service variables to approve the tickets.

Many companies are becoming fully digital to accelerate invoice processing, so everything is done on mobile devices. The problem with this is that the tickets are very hard to diligence in detail on mobile devices.

In these situations, it is common for field representatives to be told to digitally sign the tickets and not worry about the accuracy because it will be checked in the office.

I strongly discourage this behavior and system. I do not want anyone signing anything until it is checked as 100% correct, and if it cannot be confirmed, call the office and confirm before signing anything.

This reminds me of retail self-checkout systems, which were supposed to replace human cashiers. However, the technology has caused serious problems, even after many additional layers of software, hardware, and personnel were added to the floor and above, watching security cameras from multiple angles. As a result, some retailers are restricting or removing self-checkouts. A problem with a retail self-checkout may only cost a few dollars compared to thousands in our industry. AI can assist in ticket diligence but cannot replace us.

Additionally, on $100,000 plus tickets and other high financial risk items, do not sign in the field until both the office and field confirm it is correct, then approve the field ticket. If we have AI, incorporate it into the diligence process. For example, we can run our human diligence process and follow up with our AI. Then, have a human look at it again to double-check. Then consider signing.

The problem is, once the field ticket is signed, if it's incorrect, we are now fighting an uphill battle that eats up time and resources. Negotiating power is greatest before the field ticket is signed, and many tickets are not correct to the level of precision we are looking for. Plus, the office might know if the prices are correct, but they will not know if the services are correct. Both must withstand due diligence before signing.

## 3. Demand detailed line-item field tickets.

Increasingly, vendors are submitting tickets without line-item details. Their aim is to expedite approval. I understand the appeal of this approach. However, accepting such obscure field tickets leaves us with less to review and question, leading to inaccurate billing and increased costs.

The solution to this is to reject any obscure tickets. Refuse to sign and tell the vendor to fix it and resubmit.

Detailed field tickets are not only useful when determining whether we are charged correctly but also when we are looking for ways to reduce our costs through design changes, efficiency opportunities, and operational adjustments.

On complicated high-dollar tickets, to ensure every line item is checked, consider requiring the wellsite leaders to put their initials next to each line item. During the approval process, to see that each line item has initials adds value during the final review before payment is issued.

## Artificial Intelligence Invoice Companion

Everything we are reviewing can be transformed into algorithms and code, which can be incorporated into our artificial intelligence (AI) invoice auditing system to help find problems and assist in processing.

AI is not a replacement for human diligence. It is a companion. It is another check on what we are doing. There are services that we can utilize for a fee or build our own. Our Virtual Company Man (VCM) incorporates our collective wisdom to provide due diligence of the field tickets.

Every time we find an issue with the tickets, we log it in the Operational Issues List. Then, each day, we upload that data into our VCM, which advises our entire operation. With this system, we increase cash flow per dollar spent by achieving more accurate field ticket amounts. AI-supported field ticket diligence enables us to scale our human intelligence regarding field ticket error detection, across our company.

Whichever system we decide to use, confirm we can tailor it to fit our needs and build onto it when new issues are identified, or we want to test a new idea.

For example, if we are concerned with small dollar invoices below a certain threshold, which are getting less human diligence, we could build algorithms to identify red flags and have the AI go through years of historic small dollar invoices to find potential shenanigans.

That is one of the benefits of AI. Once we get it set up and programmed, we can instruct it to go back in time and look for fraud. Then, we can address those issues today.

# ——— ACTION ———
## Cost Book

Once our vendors are lined up and costs confirmed, consider building a detailed document summarizing what each service should cost. I refer to this document as the Cost Book.

This document should include line-item details so the field tickets can be checked on a line-by-line basis. The goal is to make it easy for our field team to look at tickets while cross-referencing the Cost Book to confirm we are charged correctly.

> **We must do everything we can to make our field team's job easier.**

The Cost Book can be as simple as one sheet of paper with the costs listed. This way, the wellsite leaders do not have to fumble around with a stack of papers or decipher the bid documents to determine if the tickets are correct.

If there is a mistake with costs or services, the best time to fix it is right on location, as the vendor can easily make corrections and present an updated ticket to be reviewed again.

People reading this might think building a Cost Book is unnecessary because the team already has the AFE and cost codes. The problem is that the AFE line items and cost codes lump multiple services into one number. Plus, we need to drill down to an extreme level of detail when reconciling the field tickets.

Some might think the best thing to do is to have the vendor attach the bid documents to the tickets so they can be reviewed.

This is better than nothing, but it puts control with the vendor. To independently check what the vendor is presenting, we must have our own source data. I have had several cases in which the vendor presented a ticket with supporting bid data. The problem was, it was outdated bid data, and of course the ticket was higher than it should have been.

It is not advisable to have the vendor under examination provide the backup documents from which we will audit that vendor. We are reconciling our data to their data, independent from one another.

Furthermore, bid documents can be difficult to understand and lengthy. The Cost Book cuts out all the unnecessary parts and includes things to look for when reconciling tickets. Some operators build a spreadsheet that gets sent out each week with updated costs. The point is to diligence the tickets independently from whatever the vendor puts in front of us.

During aggressive cost reductions, when we start building efficiencies into our operation, and make cost modifications, vendors may not financially recognize our hard work. It is common for them to continue to invoice us as they have in the past. Even if we have negotiated lower costs, and those costs are correct in the tickets, it is easy to make other changes to service variables to claw back revenues. This problem is a big concern during cost reduction efforts because vendors are trying to maintain their revenue and margins. As discussed previously, we should expect people to do what they are incentivized to do.

Vendors have their own price book and are diligent when it comes time to build the field ticket. Relative to operators, vendors spend much more time reviewing the field ticket than operators spend checking it.

Often, field tickets are triple-checked by multiple levels of management within vendor firms before they are presented to the operator. The field ticket is the foundation of the vendor payment process, and a customer-approved field ticket is a big deal for the service reps.

In reviewing field tickets and invoices for over two decades, I have found that at least 95% of errors on field tickets and final invoices favor the vendor. These are not honest mistakes. Anyone who has worked on the vendor side will tell you, whenever there is uncertainty while writing the ticket, the higher price or structure will be utilized, or the service variable will be included.

The person building the ticket wants to ensure everything is charged for, so they don't get in trouble. Management wants to ensure there is a cushion just in case, so they add whatever is in question. If there is uncertainty, round up; that is the mental model building field tickets. If the customer approves it, they must agree. That is the assumption.

The burden is on the operator to address uncertainty on the tickets and force the vendor to fix it. During field ticket review, we must be aggressive in fixing every little detail because it adds up. It is in the small details where the mistakes are common. We are taking massive action to reduce costs, and we need to see it on the tickets.

If we are serious about cost reduction, we should feel pain

when approving invoices—the pain of payment. Approving invoices is a stressful process, especially when ensuring they are correct. I have had many people approve tickets with zero diligence and laugh about it with their buddies. Would they act that way if it was their money?

At one point during my career, I worked with several oil company engineers who would race each other to see who could approve more invoices faster. Can you believe it?

They would let the invoicing software system build up to a hundred or so invoices and then compete to see who could click the buttons faster within the cloud-based approval systems to get their invoice inbox down to zero first.

These were invoices that mostly had approved field tickets, and the engineers were the secondary check in the office before payment. This "it's already signed—must be correct" mentality is common, and very little diligence in the office occurs once the field tickets are approved.

Even more of a reason to have a robust approval process on location because there is a good chance this is the only time the costs will be thoroughly checked before payment.

Another advantage of having a Cost Book is that it allows leadership to see the entire cost structure in one streamlined document. When looking for ideas to reduce costs or change the pricing structure, this document can be beneficial to hand out and review during meetings and discussions in the office and field. The Cost Book makes it easy to look at and address every cost mechanism line-by-line. Additionally, all the data from the Cost Book can be uploaded into our artificial intelligence VCM systems to help aid our efforts.

————————— ACTION —————————
# Field Ticket Checklist

Maximizing free cash flow (FCF), the primary objective of this book, can be boiled down to the following equation:

$$\text{FCF} = \textbf{Cash Flow from Operations} - \textbf{CAPEX}\ [39]$$

From a field ops perspective, we can clarify free cash flow by substituting standard variables with the following:

$$\textbf{Success} = (\textbf{Net Production} \times \textbf{O\&G Price}) - \textbf{Field Tickets}$$

Success equals net production multiplied by oil and gas prices minus the field ticket amounts. We cannot control the price of oil or natural gas. What we can exert a great deal of control over are the field tickets. We must ensure the field ticket amounts reflect our hard work on reducing costs.

The tickets are the quantification of our efforts. They are our unbiased report card, showcasing the results of our dedication towards our cost reduction goals. We must take the time to ensure they are correct. The best time to address mistakes or issues with vendor performance is on location when the ticket is presented.

As soon as the work is complete and the ticket is presented, everything that has occurred is fresh in everyone's mind. This is the time to make corrections or get concessions if vendor performance was less than optimal. Before the ticket is signed, we have leverage during negotiations. Consider crafting a ticket approval checklist.

# Checklist for Approving Field Tickets

1) Obtain the latest Cost Book.

2) Confirm ticket line-item prices match the Cost Book.

3) Verify price structure matches the Cost Book.
   - e.g., flat rate, bundled products, time-based, fixed fee.

4) Validate vendor services are correct.
   - e.g., time, items, equipment, quantities, personnel.

5) Confirm there are no extra or hidden charges on the ticket.

6) Did the vendor have or cause problems, delays, issues, additional direct or indirect costs, or cause any stress?
   - If yes, negotiate cost relief or reject the ticket.
   - If not, prepare to approve.

7) If the ticket is above $100,000 call or text engineers to double-check calculations before approval.

8) If major operational problems occurred, call or text office to confirm there is agreement with the cost concessions offered by the vendor during negotiations before approval.

9) If you are not 100% sure about any aspect of the checklist or there is a questionable ticket item, call office to discuss.

10) Do you recommend this vendor or any specific personnel associated with this vendor to continue with our operation?
   - If not, call the office to discuss alternatives.

**High-risk vendors must be addressed immediately.**

—————————— **ACTION** ——————————

## Request Early Payment Discounts

Most vendors are concerned with operators who slow pay, which has become common. To address this, some vendors offer a discount if the operator pays within a certain period. Prompt payment discounts are often listed on invoices. Other times, we must ask for them. A common offer is a 2% discount if paid within 10 days. Many early pay discounts go unclaimed.

Consider asking vendors for early pay discounts. Then, prioritize invoices incorporating the discounts. If a 2% savings does not sound interesting, ask for more. If vendors know we are interested in early pay discounts, during certain times of the year, they may call us with an attractive deal. I have had vendors call me when they needed liquidity quickly and offer 5% to 10% discounts if we could issue payment immediately. That's a great deal.

If paying early is not interesting and more time is needed, consider supply-chain financing. Saudi Aramco considered this method to finance $2.0B of vendor invoices per month.[40] According to the WSJ, "In a typical supply-chain finance deal, a bank or other financial institution will pay a company's supplier faster than the normal payment terms, which can range from 60 to 120 days. The supplier agrees to receive slightly less than it would get by waiting and pays the bank a fee. The company pays back the bank the full amount down the road, improving its working capital by padding out the time it gets to hold onto its cash."[41]

─────────── **ACTION** ───────────
## Adjust Cost Structure

---

Working with vendors to adjust cost structure can help reduce actual cost volatility, cost exposure, cost risk, and total incurred cost. For example, if we currently have a day rate cost structure for drilling rigs, it might be favorable to restructure the agreement to pay per foot. Another structural adjustment would be to bundle certain items while unbundling others. There is no one-size-fits-all, and the potential structural configurations are abundant.

The optimal configuration will depend on vendor prices associated with cost structures, specific services, target formation, well design, area of operations, potential risks, and historical operations data.

For example, if we regularly experience significant downtime due to weather or other hard-to-control variables, cost structures that remove or limit non-productive time (NPT) charges might be favorable.

To get a good idea of what could be optimal, consider performing a statistical analysis of the past 12 months of actual cost data segmented by service or operational basis. Dissect each service to see if costs are stable and declining, typically a good sign, or if there is cost volatility with costs trending upwards. The latter could be an opportunity to restructure costs.

Next, build detailed cost models to see how different scenarios impact service costs with current structure and vendor prices. Then, experiment with different structures to see which adds the most value based on potential vendor pricing.

Instead of only looking at worst-case and best-case situations, consider incorporating Monte Carlo simulations to look at thousands of potential scenarios.

When requesting service providers change cost structure, it is difficult to know how receptive they will be to the new structure and the associated prices. If vendors are unwilling to accommodate us, it might be time to look at supporting a new entrant into the marketplace or vertically integrating certain aspects into our company. Below are several structures to consider modeling.

## Unbundling and Bundling

Unbundling various services has become the preferred structure due to cost savings and flexibility. Services to consider unbundling include fracturing, chemicals, sand, wireline, tools, fuel, trucking, and drilling fluids. Unbundling frac services can result in significant savings in many situations, but not all.

Bundled frac services have become much more competitive compared to unbundled services. Additionally, unbundling everything may not be the optimal configuration. It might make sense to keep some aspects bundled.

Bundling certain services can add value when there is significant risk, or when having a team that continuously works together is critical, such as post-frac plug millouts.

## Fixed Fee

Depending on cost, a fixed fee structure can be an attractive deal to help stabilize costs and remove financial risk. For example, if current AFEs are $7.0 million, and key vendors agree to fix costs to consistently deliver wells at $7.0 million, most operators would consider the deal.

With this structure, most of the CAPEX risk is removed, assuming that no matter what happens, the cost is $7.0 million. In reality, it would be challenging to get all the vendors to agree to this unless a turnkey middleman got involved, which would probably take too much margin, making the structure unattractive.

However, there are opportunities to fix costs on large portions of the operation using a variety of arrangements with multiple vendors that deliver a pseudo-fixed fee structure.

For example, we can get most vendors to agree to a competitively priced, fixed cost for pad construction, conductor casing drilling and installation, surface casing drilling and installation, cement jobs, rig mob, and certain tools.

For the primary drilling rig, we could convert from a per-day cost structure to a per-foot cost structure. That would effectively fix the cost based on depth.

The rig provider would then be incentivized to find efficiency in its business to increase margins. As long as this did not impact geologic targeting, it could be a win-win. Incentivizing a vendor to be innovative with a new cost structure could result in a technological breakthrough, benefiting both parties in the long run.

For the frac crew, we could convert from a per-minute cost structure to a per-stage cost structure for equipment, horsepower, and wireline. The cost per stage would, of course, be based on whether we are conducting simul-fracs, super-fracs, or standard zipper frac operations.

This would cover a significant portion of the costs. For the remaining services, we would engage with vendors to structure a price that incentivizes our success with the appropriate terms and conditions.

Time and amounts-based cost structures often reward service providers when the operator has problems. The longer the job takes, or the more products are utilized, the more money the vendors make.

Many vendors love winter operations for this very reason. When I see bad weather on the way, it gives me stress because of all the potential problems and actions that need to occur to reduce risk. When vendors see a winter storm coming, they have the opposite reaction. They see money falling from the sky. To help address irrational incentive structures and behaviors, limits on certain items should be considered, and various provisions should be incorporated to limit exposure.

## Cost Limits

If a vendor is incentivized based on the amount of materials utilized, they might look for reasons to run more materials or not be precise when running those materials. This can happen with drilling additives, frac chemicals, millout materials, and production chemicals. Therefore, one option is to set limits on how much material over the plan will be accepted.

For example, on frac jobs, if lab testing indicates we should only run 0.5 gallons per thousand of a scale inhibitor for a total of 90,000 gallons for the job, but the blender liquid additive pump actually runs 10% above that amount, a cost limit structure would limit the amount of chemical that we would be charged for to only 3% to 5% above the correct amount. This structure incentivizes the vendor to exercise diligence on what is utilized.

## Vertical Integration

If costs are not favorable or availability is an issue, it might make sense to look at vertical integration. For example, if affordable drilling rig availability is an issue and several SCR rigs are about to be stacked, we could purchase those rigs and run them through a separate entity.

If rig providers ask $40,000/day for a high-spec rig, but we can run an adequate rig for $10,000/day, it might make economic sense. Of course, a number of additional factors need to be modeled into the decision, but it could be something to consider.

The other aspect of this dynamic is that vendors need to know that operators can and will vertically integrate services. This is important to keep vendors from grabbing excessive margin in our business and becoming parasitic.

A parasitic vendor is one which increases the cash flow it extracts from us without increasing the value it delivers. When commodity prices rise, many vendors become parasitic.

Due to the scale and risk of shale operations, there is a fine line between being profitable and unprofitable. It only takes a few undesirable events to occur to make development unattractive.

Another reason to vertically integrate certain things is to have more control. For example, lower master frac valves are a critical component, and they are often left installed for months when flowing wells up casing during production.

Rental charges more than pay for the valves. Therefore, buying them works. Additionally, overseeing inspections before putting them back in service or deciding to scrap them is critical because a failure is catastrophic. Any item that costs less to own than rent is something to consider purchasing.

By changing the cost structure, we can incentivize vendors to address problems and contribute to cost reduction efforts. Anytime an undesirable situation occurs in which a vendor benefits at our expense, we must take action, or the situation will continue and expand. Changing the cost structure to address these types of situations is an effective tactic.

If a vendor is not receptive to adjusting cost structures to prevent zero-sum game situations, then we may have no choice but to find service providers who will bring solutions. We cannot reward vendors with continued work if there are no indications of improvement that translate onto the tickets or quality of the job in terms of the value we receive.

Performing statistical analysis on the invoices will help us maximize free cash flow by dissecting how money is moving, if we are incentivizing bad behavior, and if our vendors are improving performance or becoming parasitic.

---
# ACTION
## Remove Parasitic Middlemen
---

Middlemen are pervasive throughout business. The oil and natural gas industry is no exception. From a certain point of view, everyone is a middleman in some aspects of their job. Some middlemen add value, while others are parasitic predators that must be eliminated.

Intentionally or unintentionally utilizing intermediaries comes with an increased cost. This additional cost can be small, as in a few percentage points, or over 100% above what we would pay by direct sourcing services and equipment.

The oil industry has no room for this, especially during cost reduction efforts. We cannot afford to give cash flow to a low-value intermediary. Margins are too tight.

Operators cannot afford the convenience or potential value a middleman may bring unless it is quantifiable value above the additional expense and absolutely necessary.

Every middleman says, "It's only a couple points." But it adds up. Plus, we lose control, and that further increases costs. Also, the more people we involve, the more people we have in our pockets working to extract additional money from us, and middlemen are highly sophisticated. They can outsmart us when it comes to the tickets and invoices.

In certain situations, middlemen are unavoidable, either because there is no other way to do business directly or the value they bring is so significant that it is worth the additional expense.

## Advantages of Middlemen / Intermediaries

- Expert knowledge (tech, ops, problem-solving, key info)
- Hold large inventory from multiple manufacturers
- Assume ownership risk, theft, damage, and quality issues
- Incur expense of storing and distributing equipment
- Speed, proximity, convenience, accountability
- Provide specialized transport and assume transport risk
- Insurance, MSA, meet corporate requirements covering subcontractors, consultants, and manufacturers
- Centralized, one-stop source, and bundled offerings

I look at middlemen as an opportunity to reduce costs. From this perspective, a middleman is defined as someone or something that can be removed to establish a more direct relationship. Removing middlemen removes complexity, reduces cost, and reduces risk.

The first step in minimizing the use of middlemen is identifying where they are. Once we start digging for middlemen, we may be surprised at how prevalent they are. Sometimes, it's easy to identify middlemen; other times, it's hidden beneath multiple layers and very difficult.

Uncovering middlemen will help us reduce costs, especially the parasitic, predatory type. This type of middleman is always looking to get into our business. They are not partners but cash-flow vampires looking to suck up as much cash out of our company as possible. Parasitic middlemen not only increase costs but also make it difficult to assess the personnel and equipment that will be on location.

This jeopardizes service quality and accountability, adds unnecessary layers of communication, and increases risk.

I prefer to deal directly with the manufacturer, personnel, and service providers that will be on location. I like to have a direct relationship. In other words, if I hire a specific company, I want to see that company's equipment and personnel on location, not a subcontractor.

All my vendors know this about me. I make it known so that it is clear that I do not want hidden subs on location. I want to know exactly who is going to be there and where everything is coming from. I am not interested in getting middled.

There is a reason insurance companies ask operators, "What percentage of work will be subcontracted?" It is because subcontractors are linked to higher risk.

## Disadvantages of Middlemen / Intermediaries

- Increased costs, markups, hidden fees, surprise invoices
- Lower service quality, less accountability
- Dated equipment and missing components
- Higher risk of defective items, materials, services
- Increased complexity and reduced transparency
- Loss of control, reduced oversight, inability to vet, assess, inspect aspects of the equipment, personnel, and service
- Accuracy of information, low technical knowledge
- Reduced communication and miscommunication
- Double-brokered risk and increased subcontractor risk
- Increased risk of financial shenanigans

When looking to reduce middlemen, it is not uncommon to find that over 50% of the people and equipment on location are subcontractors. If we have a subcontractor, we have an intermediary attached.

In recent years, the probability of getting middled has increased because many vendors are running lean. If they win the bid or commit to providing a service but cannot make it because they are shorthanded or equipment is booked, it has become common practice to subcontract the work they cannot service and get the equipment or items from a $3^{rd}$ party.

It is typical for much of the equipment and materials we are invoiced on to be sourced from a different vendor than to whom we contracted the work. For example, we rent equipment from our preferred "friendly" vendor, which rents the equipment from someone else. The only way this structure works is if we are overpaying or getting lower quality.

Nobody wants to turn down work, and they won't. From the service perspective, as long as they provide the service, material, or personnel at the agreed price, it doesn't matter if they subcontracted it out because they could not get to it. They do not see a problem with this, but I do. The price might be as promised, but the risk is elevated.

Another strategy utilized by vendors is to set up a business model that double-brokers or subcontracts work at a profit to them. For example, if a salesman has a good relationship with an operator or can win a bid, they now have a set income structure to generate a profit with little work by subcontracting to another vendor at a lower cost. Many vendors operate with this model, unbeknownst to the operator.

--------------- ACTION ---------------

## Minimize the Amount of Equipment
## and Personnel on Location

Anything we can do to reduce the amount of equipment and number of people physically required on location will reduce risk and cost in most situations.

For example, using a spudder rig to drill the surface section will reduce risk by minimizing the amount of equipment and number of people during the preliminary stages of drilling. The surface section can carry significant risks that are often overlooked, especially when it comes to shallow gas kicks, the most dangerous type of kicks. They can occur without warning, breakout around the rig after shut-in, and may not be able to be contained.

Drilling the surface section with a spudder rig reduces exposure since the operation requires much less equipment and personnel compared to a big rig. Spudder rig surface drilling also creates a surface borehole that is more in gage (less washout) which may allow for shallow gas and other issues to be detected more effectively compared to drilling with a primary rig.

Furthermore, drilling with a spudder allows for a better surface cement job, which could result in savings on excess cement requirements to ensure circulation to surface. Additionally, when it comes time to install the casing head and then fill the cellar with cement up to the base plate, it is easier to accomplish with the spudder rig because less equipment blocks the area. Also, if we need an intermediate string, we can do it with a spudder rig depending on depth. Some spudder rigs can drill reasonably deep.

For every operation, question the number of people on location. Does each person have to be there?

Often, when it gets slow or a vendor is looking to get more hours for workers, they will put more bodies on our location. This is usually not a good thing. One way or another, we end up paying for it.

One exercise to consider performing is to estimate the number of people required for each operation, discussing the matter with each vendor beforehand.

If we see more people on location, start asking questions. Why are there more people than planned? Are they trainees, extra hands, or is something else going on?

It is not uncommon for vendors to train people on our location and then charge for it. Find out what's going on. We may be surprised at what we learn.

Additionally, question every piece of equipment on location for drilling, completions, and production. Extra equipment crowds the location, increasing risk and cost.

If it's sitting there, we are paying for it. Ask our team what can be removed.

Some vendors use customer locations as a staging pad for their equipment. I don't like the practice for several reasons. Extra equipment shrinks the available workspace, opens us up to get charged for it, and increases liability. If this extra equipment gets damaged, the operator will be held responsible for it since it's sitting on our property. If someone gets hurt because of the unnecessary equipment, we will be involved. Additionally, people forget that the extra equipment was not requested, and it often makes its way onto the field ticket. Now, we must spend time and energy to get the invoice corrected. That is if we catch the mistake.

The vendor is incentivized to put as much equipment on our location as we will allow. It helps them from a risk perspective because they transfer the risk to us. Additionally, they benefit from getting the equipment closer to other clients in the field.

Furthermore, if the equipment or materials are on location, maybe we will use them. The logic is that the operator is more likely to use equipment and materials if they are already on location. This is why some vendors will drop off various items on location for "free." Nothing is free in this business. If someone offers a free bit run or free frac plugs, don't walk away; run. It's a red flag.

Carrying extra or backup equipment around is common and prudent in many situations. To determine the cost-benefit, consider performing a statistical analysis on situations in which we used the backup equipment or were on NPT, waiting on a part or person. Downtime associated with waiting for something or someone to arrive on location are opportunities to reduce costs by optimizing our operation.

Many technologies and processes have been developed over recent years to prevent a single point of failure from causing NPT. As part of our statistical analysis, look for downtime that can be eliminated going forward by changing technologies and removing the need to wait on a person or part.

—————————— **ACTION** ——————————

## Enhance Delivery Receival Process

Engineering designs and operations require specific items and amounts to work successfully. However, it is typical for incorrect materials, equipment, and quantities to be delivered to location. This ends up causing all sorts of problems impacting everything from daily operations to final invoicing.

If the wrong equipment is on location, there is a high probability that it will accidentally get utilized or invoiced, even when it is not utilized. Per pad, it is common to experience at least one or two mistakes regarding the wrong equipment or amounts being delivered to location. If these mistakes are not caught, risk, cost, and quality will be impacted. Delivery mistakes can easily cost millions of dollars depending on what happens after the items are delivered.

Unfortunately, over the years, I have seen everything that could be delivered incorrectly get delivered incorrectly, including drill bit sizes, cement types and amounts, floats, BHA components, wellhead parts, casing, frac chemicals, proppant, and facility components—just to name a few.

When I find the wrong equipment on location, I request it be removed immediately. If the vendor cannot get it quickly, I place it as far away as possible and cover it in caution tape, take a few pictures, and instant message it to the group as an incorrect delivery to be removed. However, we never want to get to this point. We never want the wrong equipment on our locations because that is how it gets run—costing millions.

In general, industry practice is for the service provider or truck driver to unload the items on location and bring a delivery ticket to the wellsite consultant. It is standard practice for the consultant to ask a few questions and then sign the delivery ticket without going outside to inspect the items. When running full-scale operations, the same equipment and items are typically utilized from pad to pad. In the field, everyone assumes the materials and equipment from the last well will continue to be used on the current well, and all the service providers will bring the same items. This assumption is incorrect.

People are constantly moving in and out of positions on the services and operating side of the business due to the rotational nature of work schedules and high turnover. This adds complexity, and, as a result, the wrong things are brought to location. Furthermore, as continuous optimization occurs, designs change, requiring the delivery of different things.

**Strong processes prevent problems, supported by written checklists, procedures, and digital technologies.**

There are several actions to ensure the right equipment and materials are delivered and approved. First, instruct consultants to never approve delivery without inspecting the items to ensure they are correct and undamaged. Consider adding a requirement as part of a "Delivery Receival Checklist" to take a picture of every item and upload it to the Daily Visual Report. This way, everyone can see what has been delivered and where it is located on the pad. This also forces the consultant to inspect the items. Once we have a picture or video of the items delivered, we can use AI to scan them as a double check to confirm the correct items and amounts.

Second, consider utilizing a different stamp or digital process for delivery acknowledgment documentation. Sometimes, signing the delivery paperwork can result in getting invoiced based on that document, which is incorrect. Therefore, it might be good practice to not sign delivery papers and only have a stamp (physical or digital) indicating that delivery has been acknowledged.

Third, engineers should consider including specific items and amounts, with as much detail as possible, in the drilling and completion procedures. This could be in a separate section of the written procedures to make it easy for the wellsite representatives to reference before accepting deliveries on location.

Finally, consider requiring vendors to utilize geotagging systems when making deliveries to locations. These systems include coordinates, a time stamp, pictures, and additional delivery details as part of the paperwork left with wellsite representatives when items are delivered.

Pictures included with the vendor paperwork showing the items placed on location help identify issues, should they occur. During invoice approval, I always appreciate it when the vendor uses geotagging applications and includes the data with pictures as part of the final invoice paperwork. This is especially helpful when items are delivered to remote locations, and no one is physically on location to receive them.

--- **ACTION** ---

## Scrutinize Bulk Purchases

In the process of acquiring an oil company, there comes a time when we must evaluate the sellers' inventory. While performing this task, it is always interesting to see how much newly purchased equipment ends up stacked for years, rusting away, unused. When asking why all this equipment is stacked, I have heard, "We changed our designs," "It became obsolete," "We over-ordered," "It didn't work right," "There was a problem." Then the person mumbles off, head down in embarrassment.

It's easy to be enticed to bulk-purchase items for a discount, but it is not always the best move. Anything that requires us to spend more must be scrutinized. Feedback from everyone in the office and field is critical. It is not uncommon to bulk purchase an item that has been causing problems in the field, but no one mentioned it before the order was made, because minimal diligence was done.

### Popular Bulk Discounted Items

- OCTG (Casing, Tubing), Float Equipment, Sleeves
- Line Pipe, Valves, Fittings, Controls, Accessories
- Drilling Mud Additives
- Proppant, Frac Chemicals, Guns, Plugs
- Production Chemicals, Injection Pumps
- Artificial Lift Equipment, Packers
- Vessels, Surface Production Equipment, Tanks

Some items must be purchased in massive quantities to ensure availability during large-scale development projects. The problem is that manufacturing operations in the oil and gas industry are not the same as manufacturing in most other sectors.

Subsurface variability, evolving engineering preferences, variable operational needs, continuous optimization, and technological advancements often lead to bulk-purchased items quickly becoming obsolete or undesirable. Combined with volatile commodity prices resulting in changing operational plans, bulk purchased items are often stacked in a yard in the middle of nowhere, disintegrating.

## Bulk Purchasing Downsides and Risks

- Cash outlay
- Storage space required
- Potential bulk defects
- Damage risk in storage or transportation
- Theft of items
- Cannibalism of equipment for parts
- Labor and transportation costs to move equipment around
- Decreased flexibility
- Evolving engineering requirements
- Variable operational needs
- Continuing optimization
- Technological advancements
- Changing commodity prices
- Development plan adjustments

I do not advocate carrying zero inventory. However, there are things we can do to reduce the risk inherent in bulk ordering equipment and materials.

First, restrict bulk orders to no more than one year of forecast demand. In the oil and gas industry, it is hard to forecast equipment and materials needs much further than six months. If we only bulk-order one year of items, should development plans change, designs change, or rig count be cut, we most likely will not get stuck with excess stock that goes unused.

Second, request that the vendor we order from hold the inventory until we need it. This way we do not need to increase storage space, incur additional transportation costs moving equipment from the yard to location, and we do not hold the damage risk involved in carrying and storing items.

Operating companies are not inventory management experts. Items frequently get damaged in storage and during transportation. Therefore, it is common for items to show up on location unusable.

If these items came from internal inventory and are rejected on location, we take a complete loss or incur additional costs in time and equipment. This often eats up the savings thought to have been realized in bulk purchasing or costs more money than if we purchased piecemeal.

However, if the equipment comes directly from the supplier, then we can reject the equipment and easily call for new equipment. Equipment and materials should be inspected before transport and inspected on location before they are accepted and utilized.

If the vendor, manufacturer, or distributor holds the items until they are needed on location, we can avoid this additional invisible risk. Also, if they hold it for us and we end up not needing it, it is easier to transfer the items to another client.

The final aspect to address when making large orders is defect risk. When deciding to make bulk purchases of certain items, consider enlisting the help of industry experts and 3<sup>rd</sup> party inspectors to evaluate the items. This is critical for bulk purchases of OCTG, especially casing exposed to frac pressures. Signing a contract to bulk purchase casing at a discount might seem like a victory to be celebrated, but we could be signing our company's death warrant.

Bulk-purchasing defective casing will destroy an oil company due to the clarity gap that exists. Bad casing has contributed to the economic failure of multiple companies, particularly when operating at scale. Since large batches of casing are a bulk purchase necessary when developing shale assets, we must have a thorough independent inspection and diligence process with multiple checks and balances on this critical oilfield component.

Since maximizing free cash flow with cost reduction is impossible if casing failures occur, any action to reduce casing-related risk must be seriously considered. Due to its significance and interconnected relationship with almost every business discipline, casing is repeatedly addressed from multiple viewpoints throughout this book. Because of the cost, risk, and potential impact, OCTG must be a centerpiece that receives significant attention and diligence during all stages of planning and development.

# —————— ACTION ——————
## Optimize AFE Structure

Optimizing the AFE structure will help with transparency, personnel accountability, auditing, cost control, and cost reduction. It is common for AFEs to have minimal categories that are generic, obscure, and not optimized for horizontal development.

AFE structure is a legacy relic, originally built for vertical well development and intentionally designed to hide information from competitors and non-operated partners. However, this legacy AFE structure is also excellent at obscuring cost information internally.

For modern horizontal multi-well development, an optimized AFE structure with expanded transparency will add value in a variety of ways. An optimized AFE allows the team to easily see estimated and actual costs at a more granular level.

This becomes important during cost analysis before, during, and after operations. It is especially beneficial during budget variance reviews when costs are broken down on a line-item level comparing estimated versus actual, which is typically done by cost codes originating from the structure of the AFE.

If the line items and associated cost codes are generic, with many actual costs lumped together in one cost code, understanding our cost situation at a granular level will require additional time and effort to identify problems and opportunities.

For example, a legacy "lumper" structured AFE will use "Stimulation" as a cost category which includes many different costs. Whereas a detailed transparent AFE would use 10 to 20 cost categories, splitting out stimulation costs into a more granular level for additional transparency, accountability, tracking, and control.

| Obscure "Lumper" AFE | Detailed Transparent AFE |
|---|---|
| **Intangible Example Line Items** | **Intangible Example Line Items** |
| | 9300- Frac HHP |
| | 9301- Frac Support Equipment |
| | 9302- Proppant |
| | 9303- Proppant Transport |
| | 9304- Proppant Storage |
| 9300- Stimulation Services | 9305- Fuel / Power |
| | 9306- Power Support Equipment |
| | 9307- Source Water |
| | 9308- Water Transfer, Filters |
| | 9309- Frac Chemicals |
| | 9310- Digital Services |
| | 9350- Flowback Equipment |
| | 9351- Trailer Houses & Comms |
| 9350- Equipment Rentals | 9352- Frac Tanks |
| | 9353- Frac Stack, Manifolds |
| | 9354- Inspection & Repair |
| | 9355- Lights, Trash, Septic Systems |
| **Tangible Example Line Items** | **Tangible Example Line Items** |
| | 9400- Storage Tanks |
| | 9401- Vessels (Separator, Heater) |
| | 9402- Compression Eq. & Install |
| 9400- Surface Equipment | 9403- Gas Metering |
| | 9404- Parts, pipe, valves, fittings |
| | 9405- Containment, Liner |
| | 9406- Automation, Safety, Security |

Not only can we break down generic line items to show more detail, but we can also segment one level up. The primary classifications across most operating company's cost structures are intangible and tangible expenses.

It is common for a hodgepodge of intangible costs to be grouped into two subcategories: drilling and completion. One option is to continue using intangible and tangible as the primary categories but add 4 to 6 subcategories, for additional segmentation.

## AFE Example with Detailed Subcategories

| INTANGIBLE COSTS | | | | | | |
|---|---|---|---|---|---|---|
| Subcategory Description | 100 L&L | 200 Pad | 300 Drilling | 400 Completion | 500 Facility | 600 Flowback |

## Notice we split intangible costs into six subcategories:

1) Land and Legal (L&L)
2) Surface Operations (Pad)
3) Drilling Operations (Drilling)
4) Completions Operations (Completions)
5) Facility Construction (Facility)
6) Flowback Production (Flowback)

The typical AFE only has two subcategories: drilling and completion. A significant amount of money can be spent before drilling begins. Often, these expenses are dumped into the drilling category and onto the drilling team.

It is discouraging for the drilling team when land, legal, road, pad, and all associated costs hit the drilling cost category before drilling begins. Especially when pre-spud costs exceed AFE. This is commonly referred to as "starting off behind" and does not help morale on location.

Pre-drill cost dumping establishes a culture in which no one has control or responsibility over the cost. It allows people to hide costs and accountability.

Shale operations at scale usually have a land department focused on land and legal processes, a surface operations department focused on building roads, pads, and sourcing water, and a drilling department focused on drilling operations.

The drilling team is not directly responsible for or has control over costs incurred running title or settling surface damages. Additionally, the drilling team is not typically focused on managing the construction of the pad or roads.

When these costs are included in the drilling category, it allows land and field operations to avoid the spotlight of accountability by hiding costs while simultaneously demotivating the drilling team. It also allows the drilling team to blame the other departments for cost overruns, consciously or subconsciously.

Therefore, we have the land, legal, and surface operations burying costs in the drilling subcategory, the drilling team feeling that they cannot control their costs, and in the end, no one is responsible for anything.

Clearly segmenting pre-drill costs with separate subcategories resolves most of these problems. This can be accomplished with secondary categories, subcategories, or both, as we have done in our example.

Additionally, we want the drilling team to start with a zero balance on field estimated costs for the initial daily report. Most executives would be surprised to see the reaction on location when the first daily drilling report comes out with $500,000 in the incurred field estimated cost bucket under drilling, before any drilling equipment has mobilized to location.

When the field consultants see big numbers in the field estimated cost tab for drilling costs before drilling has actually commenced, it sends the wrong message:

*This sloppy company and operation is a free-for-all.*

The thinking in the field goes like this, "People that I don't know are hitting the drilling AFE with huge costs for no apparent reason, but it definitely isn't drilling because we haven't even started drilling." The mentality is, "If it's okay to hit the drilling AFE like this, then I can hit the drilling and completion AFE with a little extra cost here and there, and they will never notice. If they do, they probably won't care anyway. These people hit the drilling AFE with anything and everything. It's a cash machine."

The field team thinks, "If the office treats the AFE like this, they cannot see what we are doing. Don't worry about the cost; it's all talk because these people do not value precision. They are messy and will never know what we do. We do what we want to do. They do not need to know, nor will they be able to figure it out."

The same goes for completion costs when facility and flowback cost are mixed in with frac costs. Especially if the facility is built before frac shows up and the facility cost hits the completion field estimated cost prior to starting the frac.

Facility design, management, and optimization are often managed by the facility department. Therefore, the associated costs should be separated from the completion category. The same goes for flowback expenses, which are usually managed by the production department. We want to separate these groups from the completions category to help with transparency and cost reduction.

In recent years, facilities have become more elaborate, and costs have gone up. Isolating these costs will help with reducing them.

Flowback costs, usually the last significant cost incurred, can easily add up to hundreds of thousands of dollars if not properly managed. Therefore, consider a separate subcategory for flowback expenses since they are usually managed by the production department.

Sometimes it can be difficult to implement these types of AFE changes due to accounting systems and other complexities. One option is to have a lumped external AFE with minimal detail and a transparent internal AFE with a finer level of segmentation.

The other option is to add additional subcategories. This is usually easier to implement because there are normally unused cost codes within most accounting systems that could be assigned to new subcategories to provide a greater level of detail and transparency.

---
## ACTION
---

# Independently Verify
# and Validate Actual Costs

---

Accounting and finance are central to our plan to maximize cash flow. We must have an accurate understanding of historical and current actual costs. Field estimates are inaccurate, and it is not uncommon for actual costs to be incorrect or exclude expenditures that should be included that were intentionally or unintentionally left out.

Accurately accounting for our total expenditures is our baseline starting point. We must know exactly where we currently stand, including all costs, no matter what happened operationally or financially. It is not rational from a free cash flow perspective to exclude certain costs because they are not typical or not a continuous, reoccurring expense.

Costs are incurred, or they are not incurred. Do not rationalize a specific expense and then remove it based on the philosophy that the expense is irregular or nonrecurring. For example, if the team got stuck during drilling or had a casing issue resulting in significant unplanned expenses, these costs should not be removed because they are non-routine.

Often, these types of costs are removed from analysis and presentations with a small "adjusted metric" footnote. I understand, and in certain cases it needs to be done. However, for what we are doing it will not help to cleanse the costs. We need to see all expenses for every well to understand what is happening. To establish a starting point, we need to visit the accounting department regularly and relentlessly.

Once operations are finished and several months have passed, actual costs should be through the accounting system. However, it may take more time before all costs are accounted for. In certain situations, it might take six months to a year to get all costs processed by the accounting group. However long it takes, please confirm all costs are captured.

In certain situations, after a well or pad is complete, I have sent out letters to all vendors to submit invoices within 30 days because we are closing the AFE on the DSU. If invoices are not submitted in a timely matter, they will not be paid.

Establishing a fixed period to close an AFE can help ensure all costs are accounted for and prevent vendors from taking advantage of open-ended AFEs.

Even with this tactic, I have had invoices submitted one year after the operation is complete. Regardless of the methods utilized, the goal is to get an accurate account of all costs incurred.

With the actuals from accounting, the next step is for engineering to audit the actuals to confirm they are correct. There are software packages to help with this, but there is no substitute for manually looking at each cost incurred to find any abnormalities.

If there is a question, pull the final invoice and field tickets. I have caught many problems with this simple method, including incorrect invoices, ghost tickets, fraud, overbilling, double billing, tax issues, and a host of other tactics vendors use to attack our free cash flow.

The point of this process is to verify and validate actual

costs to establish an accurate starting point from which to benchmark our cost reduction efforts. During past cost reduction campaigns, whenever I went into a group as an outsider to help reduce costs, I often began with a current cost audit.

For example, I would ask, "What are current well costs?" The team would often give numbers from the most recent series of AFEs. This response is a red flag. Facing actual costs can be emotionally painful, especially when difficult problems were encountered on a series of wells impacting economics, asset confidence, and team morale.

Additionally, due to time constraints and pressing daily operational issues, which always take precedence, asset teams rarely have time to review actual costs in detail. Spud-to-rig release, frac days, field estimated cost, and large ticket items are usually utilized as a proxy for actual cost performance.

Making time for detailed actual cost audits is a great starting point for every cost reduction campaign. Looking back at actual costs provides a frame of reference from which to work and gauge performance going forward.

Actual cost analysis also provides a wealth of information to help identify opportunities to improve economics and create value. One of the aspects of a detailed actual cost audit that I enjoy is pulling all the field tickets and final invoices to see the line-item detail. The data within the line items and cost structure enable statistical analysis and spur new ideas to help maximize cash flow with cost reduction.

# ACTION

## Precision AFE's

Standardized AFEs save time in terms of office hours spent on generating cost estimates, but the process of using identical AFEs can result in higher actual costs and contribute to a culture of sloppiness.

It is common to construct uniform AFEs based on lateral length. Most asset teams have an estimated cost for 5,000' LL, 10,000' LL, and 20,000' LL that are used in development models and on AFEs cranked out as part of the approval process. Change the AFE well name, and we are good to go.

However, on a 10,000' LL well, it is common for actual effective lateral length to be 9,500' to 9,900'.

It is rare to have exactly 10,000' due to legal hardlines, nudges, drilling aspects, and completion dynamics. If an AFE is calculated on a 10,000' LL with 50 stages, based on 200' stages, but only 48 stages are possible due to a more precise calculation, then well cost should be at least $100,000 less.

Most managers and engineers enjoy having this secret money cushion. This hidden back pocket money in the AFE is often used to hide cost overruns, perform unsanctioned operations, conduct science experiments, and engage in other unholy activities. The solution is to abolish generic AFEs.

Some people think that generating precision AFEs is a waste of time. I disagree. We are looking for every opportunity to manage financial risk and add accuracy to a process embedded with uncertainty.

Constructing a precise AFE that can be adjusted will not add much time once the template is built and a system is put in place that does not accept generic AFEs.

## Cost precision sends a powerful message.

The AFEs establish a quantitative benchmark. If the AFEs are inaccurate, our process of measuring cost performance is inaccurate. Generic AFEs also convey that we do not value cost diligence, which can translate into our daily field estimated costs and final invoices.

When vendors and wellsite consultants know the AFEs are generic, they may take advantage of it or not diligence operations for efficiency and tickets for precision with extra effort. At the very least, generic AFEs send a message—that cost precision is not a priority.

If you are a manager or executive, ask for a copy of the AFE template to see how costs are calculated. We might be surprised at the lack of attention to detail. Of course, asking for the calculations will usually solve the problem.

Precision AFEs increase the accuracy of our benchmark costs from which to determine actual cost performance. If additional costs are needed because of uncertainty, it is better to increase the contingency due to the transparency of that number, which usually has its own line item.

If a well or project needs a large contingency, it's better not to bury it within the AFE because it makes it harder to assess risk and make capital allocation decisions.

# 6

## PRE-SPUD ACTIONS

—

O verlooked costs frequently occur before drilling during pre-spud operations, which includes all processes and associated CAPEX incurred before spud.

Detailed daily cost monitoring typically begins when drilling starts. Therefore, pre-spud costs occur with limited oversight and minimal scrutiny. Because these costs are not closely monitored, it is common to forget that they happened, especially if we leave them off the daily reports or do not account for them in a separate cost category.

As a result, it is common to think we are below AFE based on field estimated costs, forgetting to incorporate pre-spud expenditures. Furthermore, as well count per pad goes up and pre-spud costs are spread across more wells, it is common to think the cost impact per well is minimal; therefore, minimal effort to control these cost occurs.

The mentality is that whatever these costs are, it's not a problem because we will spread them out, leading to the mindset that pre-spud costs are irrelevant. As a result, pre-spud costs have expanded across our industry, increasing to significant levels on a per-well basis, impacting economics. The lack of attention to pre-spud costs has led to operational inefficiencies and many random items and services being billed to the well. The bottom line is that pre-spud costs need increased oversight to identify opportunities to reduce them.

## Common Pre-spud Costs

- Multiple Title Opinions
- Legal and Regulatory Services, Applications, Hearings
- Technical Services, Exhibit Preparation
- Environmental, Wildlife, Cultural Investigations & Reports
- Staking, Surveying, Mapping, Plats, Signs
- Surface Use Agreement, Water Purchase Agreement
- Surface Damages Payments, Water Payments
- Water Pit Construction
- Permitting, Notification Filings
- Insurance, Overhead
- Travel, Supplies, Consultants
- Road, Wellsite Pad, and Reserve Pit Construction
- Pad Stormwater Runoff Construction Control
- Fence, Cattle Guards, and Gate Installation
- Cellar, Conductor, Mousehole, Grout, Cover Plates, Nets
- Electric Power Installation, Microgrid Setups
- Midstream and Pipeline Installations
- Offset Wells, Parent Wells, and Parent Facility Preparation
- Rig Mobilization and Rig Up

---
# ACTION
## Insource Pre-Spud Prep Work
---

Outsourcing work to gain efficiency is a popular strategy to help operating companies drill the maximum number of wells. However, outsourcing comes at a cost, often based on a time component that does not directly align with operator interests if a low-cost strategy is desired.

Title, legal, regulatory, environmental, and technical services are charged on an hourly basis with little to no operator oversight. These professional service invoices are rarely challenged due to the nature of the work, often resulting in inefficiencies and overbilling. There is simply no service provider incentive to adapt efficiencies that result in billing fewer hours, other than losing the work.

With a focus on low-cost, in an environment in which fewer rigs are running, current internal staff should have the bandwidth to insource many pre-spud services currently performed by contractors.

Instruct internal staff to take direct ownership of currently outsourced activities. Much of the work required is already performed in-house. In fact, operator staff may find that by bringing the work in-house, we have more free time because we no longer must deal with service provider personnel, inefficiencies, mistakes, politics, invoicing, workforce changes, unnecessary meetings, explanations, document gathering, and conveyance. By bringing the work in-house, we eliminate all of it.

In addition to reducing costs, staff become stronger, more knowledgeable, and more valuable to the company and to themselves. By eliminating the middleman in many cases, our company establishes a direct relationship with key individuals and governmental agencies. This minimizes uncertainty and gives more control to the operator, where it belongs.

---

## Control reduces costs.

---

For example, previously we outsourced surface use agreement (SUA) preparation and surface landowner damages negotiations to land services contractors. By insourcing, not only did we eliminate all hourly charges, travel charges, and other miscellaneous bills, but we were also able to get much better deals and establish direct relationships with surface owners, who are effectively long-term partners and neighbors to our operation.

Furthermore, we saved significant time because we no longer had to spend hours dealing with the land service contractors. Therefore, dealing with outsourced service personnel costs double— our time and their time billed to us.

Another example would be federal drilling permits. When drilling on Bureau of Land Management (BLM) or Bureau of Indian Affairs (BIA) managed land, an Application for Permit to Drill (APD) must be filed. The process is detailed and involves many steps. I have outsourced these tasks and insourced them to help reduce costs. By insourcing, we were not only able to eliminate the associated cost burden but strengthened our relationships with people working at the BLM and BIA. Removing the middleman allowed us to better understand concerns and issues from the agencies on a well-by-well basis. This allowed us to optimize our operations on federally managed land to further reduce our costs. This would not have been possible if we continued to outsource BLM and BIA work. Furthermore, we became more aware of the small details and violations that could result in costly penalties.

## Insourcing increases intelligence.

Insourcing may not be possible for every service, but it should at least be discussed regularly, especially if vendor costs are not declining. If we are outsourcing these services and we do not see a decline in the invoiced amounts, it tells me that our vendors are not getting better at their work for us. They should become more efficient over time; therefore, the billed hours should decline. If we plot service costs over the past 12 months for each service, we can get a clearer picture of what is going on. Then, decide the best course of action. At the very least, we should meet with our vendors and review our analysis of their invoices to discuss opportunities to reduce them.

Regardless, outsourcing specific tasks does add value and should be implemented for multiple aspects of pre-spud operations. However, when striving to reduce well costs, the value of insourcing must be considered and tested to see if the cost savings and associated benefits add value to our operation and objectives.

————————— **ACTION** —————————

# Establish Fixed Costs and Stop Points
# for Land and Legal Expenses

Significant land and legal expenses can dig a financial hole before actual digging begins. Land and legal work can get out of hand quickly and catch people by surprise. These costs generally receive little attention, are underestimated in the drilling AFE, and are subject to variability, making cost control difficult. Management oversight and attention to land and legal expenses are cursory, which leads to minimal pressure on internal staff and external service providers to push costs down and increase efficiencies.

When analyzing actual line-item costs, look for land and legal expenses that should not be billed to the well. This will vary by company regarding what is billed compared to what should be billed.

For example, original title opinions (OTOs), preliminary title opinions, and ownership reports might not be billed to the well depending on company policies. However, drilling title opinions (DTOs) and division order title opinions (DOTOs) probably should be billed to the drilling AFE.

Whatever is decided, we want to ensure consistency in capturing the same expenses in the estimated and actual cost records for each well.

Second, lawyers need to be actively managed, just like everything else. One way to do this is to establish hard stops with dollar amounts. For example, if we estimate $50,000 in title opinion work, ask our provider for status updates when they reach $10,000 so we can see how the work is going.

This lets the lawyers know that we are engaged and monitoring the costs. When people know they are being watched regarding costs, they will monitor costs themselves and know it will be harder to run up the costs on the client. Land and legal hourly workers are notorious for running up the hours to extract as much as possible.

Third, manage which attorneys are working on our wells. Different attorneys charge different rates. Try to find efficient, low-cost options within the firms we employ, and minimize the amount of time the high-cost senior-level attorneys bill to our wells unless it's necessary; and if it's necessary, make sure they give us a good reason for why they are charging hours to our wells.

Fourth, combat excessive hourly expenses by requesting fixed fees. A fixed fee for spacing, location exception, multiunit, pooling, and other general matters helps reduce cost and minimize variability. This is easier for legal related regulatory work for application preparation compared to title work. Depending on the complexity of the title situation, the time it takes to perform a title opinion will vary.

For example, if our acreage is covered in legacy HBP vertical wells, it will take more time. However, once we have an idea of how much time it could take, working with the title attorneys may help establish a fixed-cost plan.

Fixed cost structures also help when it comes time to shop around. Hourly rates are good to compare, but fixed costs make it easier to see who is most efficient in our area of operations.

People make mistakes, and those mistakes drive up billable hours. In a way, the hourly structure incentivizes people to learn on our dime.

The operator is paying for the attorney's mistakes, which should not be the case. It is not economic to pay for the education and mistakes of land and legal service providers because that is part of the risk they take when doing this work.

One of my biggest issues with our industry and its commonly accepted practices is that the operator pays for everyone else's inefficiencies and mistakes. Fixed fee structures help address this dynamic. The more wells we drill the more likely we will be able to find service providers that will offer fixed fee options.

Finally, shop around for land and legal services to get a sense of who is efficient and who is not. Often, people get comfortable with an attorney, and that attorney gets lazy, leading to inefficiency and high costs.

This is true for any service charged hourly. When working hard to reduce costs, it is rather discouraging when pre-spud costs come in above expectations, and we are already behind on total well cost objectives before drilling begins.

Establishing fixed cost agreements on pre-spud services will help minimize cost variability. Additionally, once we have established fixed costs, it is easier to look for opportunities to push the cost down further. Standardized fixed fee structures allow for predictable actual well costs and reduce the unfortunate pre-spud high-cost surprises.

# Pre-spud Ops to Consider for Fixed Fee Structures

- Application Preparation
- Associated Hearings
- Title Opinions
- Exhibit Preparation
- Protest Situations
- Regulatory Services
- Surface Use Agreements and Negotiations
- Master Services Agreement (MSA) Reviews
- Letter Agreement Reviews
- Cultural Investigations and Reports
- Well Survey and Staking Operations
- Well Permitting
- Camera Installation
- Water Pit Construction
- Drilling Water Transfer
- Wellsite Pad Construction
- Fence, Cattle Guards, and Gate Installation
- Cellar and Conductor Installation
- Electric Power Installation and Microgrid Setups
- Nat Gas Lines and Field Conditioning Setups
- Rig Mobilization and Rig Up

## ———————— ACTION ————————

## Construct Irregular Shaped Locations

Standard oilfield surface locations are designed as a quadrilateral with four right angles, usually resembling a square or rectangle. It is an easy shape to explain to contractors. It is also a favorable geometry for drilling and frac operations. However, this design is not always the most effective.

When in full-scale manufacturing mode, standardizing designs can help simplify operational execution but can also result in higher costs and uneconomic wells.

The surface carries just as much variability as the subsurface. This variability must be understood when deciding where to build the location, the optimal shape of the location, and construction materials utilized.

Defaulting to the same standard square is often the easiest option. It requires little thought or planning. However, this standard design can be expensive and is not always the safest or most environmentally friendly.

If our default size is 500 feet by 500 feet, equivalent to 5.7 acres plus the road, approximately 6.0 acres total, do we really need 6.0 acres, or do we need 5.5 or 4.5 acres?

Any amount less than 6.0 acres will reduce construction costs, surface use damages paid to the surface landowner, pre-frac location repair costs, post-production repair costs, general maintenance and upkeep, interim reclamation, and end-of-life location P&A costs to convert the location back to its original condition. Additionally, a smaller location can be safer as long as we do not impact operational abilities.

To be clear, we do not want to build a micro-location that is so small that it is hard to move around. That would impact our big-picture goal of reducing total well costs. To reduce drilling and completion costs, we must increase our speed, and it is hard to do that with a poor location, whether in terms of location size or location stability.

We can go too far when shrinking the size of the pad or changing the shape and impact the capital-intensive drilling and fracturing operations. However, I advocate for building a fit-for-purpose solid location rather than an oversized, unstable location that is wasteful, environmentally unfriendly, or potentially unsafe.

If we always build the same default-sized and shaped location, we may unintentionally facilitate stability problems, washouts, water retention, or other costly situations. The reason for this is that we will end up building on areas of the surface that are not optimal for an oilfield location — areas that cannot handle the traffic, equipment, or operations. With a cookie-cutter pad design strategy, our team is forced to build on unfavorable ground, regardless of surface conditions.

After a location is built, most drilling and completion personnel view every foot of the location equally. This means that if there is a place to put something, they will not take aspects of the foundation into account because it all looks the same once it is built. But it is not the same.

For example, years ago, we built 5.0 to 7.0-acre pads with cement stabilization. No issues occurred during drilling, but when it came time for fracturing operations, significant problems materialized.

For frac proppant handling, we used a containerized sand system, which relies on significant forklift operations. This boxed sand system and associated forklift put more stress on the pad compared to other sand systems.

Due to the constant movement of the forklift in the same area with sharp twists and turns, a modified high-density polyethylene (HDPE) interlocking mat floor is installed to help with stability, prevent the wheels of the forklift from digging into the pad, and prevent the creation of deep ruts in the ground. Any ruts, even small ones, can make the operation unsafe due to the amount of weight the forklift is carrying. The floor the forklift operates on is referred to as the "dance floor," and its placement on the pad is critical.

During discussions with frac service company managers, the service company wanted to place the dance floor on an area of the pad that happened to be on the fill side. They thought that area of the pad would be ideal for their rig up. We did not like it and discussed it but yielded because they wanted the equipment configured to their preference.

As a precaution, we added additional rock under the dance floor, and everything was fine until it started raining. Once that happened, the pad began to settle, impacting forklift operations to the point that we had to shut down and fix the dance floor. Pad issues occurred multiple times during the frac as the rain intensified, requiring us to repeatedly stop operations and fix the dance floor. The delays cost over $200,000 in downtime. Plus, another $100,000 to fix the pad. The additional cost burden was reconciled during frac field ticket negotiations before the ticket was signed and approved for payment.

To avoid this costly situation, we should have put the dance floor on the cut area of the pad. During planning, prior to frac mobilization, we explained the difference between cut and fill, but it did not resonate with the service company. They did not think it would matter. Later, I discovered the issue of sinking dance floors occurs across the basin and is a common problem with this sand system during heavy rain events.

When building a location, we must level the ground. Once the desired ground elevation is determined, a cut is made above the desired elevation, and that dirt is moved to fill areas that are lower than the desired ground elevation to level the pad.

Fill dirt is fluffy and needs to be compacted. Some people run a sheep's foot to help address it. Compaction takes time. Depending on the area of operation, a cement-stabilized base may be necessary. A stable location helps with efficiency and cost reduction because if the ground is a muddy mess, it slows operations down, increasing costs.

For our situation, the solution was easy: do not put the dance floor on fill dirt. Additionally, I have since adapted a new way of explaining cut and fill to team members:

*"God stabilized the cut side of the pad over millions of years. Man stabilized the fill side of the pad over the course of a few days. God is stronger than man."*

We further reduce risk and cost by minimizing areas that require significant amounts of fill. This is why always building the same default-sized and shaped location is not only more expensive but also carries more risk. We have a higher probability of encountering areas of the Earth that require significant fill, are in the path of stormwater runoff, or are naturally unstable.

---

**A strong, stable pad enables speed, which is what we need to reduce costs during drilling and fracturing operations.**

---

Building locations that take the natural topography into account, balance the cut and fill, and optimally handle stormwater, will not only reduce costs but are also safer.

Even the federal government agrees with this type of tactical cost reduction action regarding pad construction. The BLM Surface Operating Standards and Guidelines for Oil and Gas, known as The Gold Book, states:

*"Well sites should be designed to fit the landscape and minimize construction needs. In many cases, this means designing a well site that has an irregular shape, not rectangular. The site layout should be located and staked in the most level area, off narrow ridges, and set back from steep slopes, while taking into consideration the geologic target, technical, economic, and operational feasibility, spacing rules, natural resource concerns, and safety considerations. Well locations constructed on steep slopes cost more to construct, maintain, and reclaim ... Locations on steep slopes that require deep, nearly vertical cuts and steep fill slopes should be avoided..."* [1]

# Low-Cost Location Considerations

- For efficient drilling, completion, and production operations, incorporate entry, exit, and movement on and around the pad. A dependable, stable, streamlined pad enables us to increase operational speed, reducing costs.
- Evaluate natural ground stability for each segment of the proposed pad before construction.
- Identify natural stormwater runoff for the proposed pad.
- Avoid building on natural water ponding sunken areas.
- Avoid building on areas that are naturally unstable, in the path of stormwater runoff, or have safety concerns.
- Avoid redirecting the natural flow path of water.
- When staking location shape and size, incorporate the amount of fill dirt required for each area of the proposed pad. Remember, if we build it, someone will put something on it that may become unstable.
- Incorporate floodplains, erosion areas, sinkhole risk, groundwater wells, offset houses, vegetation, threatened species, and cultural resources into pad site selection.
- Take public roads and neighborhoods into consideration for travel to and from location for safety and efficiency.
- Optimize designs for ingress, egress, access roads, pipelines, utility lines, and frac source water from a cost, function, and safety perspective.
- Incorporate production facility vessels, tanks, flares, and other component spacing considerations to optimize facility designs and fulfill safety setback requirements.

## ——— ACTION ———

## Recycle and Reuse

Surface location construction materials account for over 50% of the total cost of pad construction, with the remaining cost from labor, equipment, processing, and mobilization.

Once a location has served its purpose, there is an opportunity to harvest materials from unused portions of the pad. For example, we can harvest rock from the corners of the pad and areas that do not see production truck traffic. This provides an added benefit when the site is fully reclaimed since there is less work to do. Additionally, when fully reclaiming older locations, a significant amount of material can be recovered to build new locations.

Recycling pad construction materials from offset locations can save a considerable amount of capital in materials and trucking. Aggregate rock products, gates, cattle guards, posts, and drainpipes are easily reusable and require little to no processing. Wire fencing is more challenging to reuse because the wire gets stiff over time; however, it can still be done.

Recycling location rock, including limestone, caliche, and gypsum aggregates, is the most lucrative in terms of cost reduction when replacing newly mined rock with reclaimed rock. Savings are generated not only from material reuse but also from trucking. If nearby locations are in the process of reclamation, it is logistically convenient to route those materials to the new build locations. Stockpiling reclaimed rock is another option to reduce upcoming construction costs.

—————————— ACTION ——————————
## Lean the Location

Depending on topography and time of year, there are opportunities to build extremely lean locations. These are locations that require minimal dirt work because they are sufficient for oilfield operations in their natural state.

With proper soil investigation, it can be determined that minimal dirt work is required to have a suitable location for operations. This is a situation where the location is naturally level, stable, has good drainage, and requires minimal cut and fill. Construction requires minimal amounts of materials depending on surface analysis and operational requirements.

Due to weather, the time of year is critical when implementing an extreme lean location. Oilfield locations take a beating during rainy periods, driving up costs. Having to dozer equipment in and out of location due to muddy conditions is very expensive. Furthermore, if the drilling rig becomes unstable or frac equipment starts to sink, should the issue not be easily addressable, the cost will quickly consume our contingency allocation.

Some operators build extremely lean locations and then deal with problems when they occur. Over time, they advocate that the savings more than offset the pad remediation costs and downtime due to issues with the pad, which almost always happen during bad weather. The extreme lean approach is not for everyone and certainly not for every location.

---
# ACTION
## Strike Win-Win Surface Deals
---

Surface landowners are long-term partners. We are their guests. Taking the time to get to know our surface landowners personally is a good investment. Cultivating these relationships will help us reduce our costs.

Surface landowners are often farmers and ranchers, working around our assets all day. If we have a good relationship, they can be another set of eyes on our wells that help us avoid costly issues. If we don't, they can cause problems, increasing our costs.

We want to do everything possible to ensure our surface owners are happy. When there is an opportunity for a win-win with the surface owner, we must advocate for the deal. For example, it is not uncommon for surface owners to have materials (i.e., aggregates) used to construct oilfield locations on their property. This can save us money and help the surface owner get more money. Additionally, if there is water on their property, that can be another mutually beneficial opportunity.

Providing multiple revenue streams for the surface owner can help with surface damage costs and all the other associated costs. In some cases, it may be better to keep everything separate. However, if it works out that our surface owners can also make money with pad materials, drilling water, soil farming, and frac water, it will go a long way toward building a positive relationship with this long-term partner, further reducing our costs.

---------------- ACTION ----------------

## Optimize Construction Oversight

---

When building a wellsite location, we are paying for the expert knowledge and abilities of the construction contractor. After reviewing the staked location, proposed pad site map, design, materials, safety guidelines, and agreed cost, it is the duty of the contractor to perform the job. Due to the nature of the work, minimal oversight is necessary, especially if we have a trusted vendor partner with a track record of performance.

The ideal cost structure for location dirt work is a fixed fee, where the builder manages all services. Therefore, we do not need a day-rate consultant and can eliminate this expense. Should there be a concern that the contractor is shorting loads of rock or cement during the building process, one of our field employees could arrange to be on location when materials are delivered to confirm and sign off. Additionally, an employee field supervisor could visit our locations under construction to provide drive-by oversight while performing regular duties.

Furthermore, a low-cost $200 cellular video camera could be utilized to monitor activities remotely on a mobile phone or at the office, eliminating the cost of personnel oversight. If we are not comfortable building locations without a person overseeing the work, one field consultant could manage multiple locations, spreading the day-rate cost, as opposed to sitting on one location all day, which is unnecessary and expensive.

## —————— ACTION ——————
## Require Each Vendor to Conduct
## A Low-Risk, Low-Cost Route Plan

Statistically, the most dangerous activity in the oil and gas industry is driving to and from oilfield locations. Activities with significant risk have significant costs; therefore, an opportunity exists to reduce both.

The most considerable direct transportation-related cost is drilling rig mobilization. When deciding where to build the pad, a route assessment should occur regarding the cost of getting to location. A site that is difficult to travel to will substantially drive up the cost, not only for rig mobilization but also for all other oilfield services.

Travel costs are obvious on rig mobilization invoices. They are also obvious in the mileage charged on most tickets. However, they are less obvious when additional expenses are incurred for travel dynamics and obscured when vendors present their total cost of business, spreading travel expenditures across multiple line items. When vendors incur travel costs, the operator pays for it directly or indirectly.

Furthermore, if we are running a lean, low-cost operation leveraging Just-In-Time (JIT) management strategies, logistical issues will cause delays, contributing to non-productive time (NPT). For example, if we are waiting for someone to show up on location, that time increases our costs. In many cases, this cash burn does not get captured in the daily reports because the crews move on to something else so as not to draw attention to the fact that we are waiting on a colleague or piece of equipment that got lost, stuck in the mud, stopped by a regulatory, pulled over by the police, stuck in traffic, or any one of many other situations contributing to travel time inefficiencies that end up accumulating to significant sums of money directly and indirectly, damaging our economics.

Something as simple as incomplete or vague directions to location in a written procedure or text chain can result in significant additional costs. It is typical to hear vendors comment on the directions to location, either from the written procedures or from a pin drop when using mobile apps, as contributing to delays or problems getting to location.

A travel issue on the way to location increases stress levels for our vendors running late and our people waiting. When this stress permeates into our operation, a simple lost driver or speeding ticket can snowball into a multi-million-dollar mistake in the subsurface. Travel-related inefficiency is often the first domino in a chain reaction, increasing our costs.

Transportation inefficiencies can be mitigated if each vendor independently performs a low-risk, low-cost route assessment and constructs an individual route plan for our locations. To perform the evaluation, someone in a leadership position for each of our contractors must drive several potential routes to determine the optimal path for the vehicles and equipment they are utilizing. Then, depending on the type of equipment mobilized, the time of year, the weather, and a host of other factors, an optimum route can be determined.

Rig mobilizations usually have a well-planned, permitted route due to the size of the equipment, clearance requirements, overhead powerlines, weight restrictions, bridges, railroad crossings, schools, neighborhoods, construction, and road quality. However, if the recommended route is not optimal, working with the permitting agency to optimize it is necessary.

We had a situation during pad construction in which several rock transport trucks slid off the road into the ditch, causing major delays and significant additional costs. When asking the drivers why they continued to take the same unsafe roads to location, their response was:

*"It is the permitted route; we have no choice."*

After calling their manager and the State permitting officials, explaining the situation, we changed the route. There was a safer, more efficient alternative route, which everyone agreed to. We notified all vendors to avoid taking the high-risk, high-cost road, even though it was the most direct path.

This situation could have been avoided if proper journey management had been carried out. In this case, the weather played a role in turning what appeared to be a decent road into a muddy mess. Many vendors do not perform journey management, study route options, incorporate weather-related scenarios, and test travel the route prior to the job.

There is more than one way to get to a location, depending on the point of origin for each person. Providing suggested directions can help. However, consider requiring leadership from each service company to test drive all potential routes and make a decision for their company and particular situation. Make this a requirement in writing if they want to work for us.

Vendors must know that we are serious about reducing risk and cost when it comes to travel logistics.

Letting service providers know that we pay attention to transportation costs sends a message that we analyze this area of the operation and invoices. We do not tolerate unreasonable transportation costs or delays if someone did not perform proper due diligence on travel prior to mobilizing for the job.

Many vendors see mileage, travel hours, and transportation costs as high-margin revenue streams. There is an incentive to be inefficient. Therefore, we must have systems in place to address this dynamic. Requiring each vendor to conduct a "Low-Risk, Low-Cost Route Plan" as part of our work agreement will help reduce travel risk and cost.

---

**If someone is inefficient in getting to location,
how will they work when on it?**

---

If a vendor has travel issues, runs up the miles or travel hours, or is late getting to location, it sends a message, telling us who they are as a partner. Travel inefficiencies between vendors can be used as a performance metric to determine if we want to continue doing business with them.

If travel issues continue to be a problem, we could set a fixed mileage or amount we will pay. This incentivizes the service provider to be efficient. Regardless of what method is decided, the mere fact that this area is discussed with each vendor will result in lower costs. We are giving attention and thought to an area often overlooked.

---
## ACTION
---

# Establish Electric Power
# Prior to Initiating Operations

---

Depending on proximity to existing overhead power transmission lines and available operating voltage, there is value in establishing a location connection to the grid or establishing a microgrid before initiating drilling operations.

It is common to install electric power after drilling and completion operations, when building the production facility, to have electric power for overhead lighting, heat trace, telemetry, safety systems, and artificial lift equipment. However, if we establish electric power at the beginning of operations, we can save money by reducing or eliminating light plant rentals and local generator expenses, at the very least.

Moreover, access to electric power opens the door to drilling rig and frac equipment electrification. Depending on fuel prices, usage, and total operational days, powering rigs and frac equipment with mainline or localized grid-power can deliver massive cost reductions.

Equipment maintenance costs can also be reduced because the diesel or natural gas engines are not running unless there is a disruption. Furthermore, electric power installation provides the added benefit of reduced emissions and noise pollution.

Coordinating our development program to incorporate the establishment of electric power prior to initiating operations provides multiple cost reduction opportunities, without any additional costs, just additional planning.

——————— **ACTION** ———————
## Install Gas Gathering Lines
## Prior to Rig Mobilization

While constructing the pad, consider simultaneously installing the gas gathering system as part of the initial pre-spud setup. Proactively installing the gas gathering lines in the optimal spot for drilling, completion, and production, prior to mobilization (mob), will help us when using natural gas to power equipment as part of our cost reduction efforts.

**Every cost reduction action we take prior to drilling rig and frac crew mob enables us to maximize cash flow.**

With access to natural gas, we give ourselves the option to power the drilling rig and fracturing equipment with gas. In a depressed natural gas price environment, as an industry, we must leverage every opportunity to use our own products to power equipment. There are many options to consider in the evolving and expanding use of natural gas.

Dual-fuel (Bi-fuel) systems provide the ability to run engines on a combination of diesel and natural gas, or 100% diesel as a backup option. Many drilling rigs have dual-fuel capabilities or can be retrofitted. Some systems can detect changes in the fuel quality and adjust automatically. Dual-fuel engines have an average diesel displacement rate between 40% and 85% and can run on pipeline-quality gas or wellhead "field" gas.[2]

Due to high Btu content, moisture content, and other potential impurities such as $H_2S$ and $CO_2$, field gas may need to be conditioned to prevent engine issues. Conditioning trailers or skid options are available to process field gas for drilling and fracturing operations. The conditioning unit is typically connected to the gas-gathering system. Field gas flows through a series of scrubbers to condition the gas to power the equipment.

100% natural gas-powered options also exist for both drilling and fracturing operations. With the increase in market share of electric fracturing fleets (E-Frac), field gas can be utilized to generate 100% of the electric power for fracturing equipment.

There are added benefits to eliminating diesel use for rigs and frac equipment. By eliminating diesel, we also eliminate fuel delivery truck traffic, potential spills, emissions, noise, sound barrier costs, and diesel re-fueling safety issues during operations.

With diligent planning, by including the gathering line installations with pre-spud operations, we give ourselves options to utilize field gas for drilling and fracturing, as well as remove the potential risk that gas takeaway infrastructure is not ready when production flowback begins.

Reducing or eliminating diesel fuel can reduce total fuel expenses by 40% to 90%. This is a huge cost savings that must be utilized by the low-cost operator. Any action that can reduce cost and risk while reducing our environmental impact must be seriously considered.

—————————— **ACTION** ——————————

## Fix or Cap Rig Move and Rig Up Costs

Rig mobilization (rig moving services + rig mob days) can easily be over 20% of the total cost to operate the rig. Significant money is spent during rig mob, and it is often overlooked as an area for cost reduction because the rig has not started drilling, the cost is spread over multiple wells, there is minimal oversight, and we have no detailed telemetry capturing precisely what is happening during this process.

Rig move periods are often thought of as a break between multi-well pads for office personnel to get caught up, approve invoices, and prepare paperwork for the next multi-well pad or handle other personal business before drilling begins. Some office personnel that are 24-hour on call during drilling view rig mob as a mini vacation. It is the only time they have to take a break and enjoy life outside the oilfield.

From this perspective, the longer the rig mob, the better. Inherently, there is less management attention, involvement, and oversight on this operation. This provides an opportunity.

Rig mob and rig up are considered higher-risk activities due to the amount of activity occurring simultaneously. High risk is high cost. Therefore, we must continuously look for opportunities to make operational processes safer and more efficient. This mentality, particularly for rig moving activities, will lead to sustainable cost reductions.

Controlling risk with good planning is essential during rig mob and rig up operations. Scouting the preferred route with thorough journey management combined with virtual installation, including 3D simulations of the equipment being rigged up on location, and working with field leadership from the drilling rig contractor, trucking service provider, and operator consultants will add value.

Constructing a finalized diagram with rig equipment imposed over the surface location, combined with detailed planning, will assist in preventing needless non-productive time and other inefficiencies that cost money.

We can incentivize safety and efficiency with a thoughtful drilling contract and rig moving services agreement. IADC drilling contracts usually have a mobilization subsection within the compensation section of the contract. The details in this subsection are critical when running a low-cost operation.

---

**Reducing costs with contracts is an effective, low-risk, non-invasive action.**

---

A fixed cost mobilization at a pre-determined total dollar amount or reduced mobilization day rate (preferably 50% to 85% of the full operating day rate) capped at a pre-determined number of hours or days (1 to 4 days preferably) to load, move to a new pad, rig up, and spud should be considered.

Having a contractual fixed mob cost or cap incentivizes everyone to be safe and efficient because if the drilling rig company or trucking company is not focused, they could exceed the fixed cost or capped days and lose money.

The agreed period between wells, of course, will vary

depending on distance and road conditions. The fixed cost structure or cap helps because it reduces cost uncertainty. Uncertainty almost always works against the operator because if we do not have a fixed cost or cap, both the rig company and trucking company are incentivized to take as long as possible to move equipment and rig up.

Longer mobilizations increase the risk and the cost. Please do not be confused; I am not suggesting moving equipment as fast as possible without regard for safety. In my experience, if we have two companies that are moving the same rig, and one takes four days to do it and the other takes two days, the trucking company that has figured out how to consistently move the rig 50% faster is probably utilizing better mobilization equipment, more efficient processes, and has a better team. Oversight can ensure they are safe.

The longer it takes to move the equipment, the more risk we are exposed to simply because personnel are engaged in a high-risk process for a longer period.

Additionally, the trucking company that is taking longer is probably having a difficult time with certain aspects of the operation, engaged in non-routine activities, or is having other problems. All these factors increase risk and cost to the operator, increasing the invoice and benefiting the service provider financially.

To address this, cap the trucking mobilization costs and rig moving services at a maximum fixed price. This simplifies billing and incentivizes the trucking company to be focused and efficient.

With the rig mob and the trucking cost fixed or days

capped, service providers are aligned with the operator and focused on being safe and efficient. This provides a stable foundation from which to build on.

For example, let's say our current mobilizations are fixed at around four days and $120,000 for trucking. Once we can consistently deliver on the days, we have a solid foundation. A standard from which to benchmark and make improvements.

As the rig moving service companies get to know the rig components and operator preferences, the team becomes more familiar with the nuances of the rig. We can then identify opportunities to be safer and more efficient.

There are a number of tactics to employ to enhance efficiency. Using these tactics, we can slowly chip away at opportunities to reduce move and rig up costs.

Once efficiencies are realized, lock them in by reducing the fixed mob cost, mobilization day rate, the cap on the number of hours or days for mobilization, and the fixed fee rig moving service. Improvements are a win-win when they make the operation safer, more efficient, and reduce costs for both the service provider and operator.

# 20 Tactics for Rig Move Cost Reduction

1) Benchmark service provider rig move performance.
    a. Identify who is top tier based on metrics, including:
        i.   Safety
        ii.  Rig mobilization design
        iii. Equipment and tools utilized
        iv.  Total steps
        v.   Number of loads
        vi.  Labor and manhours required
        vii. RR to spud time
        viii. Margin: cost versus invoiced amount
2) Fix or cap rig mobilization costs and rig move services based on expert knowledge and previous rig moves.
    a. Establish a basis from which to improve.
3) Incorporate rig move into pad site selection and design.
    a. Stability, size, dimensions, entrance, and roads.
4) Determine optimal route with rig movers and work with regulatory personnel on approved permitted route.
    a. Drive permitted route identifying any final issues.
    b. Discuss issues with regulatory personnel and make changes if necessary to be more efficient and safer.
5) Meet on location and virtually spot each component.
    a. Draw picture of detailed rig component layout on pad.
        i.   Distribute illustrations to the entire team.
        ii.  Discuss the pros and cons with the team on the planned layout of rig components.
        iii. Optimize layout for cost and safety.
        iv.  Communicate updates to all personnel.

6) Review rig move procedures during safety meetings.
   a. Service company rig move and RU procedures
   b. Job Safety Analysis (JSA)
   c. Past lessons learned for this specific rig
      i. Discuss potential issues
      ii. Location dynamics
      iii. Equipment pick up, transport, spotting, set down, and rig up nuances
      iv. Order, loads, timing, dependencies
7) Confirm that the equipment is clean and safe to handle.
   a. Dirty equipment is dangerous and will slow down the team during mobilization and rig up.
8) Confirm ideal rig move equipment, size, and tools are utilized, i.e., heavy haul trailers, bed trucks, winch trucks, lowboys, cranes, pickers, gin pole trucks, boom trucks, forklifts, manlifts, winch trucks, tilt utility trailers, dollies, and ramps.
   a. Proper equipment will enhance safety, reduce cost, and minimize the risk of damaging components.
   b. Calculate utilization rate for each piece of equipment.
      i. Optimize equipment selection and utilization based on findings
      ii. Eliminate underutilized equipment if possible
9) Use written checklists for each process.
   a. Improve the checklists after each mob and rig up
10) Construct a load plan.
    a. Determine the order of arrival for safety and cost.
       i. Incorporate equipment relationships and activity dependencies to streamline move.

b. For each planned day, include the number of loads, the trucks required, and the load contents for each haul.

c. Personnel requirements and total manhours

d. Customize for each rig

11) Schedule and align move rate with rig up rate.

a. Trucking equipment quicker than it can be rigged up will increase total cost.

b. Unbalanced move and rig up speed will increase risk since equipment and personnel will be waiting around.

12) Map out each rig move step in writing.

a. Calculate number of steps.

b. Include details with each step.

13) Build an ideal timeline of events.

a. Include the target time required for each step.

   i. Calculate granular minute-level targets.

     1. Pick up, travel, set down, additional handling, final position, installed

14) Eliminate empty and underutilized travel.

a. Document all travel trips with detailed load contents and number of trips / loads to move components.

b. Work to reduce total loads required to move the rig.

15) Remove redundant and unnecessary maneuvers.

16) Identify inefficient relationships between different steps during rig move and rig up.

a. Extra people on location can lead to chaotic situations.

   i. More people equals more risk and cost, but additional crews can help in certain situations.

b. Look for opportunities to reduce labor and equipment.

17) Identify and eliminate double handling.
   a. Prohibit unloading equipment in a spot that will require it to be handled again. Scattered equipment along the perimeter of the pad is a red flag.
   b. Incorporate equipment relationships and dependencies.
18) Make equipment modifications to allow for more efficient and safer mobilizations and rig up.
   a. Opportunities to leave components in place during mobilization to avoid additional rig up time
   b. Look at mast, substructure, backyard, electrical cables, and hydraulics for design optimization for rig moves
   c. Consider trailer-mounted solutions
19) Install overhead cameras on location and GPS sensors on trucks and equipment during rig move and rig up.
   a. Record all movements for playback analysis
   b. Identify inefficiencies
   c. Utilize panoramas to capture the entire wellsite
   d. Create time-lapse videos
   e. View multiple perspectives
   f. Zoom into areas of interest
   g. Study rig moves and find opportunities to reduce cost
   h. Increase safety
   i. Optimize pad designs, particularly entrance areas, by studying footage for bottlenecks and tight spots.
20) Obtain feedback from all personnel to improve rig move process, rig equipment, trucks, and tools.
   a. It is critical to get feedback from guys on the front lines of rig mobilization and rig up.

—————————— **ACTION** ——————————

## Lump Sum Total Rig Mobilization

Cost reduction work consumes significant time and resources. In certain situations, we may not have the internal resources to devote the required personnel to reduce costs in every area of the business. With this in mind, one opportunity for drilling rig mob cost reduction is to structure a lump sum total rig mobilization contract. This is a deal in which the drilling rig company takes the lead in mobilizing the rig for a lump sum, including all associated costs with the rig move.

During discussions with this type of deal, make sure to be clear on what is included in the lump sum. When I have made deals like this, I make it very clear that a lump sum rig move is not just trucking. It includes everything, especially mob days, rig labor, cleaning, and anything else required. A fixed rig move that only includes the trucking component is not a true lump sum deal. The drilling rig company is not aligned with this structure.

To leverage the brain power and expertise of the drilling rig company and the rig moving company, both must be incentivized to be more efficient.

**One of the best ways to motivate is with money.**

A lump sum structure will get their attention. For example, let's say we have been able to move our rig for an average total cost of $150,000 for trucking, rig days, extra hands, cleaning, and everything else. However, we want to be able to do it for $100,000. Additionally, we do not want to carry the risk that the cost could exceed $150,000. Therefore, we propose a total lump sum deal with the drilling rig company. We could make a deal for under $150,000, moving us towards our goal of $100,000.

Let's say we negotiate a deal for a lump sum rig move of $140,000. Now, rig leadership is incentivized to be more efficient and perform the rig move for under $140,000 to maximize their margins. They have an opportunity to enhance their financial status.

In this situation, they will put their best minds and equipment into the operation because they are incentivized to do so. Compared to other customers, for which they are paid a set day rate, they have no incentive to move the rig efficiently. In fact, it is the opposite.

No one cares more about your money than you do. However, if we structure deals in which the operator has savings and the service provider has the opportunity to make more money, it is a win-win.

Through this process, everyone will see what is possible and learn how to be more efficient. The rig and trucking company know their equipment better than anyone. Having their attention adds value during our cost reduction efforts.

———————— — **ACTION** ————————
## Split Mob

If we do not achieve the rig mobilization cost reductions we want or need, consider a more aggressive, unconventional approach. Before our next rig move, contact a group of smaller non-rig mover companies, including hotshotters, transporters, truckers, small haulers, and other companies that provide transport services that do not use large, heavy hauler rig mover equipment. Have these smaller trucking companies look at every component we need to move and provide cost estimates on moving specific pieces of equipment.

Next, get multiple quotes on a total rig move. Once we have quotes for all components, remove components that the smaller transports can move. Then, get an updated rig mob quote. Smaller transport companies can move specific equipment at a lower cost than heavy haul rig-moving vehicles.

Utilizing massive rig-moving equipment to move every component is not the most efficient strategy. These massive machines are costly to operate and often move tiny rig components that can be moved with smaller, lower cost trucks.

For example, during a rig move, drill pipe does not need to be moved with rig mover equipment. We can often use a more traditional flatbed trailer at a lower cost and haul more joints per load because the truck is lighter than the larger rig-moving machines. By hauling small components with small hotshotter trucks and large components with large heavy haul rig mover transports we can reduce total costs.

---
## ACTION
### Utilize Drones
---

The ability to obtain big-picture views, never before seen by our team, throughout oilfield activities can elevate our operational prowess to the next level in our evolution to deliver reduced risk and costs.

Utilizing drone cameras that can view, record, and take pictures will give our team different perspectives on operations, often leading to new ideas and opportunities to maximize cash flow with cost reduction. A decent entry-level drone camera can be purchased for under $500, so there is minimal cost in trying this action item within different divisions and positions across the team.

For example, before deciding on the exact location to stake a new well, build a pad, install a centralized facility, or determine the optimal route to build a pipeline, obtaining a bird's eye view of the area and topography can provide additional detail to make a fully informed decision.

Using Google Earth or similar applications combined with boots on the ground is a good standard practice for granular resource development decisions. However, integrating detailed drone footage takes our operation to the next level, further reducing costs.

During drilling and fracturing, a drone camera can be used to help optimize equipment layout and daily operations. Looking at an operation from various angles allows for advanced interpretation and new ideas. This is especially helpful for team discussions and analysis for cost reduction.

I carry a drone with me regularly. When arriving on site, I launch the drone, which autonomously surveys the operation and returns on its own. Within a few minutes, I have a bird's eye view of the entire operation and surrounding area to discuss with wellsite leaders, vendors, pumpers, and office-based team members.

**Drones enhance our intelligence,
leading to higher-quality decisions.**

Drones are effectively flying robots, and they are elevating the ability of humans to make better decisions across many industries including, construction, insurance, agriculture, utilities, security, military, media, law enforcement, and logistics. Once we see the value drones provide, we may utilize more advanced drone-incorporated systems, including drones with thermal imaging, temperature measurement, follow-me features, multi-gas detection, metal detection, bathymetry, UV corona inspection, autonomous field patrols, and emergency response services.

Furthermore, as we get more comfortable utilizing drones, we can incorporate them into our Daily Visual Reports and artificial intelligence systems, including the Virtual Company Man. AI has become proficient in automated image analysis and can help us save time and money. The following two pages include a small sample of the many drone camera opportunities in the oil and gas industry.

# Drone Camera Opportunities

| Drilling | Cost Reduction Opportunity |
|---|---|
| **Staking Location** | Position favorably for SHL, drilling nudge, landing, drainage, constructability |
| **Road and Pad** | Design incorporating topography |
| **Subsurface** | Scan subsurface for pipelines and cables |
| **Rig Mobilization** | Bird's-eye pics, video, for planning mob |
| **Rig Layout** | Engineer configuration to optimize setup |
| **Inspection** | Inspect hard to get to equipment |
| **Water Transfer** | Monitor polyline, pumps, source water |
| **Gas Leaks** | Identify gas leaks on natural gas-powered equipment, conditioning systems, CNG storage vessels, and temporary pipelines |
| **Invoicing** | Confirm and document items on location |
| Completion | Cost Reduction Opportunity |
| **Frac Layout** | Plan frac setups with detailed images |
| **Red Zone** | Inspect red zone without exposing people |
| **Frac Stack and Wellhead** | Get closeup view of frac stack and wellhead during continuous fracturing and simul-fracs to check for leaks and issues |
| **Inventory** | Get a bird's-eye view of sand, chemical, material, and equipment inventory onsite |
| **Source Water** | Calculate pond volumes, depths, remaining water, lay-flat issues, transfer pump problems, remote personnel |
| **Leak ID** | Look for leaks, areas needing mats |
| **Remote Pads** | Monitor multiple pads, monoline, equipment, frac stacks, on remote fracs |
| **Failure Avoidance** | Check temps, electrical issues, failure potential on equipment while pumping |

# Drone Camera Opportunities continued...

| Production | Cost Reduction Opportunity |
|---|---|
| **Facility Design and Construction** | Utilize bird's-eye facility images for design improvements, streamlined installation, quicker construction, knowledge transfer to field team |
| **Flowback Setup** | Reduce setup cost with bird's-eye view |
| **Vessel, Tank, and Artificial Lift Inspection** | Inspect hard to get to areas on vessels, prevent and troubleshoot problems by measuring temperatures, internal imaging of tall vessels and tanks, and monitoring lift systems using IR and thermal cameras mounted on drones to reduce R&M costs |
| **Flare Inspection** | Look into flares to identify costly issues |
| **Liquid Leak Identification** | Look for evidence indicating expensive liquid leaks including staining that is easier to identify with a bird's-eye view |
| **Gas Leak Identification** | Check for gas leaks on top of vessels and tanks, especially vertical heater treaters and VRTs that are dangerous and difficult to get on top of without a manlift |
| **Pad and Road Repair** | Identify erosion areas and stability issues that could impact assets |
| **Off Pad Inspection** | Inspect and document landowner areas for damage, risks, issues, surrounding pad |
| **Power Inspection** | Fly drones across powerlines to identify issues with poles, fuses, connections etc. to reduce downtime, risks, and repairs |
| **Gathering Systems** | Inspect pipeline right-of-way for leaks, surface erosion, and other potential issues |
| **Emergency Response** | Use field deployed drones for manual or automated response to SCADA alerts |

# ACTION

## Increase Geologic Intelligence

Identifying the most productive rock within the reservoir, landing favorably, and staying in zone are critical aspects of long-term production performance.

During cost reduction efforts, when the focus is on cost, it is easy to lose sight of the geologic element and its importance in various aspects of shale operations and development. Therefore, incorporating geology into all aspects of cost reduction and the associated decisions made is critical when working to reduce costs.

For example, when looking to improve depth versus days curves, particularly for hard-to-drill formations, it is not uncommon to look for reservoir intervals that drill faster, known as "fast rock" or "fast holing."

However, this is often at the expense of production performance. In my experience, it is better to drill the absolute best rock from a geologic perspective, even if it is difficult because if the lateral is not optimally placed, production performance will suffer every day for the life of the well.

It can be a painful reminder every time we look at production from a group of wells in which geologic importance was minimized or ignored to improve drilling days or break arbitrary drilling records.

Another option that is often discussed during cost reduction efforts is targeting fast rock and then compensating with a massive frac design.

The logic is that we get a better result with a lower cost. Although this can work in certain situations, it is not typically as good as targeting the very best rock and placing a massive frac treatment.

To ensure we do not lose sight of the importance of geology, consider using cost reduction efforts as an opportunity to enhance our geologic performance by providing more geologic intelligence and higher quality geologic information to front line personnel to help them execute operations. This will assist the team in making more profitable decisions during drilling and completion.

> **The most expensive aspect of shale development**
> **is information that we don't have,**
> **especially geologic information**
> **on the front lines**
> **when executing**
> **operations.**

Most front line personnel are provided limited, if any, geologic information to help make decisions that maximize economics. This geologic weakness must be addressed if we desire to elevate our performance to maximize free cash flow and avoid expense mistakes.

On the next page is a list of actions to emphasize geologic importance, improve subsurface understanding, and elevate operational decision-making on the front lines by leveraging increased geologic intelligence.

## Actions to Increase Frontline Geologic Intelligence

- Meet with field team to customize opportunities to enhance performance and reduce costs using geologic intelligence
- Redesign GeoProg based on feedback from the field team
- Create a geosteering focus group on IM apps for each well
- Provide offset logs with annotations to the field team
- Mark risks: depleted intervals, abnormally pressured zones, salts, injection zones, gas flows, anhydrites, and concerns
- Install cameras on shale shakers
- Build a roadmap to KOP including at a minimum: Depths, Geology, WOB (klbs), GPM, Top Drive RPM, Total RPM, Expected ROP (ft/hr), and detailed notes with geo warnings
- Mark intervals causing drill bit damage. Automate text alerts to remind team approaching depth to slowdown and make adjustments. Incorporate into Virtual Company Man
- Include all potential faults in geologic prognosis
- Create algorithm-based text alerts to automatically notify team when conditions suggest we might drill out of target interval. Incorporate into Virtual Company Man
- Build an artificial intelligence-driven geosteering model to generate AI recommendations as a second opinion
- Create decision trees to map action plan for various scenarios including variables indicating above target, below target, crossed fault, and heterogeneous in-zone
- Construct a checklist for when to drill ahead, make a target change, slow down, call geologist, and stop drilling
- Request operations team provide quant feedback on initial geologic prognosis versus actual geology encountered

# 7

## DRILLING ACTIONS

—

In the long run, all costs are variable. Price is never static, and all costs can be restructured. However, in the short run, time is the primary driver of drilling expenses in most cases. Therefore, design, efficiency, skill, and risk management are paramount when reducing drilling costs.

The structure of most drilling services is based on time. The more days, the higher the cost. 60% to 70% of drilling costs are linked to time on location. The remaining 30% to 40% are fixed, but that does not mean we will not reduce them.

## Variable and Fixed Drilling Expenses

| Variable Costs (65%) | Fixed Costs (35%) |
|---|---|
| Rig Day Rate, Fuel, Power | Land, Legal, Surface Damages |
| Mud, DD, and MWD Services | Location Construction, Permits |
| Wellsite Supervision | Cellar, Conductor, Spudder Rig |
| Downhole Equipment | Rig Mob / Demobilization |
| Surface Equipment | Casing, Wellhead, Sleeves |
| Pressure Control | Cementing, Float Equipment |
| Mud Logging, Geosteering | Casing, Thread Inspection |
| Houses, Comms, FW, Septic | CVD, Crews, Torque-Turns |
| Tanks, Containment, Trash | Cuttings, Screens, Disposal |

## Fixed Drilling Costs

Most fixed drilling costs are accounted for in land and legal, location construction, mob, spudder rig, and casing costs. Pre-spud costs were addressed in the previous chapter. However, casing is a significant fixed cost critical not only from a well cost perspective but also from a risk perspective.

Casing problems can destroy a free cash flow strategy and have contributed to the economic failure of multiple oil and gas companies, many ending in bankruptcy. Enhancing the quality, inspection, monitoring, and installation of casing is a cornerstone of a free cash flow strategy because casing problems are difficult or impossible to fix during drilling, and especially during completion.

Even though casing is super critical, it is not a sacred cow in terms of cost reduction. There are excellent opportunities to reduce casing expenses, particularly from a design perspective. Eliminating a casing string or changing the diameter will not only impact fixed drilling costs but will also impact all variable costs if we optimize the casing design to create efficiencies and reduce spud-to-rig release days.

Since the final operational drilling objective is to successfully install the production casing for fracturing, the entire shale business is anchored to the casing, literally and figuratively. The casing is the center point of a shale operation. Therefore, to make impactful reductions and add value across both drilling and completions, identifying opportunities to optimize the casing design, quality, and installation has the power to radically enhance economics while reducing risk.

—————————— ACTION ——————————
## Wellbore Design Evolution

Casing and the drilling ops involved in getting a borehole to casing point account for most drilling costs. Casing by itself is one of our largest fixed costs. If we optimize the borehole and casing design, we create value for our entire program.

Safely eliminating a string of casing can reduce direct casing costs as well as total drilling days. The savings can easily add up to over $1,000,000 per well. Most shale plays in full-scale development mode have evolved their casing design from running multiple intermediate strings with various liner configurations to a more efficient streamlined monobore long-string approach.

### Initial Design: Multiple Intermediate Strings Plus Liner

| Interval | Hole Size | Depth (TVD / MD) | Casing Size | Wt (lb/ft) | Grade |
|---|---|---|---|---|---|
| Surf. | 17.5" | 500' / 500' | 13-3/8" | 54.5 | J-55 |
| Interm.-1 | 12.25" | 5,000' / 5,000' | 9-5/8" | 36 | J-55 |
| Interm.-2 | 8.75" | 9,500' / 9,500' | 7" | 26 | P-110 |
| Prod. Liner | 6.125" | 10,000' / 9,200'-20,000' | 4-1/2" | 13.5 | P-110 |

### Evolved Design: Surface Plus Monobore Long-String

| Interval | Hole Size | Depth (TVD / MD) | Casing Size | Wt (lb/ft) | Grade |
|---|---|---|---|---|---|
| Surf. | 12.25" | 2,000' / 2,000' | 9-5/8" | 36 | J-55 |
| Prod. | 8.75" | 10,000' / 20,000' | 5-1/2" | 20 | P-110 |

Eliminating two strings of casing, a liner hanger system, and tie-back string is easier said than done. There are good reasons and regulatory requirements for running various casing configurations. Making casing design changes will impact the entire drilling operation. Optimizing the drilling mud, directional plan, BHA, drilling parameters, and a host of other variables will determine whether casing changes are possible or practical across the subsurface geologic environment for which the asset under examination is based.

Depending on reservoir depth, overburden formation dynamics, pressures, formation stability, legal regulations, and completion requirements, certain situations will not allow for the removal of a casing string. However, even though a casing string may not be able to be removed, we can optimize hole size, casing depth, and casing diameter. Discovering optimal hole size relative to our rock properties, bit technology, and drilling ROP is typically a process of trial and error.

**Initial Design: Big-hole Vertical with 5.5" Long String**

| Interval | Hole Size | Depth (TVD / MD) | Casing Size | Wt (lb/ft) | Grade |
|----------|-----------|------------------|-------------|------------|-------|
| Surf. | 17.5" | 1,000' / 1,000' | 13-3/8" | 54.5 | J-55 |
| Interm | 12.25" | 12,000' / 12,000' | 9-5/8" | 40 | P-110 |
| Prod. | 8.75" | 15,000' / 25,000' | 5-1/2" | 23 | P-110 |

**Slim-hole Design: Slim Vertical with Slim 5.0" Long String**

| Interval | Hole Size | Depth (TVD / MD) | Casing Size | Wt (lb/ft) | Grade |
|----------|-----------|------------------|-------------|------------|-------|
| Surf. | 12.25" | 1,000' / 1,000' | 9-5/8" | 40 | J-55 |
| Interm | 8.75" | 12,000' / 12,000' | 7" | 23 | P-110 |
| Prod. | 6.25" | 15,000' / 25,000' | 5" | 18 | P-110 |

Slimming down the entire casing program allows us to reduce hole size, resulting in improved drilling days. Drilling a 12,000' vertical section with a 12.25" bit takes much more time compared to 8.75" bit. This slim-hole approach is almost a 30% reduction in hole size and resulting rock volume. We are drilling and removing 30% less overburden to get to intermediate casing depth.

Additionally, we slim down the lateral section from 8.75" to 6.25". This is another significant reduction that can lead to improved drilling times. However, this approach will increase friction pressures during hydraulic fracturing. This can be mitigated with an optimized completion design.

Since fracturing costs are the most significant well expense, we want to look at casing design changes that will help reduce completion costs. A slightly larger casing diameter can significantly reduce frictional pressures, providing reduced treatment pressures and increased fracturing rates, allowing for more frac design options. Optimizing hole size and casing diameter will reduce total well cost.

The type of fracturing we are engaged in, or hope to engage in, should factor into our wellbore design. If we select simul-frac, trimul-frac, or super-frac, over conventional zipper-fracs, we can optimize our casing designs further. Determining ideal frac pressure and rate is a key factor. In general, if 4-1/2" casing can be avoided, it should be due to the additional friction and pressure, which will increase fracturing costs and reduce options. However, if we are limited on horsepower and engaged in simul-frac, it might not matter.

## Initial Casing Design: Slim Intermediate with 4.5" Liner

| Interval | Hole Size | Depth (TVD / MD) | Casing Size | Wt (lb/ft) | Grade |
|---|---|---|---|---|---|
| Surf. | 12.25" | 1,200' / 1,200' | 9-5/8" | 40 | J-55 |
| Interm | 8.75" | 11,000' / 11,000' | 7" | 26 | P-110 |
| Prod. Liner | 6.125" | 12,000' / 10,700'- 20,000' | 4-1/2" | 13.5 | P-110 |

## Evolved Design: Large Intermediate and 5.5" Long String

| Interval | Hole Size | Depth (TVD / MD) | Casing Size | Wt (lb/ft) | Grade |
|---|---|---|---|---|---|
| Surf. | 12.25" | 1,200' / 1,200' | 9-5/8" | 40 | J-55 |
| Interm. | 8.75" | 11,000' / 11,000' | 7-5/8" | 29.7 | P-110 |
| Prod. | 6.75" | 12,000' / 20,000' | 5-1/2" | 20 | P-110 |

Increasing the intermediate casing size from 7" to 7-5/8" provides the ability to drill a slightly larger horizontal hole size and set a long-string 5-1/2" production casing.

This will make a significant difference on fracturing design options, treatment pressures, and associated costs. In general, 5-1/2" casing is the preferred casing for frac due to treatment pressures, equipment availability, and completion tech. Since most wells in the U.S. run 5-1/2" casing, drilling and completion technology has evolved around this size. Although it is not always the perfect production casing, running a 5-1/2" long-string will expand the technologies and experience available to us, reducing risk and cost.

However, is it the best we can do? The answer depends on many drilling and completion factors. For the purposes of this thought experiment, let's assume our ideal lateral hole size is between 8.25" and 9.0", and we have access to run any casing that will clear that size. If we increase the ID of our production casing greater than 5-1/2", we reduce friction pressures and increase both frac and lift options. This should reduce costs and increase production. Therefore, we must investigate it.

### 5.5" Casing Standard Design

| Interval | Hole Size | Depth (TVD / MD) | Casing Size | Wt (lb/ft) | Grade |
|----------|-----------|------------------|-------------|------------|-------|
| Surf. | 12.25" | 2,000' / 2,000' | 9-5/8" | 36 | J-55 |
| Prod. | 8.75" | 10,000' / 20,000' | 5-1/2" | 20 | P-110 |

### 6.0" Casing Design for Frac Optimization

| Interval | Hole Size | Depth (TVD / MD) | Casing Size | Wt (lb/ft) | Grade |
|----------|-----------|------------------|-------------|------------|-------|
| Surf. | 12.25" | 2,000' / 2,000' | 9-5/8" | 36 | J-55 |
| Prod. | 8.75" | 10,000' / 20,000' | 6" | 24.5 | P-110 |

For frac optimization, a 6" string increases treatment rate by 25 bpm to 35 bpm. This enables a variety of optimization possibilities for cost reduction and production enhancement which we will expand on in Chapter 8: Completion Actions.

## Small Changes, Large Reductions

Small changes in drilled hole size, casing size, and casing setting depths can deliver large cost reductions for both drilling and completions. It is hard to know the exact optimal configuration, and it is not a one-size-fits-all.

When transitioning to a new casing configuration, it is nice to have the ability to run the old design just in case hole conditions are unfavorable. It is always prudent to have a preferred casing plan and a contingency plan.

## —————————— ACTION ——————————

## Consider Specialized Casing

Any casing or connection that is not generic or listed in the RedBook[1] is what I refer to as specialized. Expanding our casing options beyond the standard options can present considerable value-creation opportunities. For example, special drift or alternate drift options can provide opportunities to reduce the required drill hole size by running smaller OD casing with larger drift diameters.

Another example would be premium (metal-to-metal seal), semi-premium, integral (box directly threaded on pipe), flush (connection OD = pipe body), and semi-flush connections. These connections can reduce failure risk, increase strength, and allow for larger diameter casing to be run in smaller hole sizes or prior casing strings.

Casing grade variants, particularly P110 variants are another option. Increased yield strength, optimized steel chemistry, hardness, and a variety of other factors can be achieved with P110 variants including P110-MS, P110-RY, and P110-EC.[2] Using these options or combinations of these options with standard options can add value and reduce risk.

In certain situations, consistent success is only possible by running specialized casing and connections. Due to the significant economic impact of casing failures at scale when in manufacturing mode, it may be prudent to only run certain spec casing. Semi-premium, premium, and P110 variants (MS, RY, EC) are standard practice in certain areas of interest (AOI).

--- **ACTION** ---

## Avoid Difficult Casing Connections

Difficult to make-up casing connections can impact an operation in a variety of ways. Increased casing installation time is the most obvious but not the most impactful. Issues after the casing is installed and cemented are the most financially damaging. If there is a problem with the casing after it has been installed, there are few economic options, and depending on the failure type, very little an operator can do to remedy the situation without incurring significant costs.

When determining which connection to go with, do not fixate only on the technical. Of course, technical aspects are important, but please take the operational activity of making-up the connection on the rig floor into account during the decision process. If making-up the connection is operationally difficult, the risk of incurring a connection failure is elevated. This can happen during installation, drilling ahead, or completion, especially during fracturing, when the casing is stressed repeatedly for days or weeks.

One connection I recommend avoiding, if possible, is a short round thread casing (STC) connection. This is one of the most difficult connections to make-up. If the rig is not perfectly level, the wind is high or gusting, or personnel are not on top of their game, it will result in a problem when making up the joints on the floor. Most casing crews dread running this connection and will have severe anxiety when headed to a job where the operator is running STC connections. But the workers won't tell you that; they will say everything will be okay and that you should not worry—but you should.

Sometimes, large bundles of casing are available at a discount. Supply chain departments may see the opportunity as a great financial deal, with cursory diligence on the connection specifics. It is not uncommon for this to occur on surface casing strings because they are not exposed to fracturing pressures and deemed not as critical. However, a failure on the surface casing connections during drilling will be very expensive and can easily run into millions of dollars.

In general, heavily discounted pipe makes me nervous, especially if it is sourced overseas. Unless I know the exact mill and have confirmed the quality with an inspection, I will avoid this casing. Some operators do not diligence strings that will not be fractured or exposed to significant pressure, but there have been numerous issues during drilling in which casing strings have failed resulting in significant expenses.

Casing issues carry added risk during full-scale development and multi-well pads, in that if a problem is present on one well, the risk of other wells having the same problem is elevated. Or the career ending domino effect occurs, in that a well control event or similar situation spreads from one well on the pad to all the wells.

Once a casing concern arises, it may require rejecting large volumes of casing already in inventory. Certainly, a hard decision to make, but it is far easier to deal with a problem on the surface than to put that problem into the subsurface.

---
# ACTION
---

## Post Mill Casing Inspection

---

Since casing failures can derail a free cash flow strategy, consider reducing the risk with post-mill casing inspections.

Once we receive our casing at the local pipe yard, run it through a full independent inspection to find any quality issues before potentially putting a problem below ground. Trying to fix a casing problem in the subsurface is expensive and, in some cases, uneconomic, resulting in a P&A situation on a new well, which is a painful experience.

I perform a post-mill inspection on all new casing that will see direct fracturing pressure. If fracturing down a long string, I will inspect the long string. If it's a liner completion, I will inspect the liner and intermediate string, unless a tieback is utilized.

As for the inspection, there are a number of options. At a minimum, perform an Electro-Magnetic Inspection (EMI) and a Special End Area (SEA) inspection on the entire string, and don't forget the pups.

Marker joints, short joints, pup joints, or "pups," as they are referred to, are like stray cats. They wander around from location to location and are often scattered around the pipe yard. Therefore, they can get damaged. Also, it is common not to know where they came from. Sometimes, if a full joint is damaged, like cross threaded and backed out, it will be cut into a marker joint. Therefore, we must not forget to inspect our pups. Including images of the pups in our Daily Visual Report, incorporating when they are run, is a good idea. It is nice to have this data if we are having trouble locating them during our initial gun run during completion operations. It is not uncommon to find them off-depth or not at all.

An additional risk reduction tactic would be to place any joints that have defects between 5% to 12.5% of the nominal wall thickness at the toe of the well since we are less dependent on joints at the bottom compared to joints at the surface. A joint failing at the toe during a frac job is not as catastrophic as a joint failing in the vertical portion of the wellbore. Joints at the bottom of the string also see less stress.

I mark joints that pass inspection but have defects from 5% to 12.5% of the nominal wall thickness with two white bands on the box end and place them at the toe.

Anything below 5% is what I consider our best casing, and it is marked with one white band. Anything above 12.5% is marked with 3 red bands and kicked out as having failed inspection.[3]

Additionally, if anything does not pass the "eyeball" test, we kick it out. This is often only possible if we are engaged during our inspection at the yard before we mobilize the casing to location. I have personally kicked out a lot of casing over my career that had passed inspection but just did not look right to me. It is much harder to do this if the casing is already on location and ready to go, but sometimes it must be done.

Determine what works for you in your area of operations and put a system in place to minimize risk to help ensure success. Due to its economic significance, casing deserves an extreme amount of diligence. Systems make that possible.

# ——————— ACTION ———————

# Work with State and Federal Agencies on Prudent Operational Regulatory Exceptions

Over time, governmental regulations have increased across all industries. The oil industry is no exception. Due to the nature of our industry, regulations are added and adjusted progressively. State and Federal regulations determine many aspects of our designs. These regulations are in place for a variety of reasons that may not pertain to our operation.

Often, these adjustments and new regulations increase costs, which seem unavoidable. However, depending on the situation, regulators may offer opportunities, including variance requests, waivers, modifications, or exceptions, to mitigate the costly impact of specific requirements. Just to be clear, I am not encouraging you to avoid state and federal regulations. That is not the objective of this action item. I am highlighting a legal opportunity to mitigate the impact of regulations, which may or may not increase costs or add risk to an operation. In certain situations, it is prudent to go above and beyond regulatory requirements, and many operators do. In other situations, regulations may make an operation unsafe if implemented as written.

Since every well is unique, regulators work with operators to ensure success by granting variance requests. I encourage operators to get to know federal engineers, state engineers, and regulators. A personal relationship can help when requesting exceptions, which can improve safety and reduce costs.

―――――――――― ACTION ――――――――――
## Spudder Rig

When the spudder rig concept first came on the shale scene, it was not widely embraced. The pushback logic was that using a spudder rig does not save money. "It's a wash!" was the most common response I heard in discussions with drilling engineers about whether to use a spudder.

This is true, usually. The cost to use a spudder rig, compared to drilling surface with the primary "big rig" drilling rig, is often the same if you only look at the theoretical projected cost. However, where the spudder rig concept shines is risk reduction, cost predictability, and cost exposure.

The cost structure of a spudder rig is typically a flat rate to drill surface or to drill to a given depth and run casing. The time component is removed from the cost equation. Therefore, if we have a problem, the additional cost is almost zero or much less compared to a primary "big rig" drilling operation.

Dealing with surface drilling problems with a big rig on location can get very expensive and dangerous. The big rig is not designed to drill shallow depths as much as it is engineered to deal with deeper, more complex intervals of a horizontal drilling operation.

Spudder rig crews are focused on surface intervals and, as a result, have perfected this segment of the operation. This is often realized in reducing the risk of washing out the conductor, reducing the occurrence of lost circulation during drilling, and ensuring cement is circulated to surface, all primary concerns on the surface interval.

Spudder rig personnel are more aware of surface issues because that is their focus area, and their equipment is geared for it. As a result, excess mud, LCM, and surface cement can be reduced, contributing to lower costs. It is common to run 100% excess surface cement to ensure we circulate cement to surface, but with a spudder rig, this can be reduced for additional savings and minimized waste because there is less washout. Furthermore, if the cement falls back or we have problems circulating to surface, it is easier and less costly to identify and address it with a spudder rig on location.

To make selecting a spudder rig for surface operations more attractive, spudder rig companies are adding additional surface interval services. In some cases, spudder contracts include everything required to drill surface, except source water and casing. Furthermore, during large-scale operations, drilling consultants can oversee multiple spudder rig operations, further reducing costs.

Some operators allow spudder crews to run independently with drive-by supervision from an employee foreman present only during the casing running and cementing operation. This tactic would eliminate consultants from the surface operation, further reducing costs and having the operator take a more direct control approach, reducing risk.

The surface casing and casing head are the foundation of the well for the life of the well. Building a good, stable foundation in and around the wellhead will help with efficiency and speed during drilling, completion, and production.

With the spudder rig method, we get everything installed

and then have time to look at the wellhead and area around the wellhead to confirm we are ready to receive the big rig. This is especially important for large multi-well operations.

For example, if there is a problem with the pad or an issue with the casing head, it is far less expensive to find it during or after spudder rig operations than when the primary rig is on location. Problems become exponentially expensive once the big rig arrives.

There have been cases where a big rig drilled the surface, installed the BOPs, and then the entire assembly started sinking into the ground. This has also happened when production casing weight is transferred from the rig to the wellhead.

When the surface structure is not stable and able to support the weight, problems will manifest. This type of situation can be more likely to occur when the primary rig drills the surface segment because there is more washout, potentially more cement fallback, less time for the cement to harden before drilling resumes, and it is more difficult to identify issues.

My preference is to have the casing heads installed, have the time to clearly look at everything in the open air, and then get the area around the wellheads ready for the big rig and BOP installation. It is advantageous to see how the pad handled the spudder operation before the big rig arrives because if the pad is failing, we can fix it before the big money clock starts.

Furthermore, if the head manufacturer recommends filling the cellar with cement to the surface casing head base plate, it is easier to execute using the spudder method. This further stabilizes the wellhead and surface area. A strong, stable work environment enables speed, which is what we need.

—————————————— **ACTION** ——————————————

# Drill String Management Systems

One of the most underappreciated and overlooked aspects of a drilling operation is the string: drill pipe, drill collars, heavy weight drill pipe (HWDP), subs, and other drill string components typically provided with the rig.

These are the only components of the drilling rig that are in the subsurface. If there are problems with any other components, they are on the surface and a lot easier to address. But if there is a problem with the string, it often goes unnoticed until a failure occurs, or we end up having to purchase the entire assembly because it is damaged beyond repair (DBR).

Due to the severity of these situations, it is not uncommon for legal situations to occur as a result. However, the question to ask before getting into one of these expensive ordeals is:

*"What systems do we have in place to manage our drill string risk and cost?"*

Based on drilling history within a specific area, type of well, and geologic formation, string wear can be forecast to determine the optimal inspection and repair system for both risk and cost. Of course, anytime we are making-up joints, drilling ahead, or tripping, the team should look for issues with the drill string. This is something to discuss at safety meetings and before each trip. However, these are simple visual inspections in less-than-ideal conditions.

The primary inspections to be systemized are DS-1 inspections and subsequent repairs performed by third-party specialists. Rig contracts often address inspection requirements based on drilled footage or hours rotated. However, just because the contract requires inspections based on an arbitrary metric does not mean it is the optimal metric for which to build our drill string management system around.

If we are drilling an abrasive formation or our people love to throw in double doglegs and shallow nudges, it might make sense to perform inspections sooner or hardbanding more frequently. To maximize free cash flow, we must determine the optimal string management system for our operation.

For abrasive formations, we may end up money ahead by hardbanding 33% of the string after every well. That ends up saving considerably in DBR charges, which more than offsets the cost and reduces failure risk.

The type of hardbanding alloy and application specifics also make a difference. Additionally, DS-1 CAT-3 inspections might be a contractual requirement, but a CAT-4 inspection is not much more in cost and may find issues that a CAT-3 inspection would miss.

Furthermore, since missing a bad joint can be catastrophic, consider having an independent monitor oversee the inspection process. It's minimal cost and helps keep the inspectors diligent. Plus, it is not uncommon for DBRs to be a lucrative source of cash flow for rig companies, so there is an incentive to be aware of.

If we are not careful, there have been situations in which a string that might have been borderline in passing inspection and had excessive damage from its prior life that was not caught, gets accepted by our company. Then, after drilling only one pad, we end up having to buy the entire string: easily +$1.5 million in DBR charges. A third-party monitor can help mitigate this risk.

If we are developing assets at scale over years, the operator in many cases, ends up purchasing the entire string over time. Depending on the math and our specific operation, it might make sense to purchase our own strings and then reduce the day rate on the drilling rigs to take this into consideration. We will probably have better quality control and drill pipe joint uniformity this way.

For example, years ago, we had an issue with drill pipe after setting a cement plug to kick-off around a fish. After the cement plug job, we clogged up the BHA when we went back to drilling and had to trip out to address it. Small amounts of cement coated the inner diameter of the walls of the drill pipe, and it built up enough to be a problem. This occurred even after pumping down drill pipe wiper balls to clean the inner diameter of the string.

We rattled the joints, which helped, but we did not get all the cement off the walls. So, we purchased a dual-lens endoscope camera and a long stick to inspect the inside of every joint. What we found was very interesting. The inside of each joint was different. Some joints had an internal plastic coating (IPC), and they did not have any cement stuck on them. On many joints, the IPC was worn. There were numerous colors of IPC, which indicated they were coated at different times and probably by different providers for different reasons.

Other joints had no coating, bare steel alloy. Some joints were slightly tapered and had different internal configurations. Some had cement coated on them; some did not. The very small amount of cement sticking on the internals seemed to be based on imperfections on the inside of some of the joints that grabbed small amounts of cement that the wiper ball was not able to wipe clean. Using the camera, we marked the problem joints, and we washed just those joints thoroughly, fixing the issue.

All problems have costs attached to them. We can solve many problems cheaply using components from local hardware stores and oilfield supply shops, which is what we did in this case to build a 100 FT inspection borescope.

After looking at every inch of the inside of +25,000 feet of drill pipe, I realized that we had a Frankenstein string. You could not tell the difference externally. This situation is more common than most people think, especially during boom periods when rigs are coming out of stack, longer laterals are in demand, and strings are cobbled together to get the required extended lengths.

Developing a drill string management system will mitigate the occurrence of problems. Increasing the level of awareness on the condition of the string will help reduce risks and costs as designs continue to evolve, pushing the limits of shale technology.

# Drill String Management System Considerations

## Drill String Cost Structure Strategies

- Consider unbundling the drill string from the rig contract.
  - Compare cost between renting the drill string separately from the drilling rig to see if there are cost savings.
- Perform a cost-benefit analysis between renting the string as part of the drilling rig contract, separately renting the string from a 3rd party, and purchasing drill strings directly from the manufacturer.
- Establish relationships with string manufacturers to obtain insights on the latest drill string technology, issues, costs, and maintenance suggestions. There are many critical string dynamics that oil companies are shielded from regarding drill string information, issues, and opportunities.

## Drill String Inspection and Maintenance Strategies

- Based on historic string wear, associated repair costs, DBR costs, and failure events, calculate the optimal inspection schedule.
- Perform a cost-benefit analysis comparing a CAT-3 versus CAT-4 inspection.
  - Consider a mix of inspection types throughout the year.
- Establish a separate hardbanding schedule to protect the string from costly damage and DBRs.
  - Depending on damage, consider hardbanding a set percentage of the string (i.e., 33%) after every well or multi-well pad.

- o Test different hardbanding products and strategies to find an optimal system to minimize costly damage to the drill string, DBR occurrences, casing damage, and casing failures.
- Depending on wear, address handling practices, rig equipment inflicting damage, make-up, and transportation.
- Consider rotating string order when drilling abrasive rock.
- Hire 3rd party drill string experts and monitors to help reduce risk and costs across our fleet of rigs.

## Cost Reduction Inspection Systems

- Institute automated inspection and maintenance processes based on risk, economics, and real-world occurrences.
- Separate joints that pass inspection but are at higher risk or are close to not passing inspection and will be run last or uphole to minimize costly DBRs and pipe failures. Below is an example to consider:
  - o 3 White Bands = 80-85% remaining body wall (RBW)
  - o 1 Blue Band = hardbands <1.35 mm (dime thickness)
  - o 1 Orange Band = joints that cannot be recut
- Automate a hardbanding and drill string protection program based on data from our inspection results.
- Consider incorporating drill string management systems into our Daily Visual Report. Capturing various string dynamics with pictures daily can add significant value to our operations and office-based team.
- Request 3rd party string monitors and inspectors to incorporate photographs into their reports.

—————————— **ACTION** ——————————

# Mud Magnet Program

---

Everyone says they run magnets in the returns line, but very few people look at them until there is a problem. I hear about situations regularly, including wearing through the casing, wearing through the wellhead, incurring severe string damage, and BHA DBRs costing millions.

Recently, a friend noticed drilling mud pooling around the rig and tracked the problem to the wellhead, which had a hole below the BOPs. Fortunately, they were drilling a docile formation, with no trip gas. When the magnets were pulled, they were covered in metal shavings. The wellhead was removed and was worn through one side, including the wear bushing protector. The ordeal cost +$750,000.

> ## Problems have costs attached to them.

During drilling, wear situations are expensive. To optimize designs, improve efficiency, and reduce risk, we need a picture of what is going on below the rig floor. Monitoring metal particles in drilling fluid is a free method to help accomplish this. Running magnets does not add value unless we have a program in place to monitor them, collect the data, and incorporate the information into an action plan.

Consider implementing a process in which the magnets are pulled once per shift (every 12 hours), photographed for our

Daily Visual Report, and the metal particles are scraped, collected, washed, weighed, and examined.

Most rigs have magnets, so this costs nothing. After doing this on several wells, we will have a baseline to reference so that if there is an increase in the amount of metal shavings (grams/12-hours), or change in properties, it will give us an early warning of an issue before it becomes a costly problem.

---

**Statistical analysis and AI can transform an invisible occurrence into a FCF opportunity.**

---

Applying statistical analysis to an otherwise overlooked occurrence elevates our intelligence to see into the future and prevent undesirable situations while maximizing cash flow. Furthermore, we can incorporate this data into out AI Virtual Company Man. Using data from the magnets we can program the system to provide our team with information on what is occurring below the rig floor.

Going one step further, we can use the images of the metal from the magnets combined with computer vision to obtain a second opinion on the dynamics of the metal we are collecting. This information enhances our intelligence, helping our team make higher-quality decisions and avoid expensive mistakes.

Significant money can be saved and a lot can be learned from the metal shavings collected from mud magnets. There are also geosteering benefits in terms of minimizing metal interference with MWD equipment or collecting data on formations that have magnetic minerals.

# Use Magnets to Reduce Costs

- Establish a baseline in our area of operations using the collected amount of shavings weight per 12 hours.
- Identify trends that deviate from the baseline.
- Use data to optimize directional nudge depths and designs.
  - How aggressive and shallow is too shallow to start a nudge based on metal shavings?
  - *FYI – I prefer to avoid shallow nudges due to the risk of wearing the surface casing and wellhead. However, monitoring shavings helps reduce the risk when shallow and aggressive nudges are necessary.*
- Identify opportunities to modify well design to perform frac treatment down casing that has been drilled through. If we minimize casing wear during drilling, there are significant cost reductions in eliminating a tieback string or completely changing the well configuration and casing plan.
- Improve directional plans.
- Better select hardband materials to protect casing and string.
- Optimize mud and additives to reduce wear and friction.
- Adjust casing thickness in areas of high or low wear.
- Reduce or increase necessity to spend time pulling wear bushing for inspection based on magnet data.
- Determine need to recenter rig or top drive based on wear.
- Address lower than anticipated ROP due to frictional issues.
- Identify reasons for difficulty achieving weight transfer to bit and directional control issues when sliding.
- Identify BHA problems before DBRs or failures.

--------------------- ACTION ---------------------
# Measure Minutes

It is industry practice to measure operational performance and costs during drilling, completion, and production using a daily time frame. From a legacy perspective, the industry thinks, talks, and analyzes things using the time unit of days. However, in recent years, many vendors have transitioned to running their business and charging by the hour and, more recently, by the minute. Vendors now look at the oilfield world in terms of minutes because that is how they think, plan, charge, and maximize cash flow. Oil and gas companies, on the other hand, are often stuck in the past and continue to view the oilfield world in terms of days.

From a big-picture perspective, using days works well, as the collective wisdom is comfortable with that time unit and uses it as a frame of reference. However, when looking to improve performance and reduce costs, consider transitioning to a more granular period.

Problems reveal themselves in the second and minute level period. Fostering a culture that is watching the minutes not only helps us better audit invoices but also increases our level of operational diligence. To introduce this concept, instruct our team to capture any issues that impact operations for five minutes or more on the Operational Issues List. Most 5-minute issues are more significant than they initially appear and require us to expand our scope of investigation to prevent them from reoccurring, leading to increased costs.

# ——— ACTION ———

## Upgrade Morning Meeting

It is standard practice for large-scale operators in nearly all industries worldwide to hold a morning meeting between field leadership and office management. The structure of the morning meeting and how it is conducted sets the stage for the day's operations, reinforces priorities, conveys critical information, and connects management to the front lines. The morning meeting is essential for our free cash flow efforts and is an opportunity to reduce risk and costs.

Operational scale, in terms of the number of rigs and frac crews, influences meeting dynamics; however, it is common for most meetings to have superintendents, foreman, and wellsite leaders engage by reviewing the past 24 hours. This typically results in reciting the daily reports, sometimes verbatim. Although it is good to have people verbally confirm what they wrote, it is not the best use of our limited time.

Consider upgrading the morning meeting from whatever you are currently doing. Change keeps people engaged. Remember, 91% of workers daydream at meetings.[4] They tune out, especially when people are reciting comments from the reports, which everyone has already read.

To address this, have each wellsite leader shortlist items from their Operational Issues List. There is no reason to waste time reviewing operations that are going as expected; it's the concerns, issues, problems, and potential problems that need to be discussed.

For example, if one rig is having a problem with a vendor or procedure that we are employing across all the rigs, the morning meeting would be the place to mention it to let everyone know to look out for the issue or to poll the team to see if the problem is pervasive.

Reviewing issues and concerns at the meeting shares information across the organization and leverages group intelligence. It would be a very expensive exercise and a waste of our economies of scale not to leverage experiences and learnings across all the crews.

Unfortunately, unless there is a serious event, sharing problems rarely occurs and is often considered taboo. Near-misses, vendor issues, and wellsite leader concerns are usually not shared until after the undesirable event. People hold it back for many reasons, resulting in operations exposed to increased risk and higher costs across our shale machine.

## Reasons Workers Do Not Share Problems

- ❖ **Fear:** Lose job, current or future work. Impacts reputation.
- ❖ **Incentives:** Impacts bonus, promotion, costs, or revenue.
- ❖ **Company Culture:** It's not encouraged. Boss gets angry.
- ❖ **Systems:** A system is not in place to share.
- ❖ **Embarrassment:** Don't want to be judged by others.
- ❖ **Ego:** Belief that "Nothing goes wrong when I am here."
- ❖ **Convenience:** It is easier to say nothing.
- ❖ **Downplay:** Explain it away as not serious.
- ❖ **Mindset:** Don't want to bring problems. Cognitive biases.
- ❖ **Time:** It is not a priority. No time to get into it.

There is a culture in all industrial sectors to keep things in the field or on the factory floor. "Don't let your business get to your boss" is a common saying and mindset in addition to the thinking that certain situations are too sensitive to be shared.

If we are going to get our costs down, we must provide a continuously open environment for valuable information to be shared, especially problems. The Operational Issues List is one forum. The morning meeting is another.

Our friend, Admiral Rickover, required everyone to submit "The Pinks," but he still felt people were not open with their problems. Therefore, he held weekly meetings in groups and individually, but since there were hundreds of people it was taking too long, and he didn't want to waste time on operations that were going as planned.[5]

He wanted to focus only on problems. So, he told his secretary to call all vendors and tell them that when they report, to dispense with the pleasantries and report their problems or just say "No Problems" if everything was going as planned. So, they tried that for a while.[6]

But saying the phrase "No Problems" took too long for hundreds of people to say, so he had his secretary call all the vendors again and tell them that if they had no problems to quickly say just one word "No."[7]

If you have a mega-scale operation and a rig or frac crew is not having any problems, there is no reason to spend time discussing how everything is going great.

Building multiple systems that require people to report issues is critical, particularly when we start making changes to achieve cost reduction goals. If there is a weakness someone is hiding and we make changes on top of it, the situation could become catastrophic. What I have seen happen is that people try to fix their problems themselves without reporting them, or they hide them. If they can't fix it, then they report it, or it becomes obvious, but at that point, it is often too late. This is the reason Admiral Rickover wanted to know about problems immediately and why he put multiple systems in place to accomplish it.

## Upgraded Morning Meeting Structure

1. **Opening Safety Topic** supported by real-world examples.
2. **Reports:** from each crew.
   - Only report if there is a problem. Otherwise, state "No Problems" and move to the next rig or frac crew.
   - Current Operation (drilling ahead, tripping, mob)
   - Geology Update (in/out zone, geohazards, plan)
   - Operational Issues List: Select problems to review.
     - Discussion (field, office, next steps)
   - Questions from team on any aspect of our operations.
3. **Intelligence:** analytics, daily tips, alerts, trends, heads-up.
4. **Actions:** procedure updates, global changes, initiatives.
5. **Closing Safety Topic** supported by real-world examples.

Sometimes, people do not feel comfortable sharing problems in a group setting, which is understandable. Therefore, depending on our operational scale, a combination of the morning meeting and individual meetings is warranted.

# ——— ACTION ———
## 10X Geosteering

Drilling metrics don't mean much if we are out of zone. Horizontal wells drilled out of the target interval almost never perform up to expectations, no matter how advanced the completion is.

From a long-term production perspective, we cannot frac our way out of a poorly steered well. Therefore, having the best possible geosteering personnel, combined with the best possible geologic technology (digital and physical) for the target rock, is paramount to improving our geosteering by an order of magnitude or 10X difference.

Determining who is the best and what are the optimal tools to 10X success in terms of landing favorably and staying in zone with the smoothest possible wellbore is not easy or static. The situation is dynamic and depends on a variety of factors, including area of interest, target interval, geologic complexity, control points, seismic, and experience. Not to mention a multitude of human factors across the team but especially between staff geologists, area expert geologists, geosteerers, engineers, wellsite leaders, and directional drillers.

Within the oil industry, the geosteering position and responsibilities do not command the level of respect they deserve. Therefore, engineers and wellsite consultants often prefer to avoid getting too deeply involved in the work or technical dynamics.

Additionally, many geosteerers want to get out of the

position as soon as possible for several reasons, including difficult hours, high stress, and low recognition. Therefore, finding the best geosteerers is challenging, but it must be done to maximize free cash flow.

> **One bad geosteerer can cause significant operational damage and economic pain.**

Please see "20 Ways Drilling Out-of-Zone Increases Cost" on pages 440 through 442 and "Probing to 10X Geosteering Abilities" on pages 443 and 444.

The geosteering problem expands as operations scale since management will have an increasingly difficult time keeping an eye on all the rigs with the level of detail required. Additionally, scaling a drilling program often involves increasing geologic complexity as we step out of control areas, further compounding our geosteering challenges.

Plus, the faster we drill, the less room for error in drilling out of zone and correcting it quickly enough to minimize the percentage of the wellbore that will most likely underperform production expectations.

Based on geologic understanding of each area, geosteering performance, and well production performance, we must determine who the top geosteerers are within our company and externally regarding relief contractors. I prefer to minimize the use of external geosteerers because there is less control and accountability.

Due to the nature of the work, distractions are pervasive.

Unfortunately, for some reason it is common for contract mud loggers and geosteerers to play video games while doing their jobs—they are not deeply engaged with steering the wells.

You could say geosteering is like a video game, except that our company's future is on the line. We may have no idea this is happening and have never met the person, but he is slowly destroying our company by playing video games, surfing the web, or using social media while geosteering our wells. Contractors are detached from production performance and, therefore, do not care deeply about how their actions impact it.

However, sometimes, we have no choice but to use contract personnel to perform geosteering operations. Therefore, when using contractors, it is critical to visit their place of business and get to know them personally. This is a task for our in-house geologists, and that relationship should be managed accordingly. Under no circumstances should we outsource our entire geosteering program. Planning, landing, and staying in zone is too critical to relinquish control.

A well in-zone has less problems during all three operational phases: drilling, completion, and production. Drilling out of zone or partially out is plagued with expensive problems and additional costs throughout the life of the well.

Geosteering is a skill with many facets. Not everyone is good at it. It is not advisable to find out someone is not good at geosteering by making costly mistakes on our wells.

Drilling out of zone does not just impact well production; it can be dangerous. Therefore, it's a safety issue. Going out of zone in many horizontal plays can start a chain reaction of events leading to an incident.

## 20 Ways Drilling Out-of-Zone Increases Cost

1) Unable to initiate a fracture below max design pressures due to unfavorable mineralogy, rock properties, and frac gradient of the out-of-zone formation, rendering the well worthless. **Cost: +3.0MM.**

2) Unable to breakdown the toe through the sleeves since it's out-of-zone. Therefore, we need to get a tractor or coil tubing and perforate multiple times to find a frackable depth. **Cost: +$150K.**

3) Screenouts occur multiple times during the frac since it is more difficult to place sand out-of-zone. **Cost: +$250K.**

4) Screenout occurs on one of the first four stages which are not in the target. Therefore, the well does not have enough energy to flowback. Need to mobilize coil tubing to cleanout. **Cost: +$300K.**

5) Well slightly out-of-zone results in higher treating pressure and longer pump times to complete frac treatment, increasing fuel usage and frac minutes. **Cost: +$350K.**

6) While building the curve or in the lateral, geosteerer calls for an aggressive target change resulting in a significant dogleg. The dogleg severity (DLS) on the survey is not fully representative of reality. During the initial plug and perf run, the wireline assembly gets hung up around this feature. Guns are fished, and the plug is milled, but the casing is damaged in the process. **Cost: +$1.5MM.**

7) Due to poor geologic interpretation and steering execution

combined with challenging geology, the well is drilled with many significant doglegs along the lateral combined with sticky out-of-zone intervals. Getting casing on depth is difficult, resulting in the rig crew using heinous, barbaric methods to force the casing down. The casing fails the pre-frac pressure test, resulting in an expensive remediation operation. **Cost: +$2.0MM**.

8) Drilling out-of-zone requires sidetracks. **Cost: +$250K**. Depending on the geological situation, drilling out-of-zone on an abrasive target formation with significant dip, bound with soft formations, results in an inability to get back into zone. Therefore, we need to set a kickoff plug and redrill lateral. **Cost: +$500K**.

9) Formations above and below target interval are unstable causing BHA to get permanently stuck. **Cost: +$1.0MM**.

10) Unstable out-of-zone formations unintentionally drilled into pack-off around the string resulting in costly problems. **Cost: +$200K**.

11) Drilling out-of-zone on hard rock target results in having to TOOH for a more aggressive BHA to get back in zone. **Cost: +$150K**.

12) Getting geologically lost results in multiple aggressive undulations impacting weight transfer, steerability, and ROP. **Cost: +$400K**.

13) Zones above and below target cause mud losses or change chemistry of mud system resulting in the necessity of additional additives or increased diesel. **Cost: +$300K**.

14) Mobile clays out-of-zone flowback to surface plugging wellhead. Production tree needs to be cut off under pressure. **Cost: +$500K.**

15) Due to crossflow from being out-of-zone, coil or stick pipe gets stuck during post-frac plug millout, resulting in a fishing operation. **Cost: +$1.0MM.**

16) Drilling out-of-zone provides less energy to flow during plug millout resulting in difficulty. **Cost: +$200K.**

17) Complex well path due to getting lost and searching for the target multiple times along the lateral results in an inability to get coil to bottom. Thereby requiring a snubbing unit. **Cost: +$150K.**

18) Formations out-of-zone flow increased water amounts, scaling, corrosion, or higher $H_2S$ for the life of the well, resulting in higher disposal and chemical costs. We also need to modify the facility design to handle the unexpected production dynamics. **Cost: +$250K.**

19) Drilling out-of-zone causes a pack-off event, loss of circulation, stuck BHA. The well becomes underbalanced. Issues on surface result in a loss of well control incident. **Cost: +$20.0MM.**

20) Drilling out-of-zone increases stress levels on the entire team including drilling, completions, production, and management personnel resulting in less attention on general operations, efficiency, cost reduction, and maximizing free cash flow. **Cost: Priceless.**

# Probing to 10X Geosteering Abilities

- Who are our best geosteerers?
  - What metrics are we using to determine this?
  - How do we recognize and reward individual geosteering performance?
  - How are we helping mediocre steerers improve?
- Who are the best geosteerers that don't work for us?
  - What is our plan to hire them?
  - Do we have a list of top draft picks?
- What are we doing to improve our planning, prospect models, geoprogs, and pre-spud geosteering models?
  - Do we have any time constraints?
- What training are we doing internally or providing externally to improve our geosteering ability?
- Are we using simulators to practice and train?
  - What is our geosteering simulator training schedule?
- Do we watch drilling playback to capture learnings?
  - Is geosteering incorporated into this?
- What is our seismic data situation?
- What digital and physical technologies should we try?
- How are we enhancing our mudlogging program?
- How are we using artificial intelligence (AI) to assist our geosteering program?
- How do our human geosteerers compare to our AI geosteering performance?

- What are we doing to improve our AI?

- What is our process to code our learnings into our AI?

- How are we incorporating our targeting systems into our automated alarms notification system?

- What is our process for taking drilling parameters (ROP, WOB, RPM, etc.), MWD data, mud logs, gas shows, and surveys to determine the most likely interpretation and geologic projection forward?
    - Where are we in terms of automating this process?

- How are we advancing our geosteering decision communication across the team?

- How are we improving style: quantitative, qualitative, target lines, (waypoints + range), (target lines + range), corridors, etc?

- What are our direct competitors doing?
    - Rank our performance compared to our competitors.
    - What are they doing differently?

- Who are the top geosteering firms in our AOI?
    - What is their competitive advantage?
    - What software and technology do they use?

- Should we have a 3rd party geosteering firm run a 100% independent assessment alongside our geosteering operation as a second opinion on challenging geologic situations?

- What systems do we have in place to interact with our operator competitors and contract geosteerers?

## —————— ACTION ——————

## Leverage Advanced Alarms

Detailed alarm notifications play a critical role in many industries. For decades, they have had a heavy presence on the production side of the business, allowing the industry to monitor remote locations without requiring people to be physically stationed at every producing well. Can you imagine how expensive that would be?

On the drilling side of the business, detailed alarms have been around, but historically, they have not added much value unless we are physically in the driller's chair, in front of the EDR, or have the mobile app open. Additionally, we cannot configure advanced alarms on legacy systems.

To get the most out of alarms, the system must have several features. First, the alarms must have text, email, and phone notification options. My preferred notification method is text because of its reach and sense of urgency. However, if someone is a heavy sleeper, an automated phone call might be necessary to wake them up.

The alarm system must be able to notify an unlimited number of people anywhere in the world. Additionally, the system must include options for various notification strategies, notify multiple people until the alarm is acknowledged, and have a tiered escalation alarm system capability.

Second, the alarm system must be fully customizable so that users can code an unlimited number of alarms using multiple variables, depending on what is happening in real-time. A no-code or low-code system with an intuitive user interface is preferable because users may need to build an alarm quickly for a given situation and push it out to the team within minutes.

For example, while drilling an abrasive formation with an average ROP in rotation of 25 to 50 ft/hr and 5 to 10 ft/hr while sliding, we noticed that under certain conditions during a slide, we could get hung up and potentially stuck. If instantaneous-ROP or ROP-Fast dropped below a certain point for more than 60 seconds, and gravity toolface (GTF) did not change during this same period of 60 seconds, we wanted to pick up off bottom and reset to ensure we were not hung up.

> **Drilling alarms must be configured to identify AOI specific drilling dysfunctions, inefficiencies, high-risk situations, abnormalities, and cost reduction opportunities.**

The longer we would stay on bottom in a slide under these conditions, with instantaneous-ROP below 5 ft/hr and minimal change in gravity toolface, the higher the probability we would get permanently planted. Therefore, having an alarm system that could be set up to send a reminder to pick up off bottom under these parameters would add significant value when we were sliding, potentially saving +$1.0 million if we got stuck and had to sidetrack.

We want a system that can send a mobile phone alarm (text or call) to the rig drillers, directional drillers, MWD hands, wellsite leaders, and engineers when this type of critical situation occurs. If our current system cannot setup an alarm for a simple situation like this—during a slide, if X occurs and Y occurs for a specified time period, contact a group of people every 10 seconds until the system registers a pickup off bottom—then it's time to find or build a system that can.

Finally, the system must have the ability to turn on and off a group of alarms (+100 alarms per group) for different situations (i.e., sliding, tripping, running casing, etc.). For example, if we are doing a negative pressure test after bumping the plug on a cement job, we want to have the ability to turn on a group of preset alarms for negative tests. When these alarms are turned on, a text is sent out to let the team know that we are doing the test now. Then, once the negative test is occurring, the system monitors flow and pressure. If all are zero for 30 minutes or whatever the required period is, the system sends a text confirming a successful negative test.

If the test fails, the system sends a text to everyone indicating that the test failed. A system like this would have prevented the $67 billion, 11-death accident on Deepwater Horizon[8], that is, of course, depending on whom the text alarms went out to. A failed negative test on an offshore exploration project probably needs the alarms to go all the way up to the highest possible operational level.

Whenever there is a near-miss, high-cost incident, or any undesirable event during drilling, completion, or production, identify how we can program our systems to automatically alert our team to take action to prevent a similar situation from reoccurring.

—————— **ACTION** ——————
## Roving Company Man

One wellsite consultant can manage multiple drilling rigs depending on comfort level. This is possible by leveraging technology, team experience, and a variety of factors we will now discuss.

Historically, for land-based U.S. operations, it was common to have one company man living on location and managing operations 24 hours a day. That same person often oversaw drilling and completions.

As shale operations evolved and became more complicated with increased risk and cost, the industry added night consultants and split operations into multiple phases (i.e., location construction, setups, drilling, toe preps, frac, millouts, facility construction, and flowback).

For drilling operations, wellsite supervision evolved into at least one day consultant and one night consultant, both living on location attached to each drilling rig. As the importance of the position increased, demand for consultants increased, and day rates followed to the point that the associated cost is now a significant line item on the AFE.

As monitoring and automation technology have progressed, some operators have addressed the wellsite supervision cost line item by allowing one wellsite consultant to oversee multiple drilling rigs, functioning as a roving wellsite leader.

This allows costs to be reduced by at least 50% when the

day rate is split between two rigs, which is huge in terms of cost reduction. If we have strong wellsite consultants, rig managers, drillers, and directional drillers, this action is something to consider. However, it depends on the area and geology. If there is subsurface complexity and challenges, it might not be a good idea. Therefore, each situation needs to be evaluated on a case-by-case basis.

Leveraging modern technology can help alleviate the burden on the wellsite consultants by monitoring drilling parameters with advanced alarms, adding sensors to monitor additional equipment, using AI to write daily reports automatically, and placing cameras across the operation with the ability to control the visual field and monitor sound. These items are part of our Virtual Company Man discussed earlier.

---

**Advanced technology enhances our general intelligence, enabling us to make higher-quality decisions, and do more with less.**

---

Additionally, a 24-hour command center that monitors operations can help oversee things when the wellsite consultant is between rigs. If we are twinning rigs, this tactic is more easily executed at scale. If we are running multiple spudder rigs, testing this tactic on spudder operations is a good starting point to see how it goes and assess comfort level.

A strong wellsite consultant adds significant value above the day rate cost, especially when it comes to cost reduction efforts. Spend quality time selecting wellsite consultants, getting to know their strengths and weaknesses, and working with them on challenging operations. Strengthening relationships with our field team goes a long way towards achieving massive cost reduction results.

While strong wellsite consultants add value, weak or untrustworthy consultants destroy it. Therefore, it is not a given that a company man, by default, adds value. The person must show value to make economic sense of the expense.

One of my pet peeves is to show up on location unannounced and find the wellsite consultants watching television, surfing the internet, or handling personal business. Wellsite leaders running a side business from location is not uncommon. Many owners, executives, and managers in this industry see behavior like this or have expensive mistakes on location that could have been prevented and think:

*"Why am I paying for this person?"*

Wellsite consultants should be high-graded, just like everyone else, and the top consultants asked to take on more responsibility to help with cost reduction efforts.

Most operators are currently sticking with two or more consultants and have them take on expanded roles during drilling and completion. The actions in this book require the wellsite consultant's role to provide more value to the operation. However, some operators are going a step further and have the wellsite consultants replace directional drillers, flowback hands, frac stack operators, technicians, tool reps, and mud engineer functions, which we will discuss next.

---------- **ACTION** ----------

# Drive-By Mud Management

---

Traditionally, the oil industry has two mud engineers living on location, managing the mud throughout the drilling process. Some companies have gone down to one full-time mud engineer. However, both situations often result in significant additional costs. Not only do we have the high day rate and the added cost of housing, water, power, and septic, but we also end up with a much higher mud materials bill.

One might think that with a full-time mud engineer we should have a lower materials bill, but it is the opposite. With a full-time mud engineer on location, the engineer wants to get the mud perfect, which comes at a price.

There have been cases in which an operator switched from full-time on-location mud engineers to drive-by mud engineers and reduced their mud cost by $500,000 to $750,000 per well, a huge savings from a small action.

When the mud engineers live on location, they often feel like they should be doing something to justify their presence and keep busy. This results in what seems like minor tweaks to the mud system. But over time, those tweaks add up, and if we have a deep (TVD) extended lateral with spud to TD of 60 to 90 days, as in the more challenging horizontal shale plays, the mud bill will get very high.

Consider switching to drive-by mud engineering and have the wellsite leader take a more active role in managing the mud.

Another option is to incorporate automated mud skids

which monitor mud properties in real-time.

Drive-by mud engineering combined with advanced drilling alarms, is an effective low-cost action that can deliver significant cost reduction results, especially if we have strong wellsite consultants.

To ensure success, we want to provide our drive-by mud engineers with full access to our remote monitoring systems, include them on our mobile application team communications threads, and quickly communicate issues directly by text and phone calls. Furthermore, establishing a daily call with our mud engineers, wellsite consultants, and office engineers can add value by ensuring we do not have any problems or make any mistakes regarding the daily instructions that our mud engineers provide.

> **Increasing communication is often key to ensuring our cost reduction actions are successful.**

With increased communication, our mud engineers know everything happening on location without having to be on location. Some operators have gone a step further and vertically integrated the entire drilling fluids service. Therefore, they directly source mud additives and employ their own mud engineers either as contractors or company employees. Regardless of what is decided, focusing on this aspect of the drilling operation can provide significant cost reduction opportunities while improving performance and strengthening our team.

—————————— **ACTION** ——————————

## Mud Losses
## Special Interest Group

If we have a highly complex problem that requires detailed technical expertise involving multiple disciplines, consider forming a special interest group (SIG) to help solve it. In the realm of cost reduction, a SIG is a group of people focused on cost reduction solutions for a specific problem.

The potential savings from addressing the complex issue of mud losses and additive usage, especially diesel for OBM, justifies forming a SIG if we suffer from this problem.

The primary objective of a SIG is to produce solutions, hopefully game-changing ones. Therefore, we want to form a group with people who are determined to deliver results.

## Mud Losses Special Interest Group Members

- **Drilling Engineer:** Lead SIG with cost accountability
- **Wellsite Consultants:** Lead initiative on location
- **Geologist:** Provide geological insights on losses
- **Mud Engineers:** Improve accuracy, detailed uses & losses
- **Rig Hands:** Track actions involving diesel, additives, mud
- **Fluid BD Manager:** Help with solutions & improvements
- **Automation Rep:** Add sensors, other monitoring systems
- **Digital Tech:** Build app to track all aspects of mud system
- **Camera Tech:** Install cameras in critical areas
- **Solids Control Rep:** Equip. adjustments, changes needed
- **Diesel Supplier:** Help with tracking diesel movement

When working to reduce costs, it can be rather painful to get to the end of a well with a good depth versus days (DVD) performance only to come up significantly short on mud volumes or have undesirable diesel usage for oil-based mud (OBM), resulting in a significant cost overrun.

Increasing diesel usage is a standard solution for a variety of drilling problems, and it often must be done to keep up with losses. However, if we can swap OBM for a more economic option, that is the ideal solution.

In many situations, there is not a good alternative to OBM, but we can eliminate OBM usage during certain sections of our wells. Through analysis, we may find that OBM is only needed in the curve or lateral. If we can stay on WBM to KOP, considerable savings are achievable. In some cases, easily over $100,000 per well if significant losses are experienced until a good mud cake is built up.

Furthermore, through subsurface analysis, we may find that we only need OBM in specific areas or situations.

> **Scaling operations across vast areas amplifies efficiencies and inefficiencies.**

When scaling drilling programs in full field development manufacturing mode, it is a common practice to employ the same designs across vast areas, even though the subsurface environment is changing. The optimal drilling fluids approach may not scale economically across our asset, especially when subsurface heterogeneity, faulting, and depletion exist.

When utilizing high-dollar drilling mud, we must look at every possible loss, no matter how small, including seepage, shakers, centrifuge, pits, cleaning, tripping, evaporation, displacement, transportation, and theft. Identifying a small mud loss venue might lead to the solution to our high cost drilling fluids and diesel invoices.

There is a good chance the solution will come as a revelation that nobody expected. That is the fun part. For example, what might seem like a well problem could turn out to be a human theft problem of diesel and OBM, both more valuable than oil and easier to monetize.

If the rig runs on diesel and the mud system uses diesel, it is easy to skim diesel—a common occurrence in some areas. Theft can occur anywhere along the supply chain. There are even cases where farmers and townspeople would stop by the rigs to fuel vehicles and farming equipment.

Uneconomic mud losses are often the result of a combination of factors, small individually but significant collectively, resulting in massive invoices damaging our company. The first step to solving the problem is to accurately account for uses and losses, including The Five Ws and One H of mud losses: Who, What, When, Where, Why, and How.

Drilling mud accounting is as critical as financial accounting, maybe even more so, as drilling mud is central to operational safety and success, serving in multiple capacities, including the primary flow barrier and wellbore stabilizer. Therefore, look to make incremental changes that add value over weeks or months instead of taking single-day drastic actions or changes that could have unintended consequences.

# Mud Losses Cost Reduction Actions

- Eliminate OBM in all or part of the wellbore.
  - Look for lower-cost mud alternatives and additives.
- Compare the calculated top of cement to actual with CBL to determine washout and subsurface volume accuracy.
  - Increase cement spacer to recover more OBM.
    - Address spacer design, pump rates, contamination, and personnel monitoring during mud recovery.
- Look at relationship between each drilling rig and mud related costs. Some rigs have better mud handling systems.
- Enhance tracking of additives and diesel usage and losses.
  - Install additional sensors on equipment.
  - Build a digital app for mud and diesel usage and losses to identify opportunities, hidden losses, and accuracy.
    - Mud loss ratio (MLR), volumes, reconciliation, additions, losses, and other factors (i.e., ROP, sweeps, rig processes, personnel, weather)
    - Identify relationships, correlations, and causes.
- Document every action involving the mud system.
  - Mud transfers, diesel usage, sand trap dumps, etc.
- Install cameras across locations including pits and shakers.
- Develop a system to check half rounds for excessive mud losses and transfer back into the active pits.
- Document shaker deck angles and screen wear regularly.
- Analyze how solids control equipment is operated.
  - Develop systems for operating each piece of equipment to improve performance and reduce mud costs.

--------------------- ACTION ---------------------

# Phase Out Housing

---

There is no reason the team must live on the wellsite location to perform oilfield work in the United States. The success of shale has attracted many businesses that specialize in providing living accommodations, including hotels, man camps, trailer parks, apartments, vacation rentals, short-term room rentals, camping, and traditional housing. We should support these businesses, markets, and communities by phasing out all on-location accommodations.

This action is good for the industry and for the United States because it encourages population migration into areas where we do business and want to see flourish.

If someone lives close to the rig, they should have the freedom to go home to be with family and not be required to stay on location. By phasing out on-location housing, we support families moving into communities in our area of operations, further improving focus and work quality.

I prefer not to have anyone living on site, if possible, primarily due to safety concerns.

It is safer and psychologically better for the team to get a break from the rig and take rest away from the risk, commotion, noise, and stress of the workplace.

The more people who live on location, the higher the risk and cost. When people live at the job site, we get an increase in non-work-related problems, incidents, and personnel issues occurring in close proximity to the operation, which I prefer to hold at a high degree of professionalism. People living on location is an unnecessary distraction.

For the operator, there are many benefits to phasing out on-location housing. First, the more houses we remove from location, the more room we have to work and move around on the pad. Second, we have less truck traffic because we can reduce non-potable water deliveries, septic hauling, housing maintenance visits, and a few other services.

Third, we do not have to spend time, attention, and resources dealing with housing problems, including water, heat, air conditioning, plumbing, sewage grinders, television, cooking, and all other people-related issues, when we should be 100% focused on drilling. When everyone is living onsite, we spend a considerable amount of time dealing with all the problems of running a small town. Who wants that? I don't. Eliminate the housing, and we eliminate the problems.

Fourth, we will have fewer people and vehicles on location to worry about, further reducing risk and cost. I can remember several instances during my career where investors were visiting a drilling operation with me and would ask in a half-joking manner if I was in the oil business or used car business because of all the cars and trucks parked on location.

It is embarrassing to have an investor say that. I even had a rig whose hands each had a personal vehicle on location and then decided it was Take-A-Harley-To-Work-Month and started arriving on hogs, which is cool, but it looked like a motorcycle gang and a used car lot combined. It was flat-out crazy on location, a huge distraction from work. Eliminate the housing, and we cut the vehicles parked on location in half, further reducing incidents, especially equipment and personal vehicle collisions on the pad.

Fifth, we will have fewer people visiting. When everyone lives on location, we get more visitors and many of them should not be there. With people living on location, we blur the lines between work and play, as in things that should not happen on an industrial job site end up happening. One of those things is having family and friends on location.

Even though everyone knows it is against the rules, it happens all the time, especially on weekends when there are fewer eyes watching, and people let it slide. I even had several cases where a wellsite leader thought it was acceptable to have family and friends stay overnight in the trailer. The only reason I caught it is because I do field visits when concerned about operations, regardless of the day of the week, and I was visiting an offset frac job, so I stopped by a nearby rig site.

This was on a Saturday evening, when nobody expects it. When I confronted the wellsite leader about his family staying in the trailer, he said all the major operators allowed him to do this. His wife told me that when he works for XYZ Corporation, a supermajor, they allow it.

If true, I highly doubt they know what he is doing because that is a liability nightmare. Even if we are strict with visitors, what happens is a bunch of family and friends congregate around the entrance to location right around shift change. A congested entrance is the last thing we want.

Sixth, by eliminating the trailers, we improve ESG metrics from a safety and environmental perspective. A lot of incidents happen in the trailer houses, including slip-and-falls, cuts and burns in the kitchens, fires, damages, electrocutions, shower incidents, water contamination, choking, drugs, pills, medical health events, horseplay, physical altercations, pranks, and other undesirable situations.

Finally, housing and related services generate a lot of individual invoices. These are not massive invoices, but they are nuanced and require a lot of error-checking and diligence before approval. Over a period, they add up, especially the non-potable water deliveries and sewage handling. If we can replace all the houses with a nice safety trailer, it will streamline the operation and associated invoices.

Some individuals and service companies on the drilling side of the business fight to have their people sleep on location, and they will provide many reasons for this. However, on the fracturing side of the business, no one wants to sleep on location, and no service company wants their people staying on location after their shift is up.

It's the same location, with frac work that is arguably more intense, and no one is sleeping on location. There is a tradition in the oil industry that drilling people have grown accustomed to, which is to provide drilling with accommodations but not provide accommodations for completion people. Drilling people want to sleep on location, and completion people do not. Nobody wants to change.

For example, while working on a shale asset in Wyoming years ago, due to weather concerns and remote locations, we provided housing on location for completion wellsite leaders. The trailers were brand new and very nice. After several years, one of the superintendents approving invoices noticed that the completions consultants were charging day rates and hotel charges on their invoices. This came as a shock because we were paying for houses on location. After an investigation, it was found that nobody was sleeping in the houses. The consultants were driving into town, even during massive snowstorms, and staying in hotels.

Once this was found out, everyone was so angry that the superintendent had to punish or release all company men. Since the invoices were approved, the company men all thought it was okay to charge for hotels even though we provided accommodations on location.

Therefore, the decision was to remove all the houses. We did not have a single house on location, not even a small office. The consultants were forced to work out of their pickups. Each person outfitted their work truck with a small office. I feel that they did a better job—it forced leadership to be outside and engaged with the operation more than before because there was no office to sit in. We did this during the entire year in sub-zero weather. The consultants had to handle their own accommodations and we did not reimburse them for it. Some stayed in the hotel of their choice, or purchased their own trailers and parked them offsite.

Allowing the team to leave the worksite and have the personal freedom to handle their own living situation supports the spirit of America and is a win-win. Location risk and safety are improved, focus on the operation is enhanced, work quality is increased, we support local communities and families, reduce the processing of invoices, minimize environmental impact, and reduce costs.

---

# ACTION

## Two Forklifts

---

"Two Forklifts" is my oilfield metaphor for the management philosophy called "Theory of Constraints."[9] The telehandler reach-type telescopic forklift, often referred to as the "SkyTrak" or "Forklift," is the most relied upon mechanical workhorse of oilfield operations. It is directly or indirectly involved in almost every drilling action. Therefore, if the forklift is down, the drilling operation is down.

The Theory of Constraints applied to drilling suggests that each drilling activity has at least one constraining factor restricting performance. According to the theory, if we identify the primary constraint or bottleneck and improve it or remove it, we improve the efficiency of the entire operation.[10] Once the current constraint is removed, we look for the next constraint in the system and so on until we reach our cost reduction goal.

For each drilling state, we typically have one primary limiting factor. Identifying the constraint and listing it in the Operational Issues List each day may help the team focus on addressing it. For example, if the forklift has a mechanical problem or is being used and another activity is waiting on it, that is the constraint limiting our performance at that moment. Since the forklift is a central piece of equipment known to have problems, to remove the constraint, it might make sense to rent two forklifts. This action removes the bottleneck. During each operation, look for the primary constraint. Identify potential actions that could improve or remove it.

--- **ACTION** ---
## 360-Rig Review

It is common to analyze rig costs by focusing on day rate. However, a lot of what I like to call "baggage costs" are missed when looking at that metric in isolation, and some rigs come with a lot of excess baggage that we cannot simply shake loose.

The rig and its capabilities are a central factor constraining cost reduction. For example, if the rig has a maximum safe tripping speed of 350 feet per minute while moving the drawworks with the rig controls, we are going to be limited by that constraint. Combined with a few other constraints, we can estimate the theoretical maximum drill string trip speed per hour by activity. Therefore, no matter what we do, we cannot go faster with the current system.

In a similar fashion, each rig has a number of non-obvious inherent constraints or "baggage" limiting our cost reduction ability, including mobilization costs, time costs, mud costs, fuel costs, rental equipment costs, BOP costs, and drill string damage costs. To identify all constraints in an attempt to release some baggage, we will perform a 360-rig review.

Most are familiar with the term "360-performance review" which is when an individual is evaluated by the people who work around them.[11] We will do the same thing but replace the person with a drilling rig. In many ways, a drilling rig is like a person with physical, interpersonal, and intellectual abilities and limitations.

We want to obtain multiple points of view on the rig from

everyone and everything that interacts with it. Additionally, we will do a negative-360 by searching for the most critical feedback on the rig to find opportunities to reduce risk and cost.

The goal is to find and improve or remove all the constraints the rig inflicts on the operation. If the constraints are too costly or cannot be improved to reach our goals, then we must release the rig.

Drilling metrics help identify the more apparent constraints a rig perpetrates. Therefore, we want a detailed quantitative analysis of all aspects of rig performance.

If we break down each rig activity and then calculate its theoretical maximum achievable performance and the associated cost, we can compare current performance to the theoretical optimum. Therefore, we are not only comparing performance between rigs, crews, or over-time but also with the maximum possible performance and lowest possible cost based on constraints. During this process, the goal is to get to a point or find that the rig or some aspect of the rig or crews are no longer the limiting factor constraining performance.

Once we have a frame of reference based on the metrics, we can turn our attention to the invoices for everything associated with the drilling rig operation.

Invoice analysis centered around the rig can go in many different directions depending on the specific situation. An excellent place to start is to look at every invoice received from the drilling rig provider or associated subsidiary.

For most rigs, the day rate is not the only cost incurred by the rig provider. There are often many additional costs, including equipment damage costs, string DBR costs, OBM fees, additional rig crew costs, additional crew member costs, miscellaneous 3$^{rd}$ party rebills, BOP costs, and required chemical costs. A lot of these costs are small, but they add up.

Drill string damage costs can be significant, especially if we are contractually required to pay replacement value, which is often brand-new cost. Therefore, if we DBR half a string of drill pipe, even though it is an older string that was used for years before coming to us, they will charge full value. Some rig companies are notorious for doing this.

Therefore, we rent a rig with an older drill string and are exposed to the risk of an older string. Then, as it is approaching the end of its life and has been fully depreciated, we are forced to purchase a new string at full price to replace it, but we must give the old string and the new string to the rig provider, so they can rent it back to us, maybe, because most likely we will get another old string. What a great business model.

This risk can be addressed with more thorough inspections and string management, as previously discussed. However, it might be the case that no matter what we do, the rig provider continues to hit us with excessive string damage costs. If that is the case, it might be time to release that rig. A string at the end of its life is a serious risk. We are overexposed to string failure while at the same time being charged joint-by-joint to buy the rig provider a brand-new string.

Another aspect of risk transfer is if the rig provider requires its team to live on the wellsite. Usually, they provide a bunkhouse, but the operator pays for water, septic, and power. This might not seem like much, but if we compare rigs in which crews live offsite to those that live onsite, the water and septic costs can easily be double or triple because the water is constantly running for washing machines and other things. As discussed previously, there is no reason to live on location, and many rig providers agree with this philosophy. If our rig provider refuses to have the crews stay offsite, from a safety, liability, risk, and cost perspective, maybe it's time to find another provider.

Fuel usage is another significant rig cost to address. Based on rig activity, calculate estimated theoretical fuel usage daily. Compare this number to actual fuel usage to see if there is an inefficiency with fuel consumption that can easily be fixed with minor adjustments or if something else is going on. In the bigger picture, diesel usage should be replaced with natural gas or grid power as much as possible.

Moving on to drilling fluids, some rigs have mud handling systems that are more costly to operate. Therefore, no matter what we do, on those rigs, our mud bill is higher because we must run more additives to get the desired mud properties; mud loss costs are higher because we lose mud in the rig system or on cleanup, and during open hole displacement.

If this is the case, the rig may not be optimally configured to handle mud in a low-cost low-risk manner. If a rig suffers from a poorly designed, engineered, and built fluids handling or pit system, it is too much baggage to deal with, and the additional cost is significant, easily more than $50,000/well.

Regarding rig control systems and workspaces, some rigs are more human-friendly than others. If the rig has something about it that makes it less favorable for people to work in the driller's cabin or on the floor, that issue will surface in a number of ways. First, we may see it in different drilling metrics. Second, it may show itself during geosteering when making target changes or sliding. If something about the rig makes it harder for the directional drillers to do their job, it can be very costly for both well costs and production performance.

Finally, some rigs are more dangerous than others. If a rig is poorly engineered, cumbersome, unpredictable, or has a bunch of "gremlins," as the field would say, it is probably not a pro-human rig. This is a problem that must be known by the operator and fixed.

Once we have analyzed all invoices associated with the rig, as part of our 360-rig review, we need to meet individually with each member of the team that interacts with the rig. It is best for someone from the office to conduct these meetings, and it will probably take more than one visit for people to get comfortable and share critical feedback.

At a bare minimum, it would be good to meet with the rig managers, drillers, directional drillers, MWD hands, mud engineers, rig crew, casing crew hands, and wellsite leaders. These meetings are an opportunity to get feedback on different aspects of the rig from different perspectives.

We are looking for bottlenecks, rig issues, risks, hidden costs, and ideas to improve based on things people see and struggling with. Sometimes, the meetings add no value; other times, massive opportunities are uncovered. It depends on the people and interest level. If we conduct a 360-rig review, and nothing beneficial comes from it, do we really want to keep that rig and team working for us?

---
## ACTION
## Delete Diesel
---

When diesel prices are high and utilized to power drilling rigs, completions operations, and as the base fluid for mud systems, diesel invoices become painful and relentless.

With natural gas and electricity available on the pad or in close proximity, it does not make economic sense to continue to utilize diesel as the primary energy source for a stationary industrial operation, particularly for full-scale multi-well pad development. Therefore, we must identify every opportunity to reduce or eliminate the use of diesel.

## Electric Grid Power

Highline electric power has served as an option to power drilling rigs for decades. Industry interest waned throughout the years due to difficulty implementing it at scale, combined with competitive diesel prices. Therefore, it never caught on, other than in specific areas or situations (e.g., noise limits, emissions limits, or other situations preventing diesel usage).

However, with the growth of independent microgrids, mobile power options, quick setup natural gas-powered generators, overhead utility power lines, electric infrastructure already in place across many basins, expansion in the number of wells drilled per pad, and an elevated interest in improving ESG metrics, many operators and rig providers have converted diesel-powered drilling rigs to operate on electric power. Most AC rigs can be converted to grid power without issue.

If electric power is interrupted, most setups have diesel-electric generators as backup. With this configuration, we get power source redundancy, further reducing risk and downtime potential. More importantly, some operators are projecting a +50% reduction in energy costs by switching to fully electric drilling rigs.

## Dual Fuel and Natural Gas Engines

If there are concerns with electricity logistics, working with electric utility companies, or availability of electric power in relation to the drill schedule, utilizing natural gas to replace as much diesel as possible is a good plan of action. Dual fuel or natural gas engines have the potential to use field gas, liquefied natural gas, or compressed natural gas.

Dual fuel engines operate using a mix of diesel and natural gas and can run on diesel if natural gas is unavailable. Therefore, rigs with dual fuel capability can seamlessly integrate into the drill schedule if a pad does not have access to field gas or if there is a problem with gas composition. Some dual fuel engines have the potential to run on landfill gas, biodiesel, or other fuels, providing additional options to reduce costs.

If field gas compatibility is a concern, there are service providers that specialize in providing field gas conditioning trailers to address this issue. If source gas proximity is a concern, temporary gas lines can be set up within a day, so there is no reason not to have access to gas in most cases.

Additionally, if field gas is not available, compressed natural gas can be delivered. Virtual CNG pipelines are expanding the reach of natural gas, connecting sources and uses to provide natural gas to multiple industries, including drilling rigs via truck transport.

Furthermore, the ability to take field gas and condition it for use in generators that charge batteries as a power source is another technology gaining interest to eliminate diesel.

With so many options and new natural gas technology introduced regularly, there is no reason natural gas cannot replace diesel as the power source for drilling rigs.

Calculating the cost-benefit of natural gas compared to diesel will vary by area and commodity price, but research suggests up to a 50% reduction in costs. Plus, if we can reduce or eliminate diesel to power the rig, our invoice tracking and processing will become far easier since all diesel use would be for oil-based mud.

## Produced-Brine-Based Mud

With our goal to eliminate diesel usage, the final area to address is drilling fluid. Oil-based mud (OBM) is probably one of the most popular mud systems in the United States. Since diesel is used as the base fluid, diesel is added to the system to address losses and maintain mud properties at significant cost. For some wells, large amounts of diesel must be added to the system, costing +$300,000 per well in mud diesel.

When we use massive amounts of diesel in the mud system and to power the rig, it is difficult to maintain strong controls over usage to run a tight ship, in terms of cost control. It is easy to make a mistake with the amount of diesel being trucked, loaded, unloaded, coded to OBM, or to the rig.

This further complicates tracking and accuracy. With all this diesel moving around, it is easy for losses to occur due to errors, trickle rates, or theft. For example, if a company man is low on fuel and needs to get to town, he may ask the derrick man to fuel him up, and for that favor, he will allow the derrick man to fuel himself up too—how nice.

Once that happens, not unlike "borrowing" office supplies, it becomes acceptable, with every hand fueling up trucks and in-bed tanks. Then one of the hands, who is a local, has family and friends stop by with large portable farm skids, because a poor farmer or landowner needs some diesel in a winter storm or some other emergency, and we can't forget about them when everyone else is enjoying. Now we are losing massive amounts of diesel. At the same time, everyone in the office is scratching their heads at our diesel usage. Our office engineers calculate a variety of metrics to figure out why we are running more diesel, oblivious to the truth.

If diesel is replaced with brine or produced water, we eliminate these types of problems. In some areas, OBM can easily be eliminated, but in others, it has been more difficult due to swelling clays, instability, lubricity, and cleaning ability.

Few operators want to experiment with brine systems. Therefore, we have less data and experience to rely upon, and nobody wants to be the guinea pig. However, if the savings are significant, it might be worth it as there are additional savings with saltwater-based mud, including increased ROP and reduced contamination risk.

---
## —— ACTION ——
## BHA Special Interest Group
---

The bottomhole assembly (BHA), including bit, steering unit (RSS or mud motor), MWD package, stabilizers, reamer, agitator, collars, heavy-weights, jars, and any other jewelry besides drill pipe, are focus items for the BHA Special Interest Group (SIG). The potential savings and complexity from optimizing the BHA justifies forming a SIG.

With a focused effort, we can deliver game-changing cost reduction solutions. A slight adjustment to the BHA has the potential to provide significant value. Below is a list of people to consider including in the group.

## BHA Special Interest Group Members

- **Drilling Engineer:** Lead with cost accountability
- **Wellsite Consultants:** Lead initiative on location
- **Geologist:** Provide geologic insights
- **Directional Drillers:** Optimize configuration, operations
- **Drill Bit Engineers:** Bit design, selection, execution
- **Diamond Engineers:** Engineer PDC cutter design
- **MWD Experts:** Optimize readings, accuracy, QA/QC
- **Digital Tech:** Build models to support the effort
- **AI/ML Engineer:** Process data, find insights, build AI
- **PDM, RSS Techs:** Performance, selection, optimization
- **BHA Inspector:** Improve BHA reliability & performance
- **Tool Reps:** Help provide solutions and improvements

For many cost reduction campaigns, optimizing the BHA has the potential to provide significant value in terms of total drilling cost reduction since it can reduce the number of drilling days, reduce the number of trips, reduce BHA rents, minimize DBR costs, reduce risk, improve wellbore trajectory, and enhance production performance.

The BHA is the tip of the spear. Therefore, the engineer leading this SIG will have a considerable amount of work and responsibility. The team lead must be interested in improvement and held accountable if the SIG does not deliver results.

Incremental changes can add considerable value and lower risk compared to doing something drastic. We do not have to pull a rabbit out of a hat, so to speak, even though some may expect it. That often requires pushing operations into the upper right-hand corner of the risk matrix.

## Drill Bits

Bit technology advances on a weekly basis, and small changes in bit design can provide a broad spectrum of cost reduction opportunities based on geology and drilling challenges encountered. Bit technology determines not only ROP but also directional control, BHA jewelry necessities, required trips, drill string damage levels, and casing running performance. An unoptimized bit can make it very difficult to smoothly run casing to bottom, resulting in further problems, if a low depth of cut or high speeds creates bumps, ridges, or undulations.

When performing research on bit selection, bit records

reveal very little in terms of detailed designs since they typically only show bit size, manufacturer, bit type, nozzles, and run details. They don't show, nor do vendor marketing spec sheets show, many of the details that matter, including diamond cutter type, table thickness, substrate thickness, cylinder specs, density, shape, back-rake attack angles, chamfers, thermal resistance, abrasion resistance, impact resistance, leaching process, or diamond manufacturer. The bit serial number does not have this information either.

Developing a strong relationship with the bit manufacturers and the diamond cutter manufacturers is critical for cost reduction. These are often two different entities. The bit manufacturers may design and produce the Polycrystalline Diamond Compact (PDC) bits, but they do not necessarily engineer or manufacture the diamond cutters, one of the most essential components of the bit.

Diamond cutter details are extremely important in BHA analysis because the diamond cutter tables (tiny components on the tips of the PDC cutters) are the only part of the BHA that is in direct contact with raw undrilled Earth, drilling (primarily by shearing) the rock, creating the borehole.

Drill bit and BHA selection and optimization are based on many variables of which we have detailed records. However, diamond cutter details are often not included in the analysis because the data is not readily available or the details are unknown. I have had several cases where a preferred drill bit stopped performing as expected, and it is believed this was due to undisclosed adjustments with the cutters, even though it was blamed on subsurface geology at the time.

Just as we are working to increase efficiency and reduce costs, so are the manufacturers of all the products we are using. If we make changes that we know about, and our vendors are making changes that we do not know about, it is possible and potentially highly likely that we will drive ourselves crazy trying to figure out what is going on.

Since the diamond tables are critical in the drilling process, we need to be hypervigilant regarding every aspect of the cutters. Therefore, we must engage with the manufacturers of every component of the bit, especially the diamond cutters.

The diamond table is the technology we depend on to maximize cost reduction at the Bit-On-Bottom-Interface (BOBI), removing each grain of rock, the obstacle causing our primary cost constraint during drilling.

Most bits, cutters, and tables are designed to have broad appeal and utility across the globe. The manufacturers must do that to make their business economic when manufacturing their off-the-shelf inventory. The bits are not designed for a specific area or target formation, with geologic properties unique to an individual operator's landing preferences. It is common for two offset operators to permit the same shale reservoir but intentionally target and drill vastly different beds within that target.

Therefore, if we are looking for game-changing results at the rock interface, we must consider developing custom bits and custom cutters.

With a drilling program in place, most manufacturers will work with us to customize the bit and cutters for minimal or no additional cost. Pursuing a custom approach for this aspect of the operation elevates our team to be more engaged with bit technology, cutter technology, quality, performance, and the manufacturing process. Additionally, we become more aware of the reconditioning and repair process, increasing the level of sophistication we are operating at.

Furthermore, when we have put the work in and developed our own customized bit and cutter technology designed specifically for what we are targeting, and our performance has surpassed competitors, it tells a good story from a shareholder investment value creation perspective.

I would rather invest with someone who understands the intricate dynamics of diamond table technology for the interval they are targeting over someone who uses off-the-shelf products and is jumping from vendor to vendor and bit to bit, depending on which salesman has good jibber-jabber or propaganda materials.

## Steering Unit

From an economic perspective, running rotary steerable systems (RSS) versus conventional mud motor configurations depend on a variety of factors. In certain cases, there is value to running RSS in the vertical and mud motors in the lateral or vice versa. In some areas, to be competitive in terms of performance and cost, RSS is a must.

In other areas, RSS is uneconomic or unnecessary. If RSS is not needed, from a risk perspective, it does not add value to have an additional $1.5MM to $2.5MM at risk in the subsurface. That is significant additional money in the hole and associated risk. Of course, as previously discussed, LIH insurance is something to consider, but if RSS is not providing superior performance, why use it?

In general, if we are drilling fast rock with geology that is more homogeneous with minimal faulting, RSS may be the best option. If we are drilling slow abrasive rock that is more heterogeneous with faulting, requiring many bit trips, conventional mud motors may be more economic. However, if there are weight transfer issues, control issues when sliding (steerability), or some other issue making sliding difficult, RSS may be the best option.

Utilizing RSS instead of conventional tools has provided significant savings to many operators, delivering increased ROP, reduced trips, one-run achievements, record spud to TD times, complex step-outs, and reduced BHA failures.

However, if we replace mud motors with RSS to solve problems, but RSS introduces new problems or limits performance (e.g., reduced RPM, etc.), then the RSS value proposition may not be there, especially with the increased economic risk and complexity.

The most economical approach may be to use both RSS and conventional technologies. Run mud motors until or if there is a point further out in the lateral with challenges requiring a switch to RSS, including azimuthal walk, or sliding performance.

Some operators that were previously diehard RSS fans for 100% of the well have switched back to mud motors because of quality issues and cost. At the end of the day, there are pros and cons to each technology that vary by application.

## Additional Components

Stabilizers, reamers, agitators, and other BHA components can provide significant cost reduction value if they are added or removed. Understanding how these components interact with the bit, steering unit, and MWD is critical to maximize performance.

For example, a group of managers and engineers at a public oil and gas company wanted to build their own company, so they raised a bunch of private equity to purchase a large producing asset with upside potential. Since the principals were not experts in detailed operational dynamics, they hired an engineering firm to assist in drilling the first well.

The engineering firm recommended their favorite directional company to handle all aspects of the directional drilling operation. While drilling the curve, they had trouble and accidentally created an unfavorable geometry, resulting in the inability to drill more than 2,000 ft of a planned 10,000 ft lateral. Since they were already over budget and could not make any more footage, they called TD. The well produced but did not meet expectations due to the short lateral.

After a post-well analysis, it was discovered that the directional company did not have much experience drilling this formation in this specific area of the basin and did not install the optimal stabilizer package for the curve assembly. This resulted in difficulties when building the curve, and the directional driller most likely threw them a double dogleg that was not visible on the directional survey.

Since they felt that the problem had been identified, their

primary investor agreed to let them drill ten additional wells. Unfortunately, on the second well, while in a slide, they got hung up and stuck. There were a bunch of complications, and they ended up losing half the lateral. The investigation determined that the stabilizer they had on the lateral BHA was not optimal for the conditions they were experiencing.

Based on what was happening in the lateral, they should have run a slick BHA with no stabilization, or as soon as it appeared that the stabilizer was jamming into the side of the wellbore when they were in a slide, they should have POOH and removed the stabilizer to reduce the max outer diameter of the BHA and the related friction.

There is an endless configuration of BHA components and combinations. Most areas in the U.S. have a standardized semi-optimized approach based on hundreds of runs. Taking a BHA that works in one area and applying it to another area, regardless of what is happening in real-time, is an expensive exercise. Leveraging data and experience that many others have already paid for is prudent and financially intelligent.

Regarding the team that raised capital to build their own company, their primary investor was so upset after two failed wells, both due to BHA stabilizer issues, that the investor fired the entire management team and took direct control of +500 producing wells. When dealing directly with investors on the smaller end of the spectrum, unlike a large investor or public company operation, there is little room for error, especially on the initial wells. Studying the successes and failures of others and leveraging experience in a particular area is much less expensive than a trial-and-error approach.

---
# ACTION
## Yard Prep
---

Operations, processes, assemblage, or any actions that can be prepared in a controlled environment before arriving on location will reduce cost and risk. The most expensive place to prepare equipment or troubleshoot issues is on location. Anytime I see vendors working on equipment during drilling, completion, or production operations in the field on the wellsite location, I ask myself:

**"Could this have been done before coming to location?"**

During drilling, the highest risk and highest cost area to perform work is on the rig floor. If we have no option other than to prepare items on the rig floor, then that is what we must do, but it's expensive and has elevated risk.

Working on the rig floor always presents the possibility to drop something into the well, to have a problem with a tool, to realize something is not working, or to damage a critical piece of equipment. All the while, the entire operation is stopped and waiting for the issue to be resolved.

For example, if we cross-thread float equipment, realize we have the wrong float threads (e.g., BTC instead of LTC), or have trouble working with tools on the floor, there will be NPT while we wait. This could easily burn several hours or days.

However, if we make-up the floats at the pipe yard or with

a mobile bucking unit in advance, and there is a problem, there is plenty of time to resolve it. Another cost reduction option is to make-up the drilling BHAs on the ground in advance instead of picking up the components piecemeal and assembling them on the rig floor while the drilling operation is stopped.

Analyze every action we take on location to determine if there is value in performing that action in the yard before coming to location. For example, I have had a good experience with cleaning, visually inspecting threads, and full-length drifting (CVD) casing at the pipe yard rather than doing it on location. In the pipe yard, it can be performed in a controlled environment, there is no mess made on location, there is less stress, and we have fewer people on location.

When I do CVD in the yard, there is the added bonus of having the thread rep there to inspect the casing for issues. We have caught several casing issues doing this and have had time to speak with the manufacturer to fix the threads or swap out the joints in question.

If we find a problem during the CVD on location, there is so much pressure to rationalize that it will be okay because we are under time constraints to run what we have. Plus, no one is really looking too hard at the casing on location to find problems, especially if there is bad weather or other distractions. Generally, everyone is just happy the casing is there, and the team is ready to get it in the well and move on.

Furthermore, when we do the CVD in the pipe yard, the CVD crew can be more thorough. They can work under shade or out of the weather, take a break for lunch at a restaurant, access a nice bathroom, discuss issues they find with managers at the pipe yard—as a result, we receive a higher quality inspection. When performing CVD on remote locations, it's a much tougher environment to identify intricate issues in.

For example, an offset operator was running a 30,000 FT premium string of casing with CVD on location. As they picked up joints onto the rig floor to run, a large chunk of metal slid out the bottom of the casing. The metal looked like it was in the casing from the factory. This immediately stopped the entire operation as there were many questions.

How did they not catch this during the drift process? Why is this metal in the casing? After inspecting the casing that remained on the racks, they found several additional joints with sharp chunks of metal. I am not sure how it was resolved, but they were down for several days trying to determine the best course of action. Anytime there is the potential to prepare something in the yard, it is worth considering.

Additionally, vendors sometimes take advantage of the operator by intentionally preparing equipment, materials, or services on location because they can charge for that time, and they transfer the work-related risk to the operator.

When work occurs on location, there is elevated risk, and it's not the ideal environment, especially if something unexpected happens or is needed to get the job done correctly and it's not available on location. When this occurs, and it's common, people will make do, which is not ideal.

Therefore, encourage engineers to tell sales reps that performing tasks that can be done at their yard, on our location, is frowned upon and unacceptable. Encourage wellsite leaders to look for this behavior and report it.

# ACTION

## Zero Deadheads

Throughout drilling, completion, and facility construction, items are hauled to and from location by multiple vendors. Each entity uses their preferred third-party hotshot company. Therefore, in most cases, a transport comes to location to pick something up with an empty trailer. When dropping off an item on location, backhauls are often empty. In the trucking world, deadhead is a term for driving with an empty load. Anytime a trucker is empty, they are deadheading.[12] To reduce costs, our goal is to eliminate deadheading.

As the operator, the first thing we need to do is to get all our vender-preferred third-party haulers on our MSA as approved vendors. The next step is to start a group chat with all vendors hauling items to and from location. Whenever someone is going to haul items, they notify the group, and we can utilize the situation to add items that we need to haul so that there are minimal deadhead situations.

Depending on what is hauled and the size of the trailer, we can determine the optimal number of items a transport should have. An efficient hotshot might be when we load and deliver three items and backhaul three items. Inefficiency would be when we load and deliver one item and backhaul zero items. The overall goal is to reduce trucking costs. As the operator, when calling for a hotshot, unless it is urgent, it would make economic sense to wait until we have at least one item to haul or backhaul.

## ——————— ACTION ———————

# Total Depth Process

Engineering the process of TD'ing a well can be one of the widest-ranging cost reduction opportunities available because it affects the economics across five operational phases: drilling, pre-frac toe prep, plug-and-perf fracturing, post-frac plug millout, and production. The decisions and actions when drilling the final 500 to 1,000 FT before calling total depth (TD), running casing, and pumping cement are what I consider the "Total Depth Process."

During the final 1,000 FT to TD, geology, drilling, and completion personnel must be fully engaged, as decisions at this point carry a significant cost impact on the entire operation. As ops teams push to finish drilling without tripping for BHA wear, drilling dysfunctions occur. It is common during the last 500 FT to have directional control issues, difficult ROP, and BHA problems, combined with an urgent desire to finish the well and reach TD as fast as possible.

As a result, in the morning meeting, someone will say, "Let's rotate it out!" Everyone giggles, relieved that we don't need to worry about geosteering this well anymore. Then, we go into rotation to TD as fast as possible. The problem is, very often, we drill out of the best rock, sometimes dramatically.

Drilling out of the best rock at the toe is the worst possible place to drill out of zone from a cost and performance perspective. It makes it harder to get the casing down to bottom and get a good cement job if the rock is unstable or mobile outside the target interval. If the wellbore starts to pack off at the toe, impacting cement job quality, it jeopardizes production performance of the entire well. We are money ahead TD'ing the well early compared to rotating out of zone at the toe.

On the final 500 FT, we must be in the very best rock because the process of initiating a frac will be ideal, and total well production has a higher probability of meeting expectations. We need both to reduce costs. When we drill out of zone at the toe, it can be difficult or impossible to initiate a frac below max pressure. If we cannot inject at a reasonable rate out the toe sleeves, risk and cost go up, sometimes dramatically, depending on what happens next.

When we cannot inject at 10 to 20 bpm below max pressure during toe breakdown, we end up riding the lightning (aka pumping at max pressure) or sometimes slightly above, and people may get rough trying to get the formation to break.

This puts stress on the casing and wellhead, which has the potential for damage. If we try to get acid down and it locks up, we have other problems, or we get acid down there, and because it is out of zone, it locks up when acid hits the rock.

Now we must get coil or a tractor to perforate and try to find a depth we can frac. Right off the bat, a simple operation has become very complex, with room for many additional problems. Drilling out of the best rock at TD can start a chain reaction of events, sending our costs into the stratosphere. Few companies can run a low-cost operation at scale with systemic problems initiating fracs out of the sleeves.

Let's say the toe is not in the best rock, but it is frackable. Once we initiate a fracture, we have created a weaker level in the formation. This can cause additional stages to frac at that same depth. Therefore, it's ideal to TD in the very best rock to weaken the target with the first five frac stages in the best rock to carry that preferred interval along the wellbore. A few feet can make a big difference in some horizontal plays when it comes to production.

Additionally, if we screenout early because we are out of zone, often the well does not have enough energy to flowback on its own, so we need coil. Lack of energy at the toe also impacts us when it comes time to millout frac plugs, arguably one of the highest risk standard completion activities.

Within the process of milling plugs, when we are at TD and preparing to pull out with coil or stick pipe, that action carries the highest risk of getting stuck. If there is no energy in front of us to carry debris behind us, it compounds the problem.

Finally, on the production side, when we have the well TD'ed in the best rock, that energy helps sweep all the liquids and sand held up along the lateral out of the well. The very best producers that flow naturally for more extended periods and generally have better production profiles are sometimes correlated with rock quality and gas shows at the toe.

As mentioned previously, sometimes it is better to TD early than to try to take a well all the way to planned TD. There are many factors, but risk, reward, and probability must be at the top of the list.

Quantifying these three variables is challenging but not impossible. For example, a friend of mine, who is an excellent drilling engineer, recently mentioned a situation in which they were 300 FT from TD on a +30,000 MD well with TVD

+15,000 FT and they were debating whether to trip the bit to get the last 300 FT.

When I heard that, I thought they were crazy to even consider tripping. For me, from a risk management perspective, call TD. I might ask one question to estimate probability, "What problems have we had on this well?" If any problems were mentioned, it's a no brainer to call TD 300 FT early, especially if we are in the best rock at current depths. What did they decide to do? They ran point-forward economics and, of course, determined the value exceeds the cost and decided to trip the bit.

While drilling the final 300 FT, they got stuck and lost the entire well. As a midsize operator, that would be devastating. Good luck explaining the decision-making process to your investors. Only big dogs can make decisions like that and continue to operate because, eventually, you run out of money.

It does not matter what the point forward economics indicates. To put it simply, they are risking 14,700 FT for 300 FT on a difficult well. The risk-reward is not good, and the probability is not good. You must understand all three variables to make a good decision.

Point forward economics looks at the estimated value of the final 300 FT compared to the cost to trip the bit and drill. It's good to know that, but it focuses solely on reward value, ignoring risk and probability. Often, what happens in these situations is the reservoir manager asks the drilling manager in front of everyone at the morning meeting if we can get the final 300 FT. What drilling person is going to say "no" in front of everyone? Nobody. Never ask a barber if you need a haircut.[13]

The answer will always be "yes."

People do what they want to do and set up questions like that because they know what the answer will be. Of course, it's not their money at risk. Most likely, these managers are interviewing for executive-level jobs at another company and not fully engaged anyway, so who really cares? Well, someone must take ownership, and I think it should be the person reading this book.

Once we decide to TD, we must have a detailed process to safely trip out, run casing, pump cement, recover our mud, and walk. When writing the procedures for these steps, try to be as detailed as possible because each drilling consultant has a slightly different preference on how they want to do things. As independent contractors, they have the latitude to conduct the work as they see fit unless we explain how we prefer to do it.

From a cost perspective, consider including how we want to calculate hole washout to determine cement volumes and drilling fluid recovery spacer volumes. Significant money can be lost if these calculations are incorrect. It pays to perform statistical analysis on this metric from previous wells in the area to have a frame of reference for percentage excess. Also, construct a checklist for the casing running process, including the thread rep, torque-turns, and casing crew on the more technical aspects of the operation. Additionally, consider crafting a checklist for the wellsite leader to QA/QC the cementing process. There are many things that can go wrong during this phase of the operation, which dramatically impact the cost of the remaining phases. We must do everything possible to put the completion team in a position to succeed.

# Never Stop Looking for Drilling Cost Reduction

Mindset is more powerful than any individual cost reduction action. However, it cannot be the mindset of just one person or even a few people. The team and each member must believe that costs can be reduced and are interested enough to take massive action, supported and driven by executive leadership, particularly the CEO.

Preventing problems and managing risk provides the foundation for our cost reduction efforts. With everything we have reviewed together when we hear about a problem, we should think about the corresponding costs. Problems equal costs, and risks eventually equal problems, which equal costs.

Therefore, if we reduce risk or prevent problems, we reduce costs, even though we may not reduce the AFE or forecast public cost numbers. The actual costs will be lower and more predictable. Attributes that can be more impactful to increasing our company's realized long-term free cash flow.

Stable, predictable costs provide the foundation for our cost reduction initiative. If we have stable, predictable costs, it may be tempting to not change anything to avoid introducing risk and problems.

However, if we do nothing, costs go up, just from inflation alone, not to mention all our vendors working to increase their cash flow. Therefore, we have no choice but to act.

We have no choice but to never stop looking for cost reduction. In the spirit of achieving 100X returns and maximizing cash flow with cost reduction, here are "100 More Drilling Cost Reduction Actions" in no particular order.

ONE GOOD IDEA

EXECUTED AT SCALE

CAN BE A

GAME-CHANGER

# 100 More Drilling Cost Reduction Actions

1. Perform asset trades to unlock longer lateral development and economies of scale. Establish strongholds.

2. Map assets with cost metrics (e.g., CAPEX/EUR).

3. Sell or swap higher drilling risk and cost areas for more economic areas relative to our company and objectives.

4. Identify areas for mega-scale development projects.

5. Use artificial intelligence and machine learning (AI/ML) to assist with development strategy optimization.

6. Institute a drilling pause. If costs are overwhelming or wells are not yielding predictable production performance, drop all drilling rigs as rig contracts expire. Give the team time to reassess and reorganize. With a fresh plan of action, restart drilling operations.

7. As more wells are drilled per rig per year, proactively cut lower-performing rigs to honor budgets, respect development plans, and prevent drilling areas that have not been fully vetted from a subsurface perspective.

8. Standardize drilling data capture, processing, and usage.

9. Identify and stop end-of-life dumping practices. When equipment is at the end of its useful life or close to being considered DBR by 3rd party inspectors, it is not uncommon for vendors to identify an operator to dump it on—charging full replacement value to the operator.

10. Build a 24-hr Operations Command Center.

11. Twin rigs for economies of scale and competitive spirit. [14]

12. Identify and reduce invisible lost time.

13. Determine optimal placement of reamer behind bit to pull slick, without tight spots when drilling vertical to KOP, curve, lateral, and run casing smoothly to TD.

14. Utilize a mobile bucking unit. [15]

15. Managed pressure drill (MPD) to increase ROP, lower mud weights, reduce losses, enhance hole cleaning, prevent pack-offs, prevent stuck events, and other NPT. Assist in our efforts to eliminate casing strings. [16]

16. While drilling, call out mobile MPD units for specific hole segments or challenges, then release. Low cost mobile MPD can rig up / rig down in less than 1-hour. [17]

17. Reduce ellipse of uncertainty with upgraded process.

18. Administer diesel fuel enhancers for drilling rig gensets. [18]

19. Pull and reuse cellar tinhorns.

20. Build custom digital applications.

21. Unbundle drilling mud system products, services, and testing. Perform independent testing to reduce bias and optimize mud system for performance and cost.

22. Request vendor discounts for multiple drill bit runs.

23. Offline cementing. [19]

24. Eliminate casing pup joints with short regular joints.

25. Automate casing runs using top drive integrated push-button make-up and cross-thread detection systems. [20]

26. Ban all social media on location (e.g., TikTok, Instagram).

27. Establish drilling incentives with midstream entities to cover part of the drilling and completion CAPEX in exchange for the midstream dedication.

28. Eliminate jars from BHA based on use and value-added.

29. Roundtrip truck transports handling OBM.

30. Add an anti-stick-slip tool to the BHA for increased ROP, longer runs, and reduced damage from less vibration, friction, and weight on bit.[21]

31. Reduce mud motor bend angles.

32. Coordinate or substitute supervak use with vacuum trucks.

33. Initiate a geochemical program to improve landing and targeting of the optimal rock. Help avoid costs associated with drilling out of zone.

34. Automate sliding.[22]

35. Install dual agitators to increase ROP.[23]

36. Identify operations to reduce or remove human exposure.

37. Eliminate trash hauling with a self-contained compactor.

38. Install a centrally located natural gas-fueled microgrid to power multiple drilling rigs and equipment.[24]

39. Eliminate all house cleaning services. Require personnel to clean their own space.

40. Improve arrangement of tools and equipment on rig floor.

41. Transition rental equipment from a cost-per-day structure to a fixed price per well or per pad.

42. Develop a scorecard to benchmark vendors across all services—rank and release underperformers.

43. Batch drill sections on multi-well pads.

44. Drill vertical, curve, and lateral with single BHA and no trips using downhole adjustable motor bend setting tech. [25]

45. Use downhole adjustable BHAs to lower WOB needed, reduce bit wear, and increase ROP. [26]

46. Use shock & vibration data to improve BHA selection. [27]

47. Craft a recipe (checklist) for systematically getting BHAs unstuck. A checklist for getting free will help prevent a team member from making a mistake or panicking during a high stress stuck pipe event.

48. In certain areas or sections, prohibit the use of RSS.

49. Consider foam cement for areas with circulation issues or situations in which formation loads can deform the casing.

50. Eliminate pad trenching with linebackers and pipes.

51. Utilize active drilling rig engine management systems. [28]

52. Install sensors across drilling rig and BHA to gather additional data to help improve drilling performance. [29]

53. Leverage proactive sweeps to increase subsurface knowledge and prevent or diagnose hidden problems.

54. Establish a weekly deep dive vendor invoice account reconciliation program that focuses on a different vendor each week. Discover erroneous charges, invoicing issues, irregularities, fraud, pricing structural concerns, and opportunities to adjust pricing.

55. Replace OBM with lower-cost next-gen WBM systems, including produced water-based drilling mud technology.[30]

56. Structure a performance-based drilling contract with rig providers to engage contractors and improve economics.

57. Eliminate drill collars and HWPD as much as possible.

58. Simplify designs to reduce operational execution complexity, associated problems, and costs.

59. Use AI to predict and provide alerts on the probability of any undesirable event.

60. Run hybrid drill bits in the curve, lateral, and to get back in zone when PDC cannot perform efficiently.

61. Combined hybrid bits with RSS technology to address slow ROP rock and issues when sliding.[31]

62. Avoid needing to rotate casing to bottom. Instead, use vibration technology to get casing to TD.[32]

63. Vibrate casing during cement job to address poor OBM recovery, poor cement isolation, and issues lifting cement to desired depths.[33]

64. Use a casing floatation sub to reduce drag force when running casing.[34]

65. Have vendors provide fully assembled steering units and MWD delivered to location ready to run downhole.

66. Use surface circulated cement to strengthen weak areas on our location pad.

67. Establish procedures to confirm backup BHAs are identical to primary BHA.

68. Modify spudder rig setup to extend spudder drilling operation to drill deeper, potentially to kick-off point.

69. Consider designs and technologies that slicken up the BHA to reduce hooks and probability of getting stuck.

70. Perfect and automate go-to-bottom and come-off-bottom process during drilling.[35]

71. Prevent rock instability with WBM nano-tech additives.[36]

72. Run in-bit sensors to map drilling dysfunctions at the drill bit for programing into the Virtual Company Man.[37]

73. Install vibration mitigation tools to address high-frequency torsional oscillation (HFTO), improve ROP, reduce bit wear, protect BHA, and reduce string damage.[38]

74. During casing run, establish an autofill process.

75. In-string reamer or post-TD mill run to remove tight spots before running casing to streamline running process, reduce tight spots, reduce side loads, and minimize casing ovality risk and casing failures during completion.[39]

76. Analyze past high-cost operational events. Then develop automated alarms to proactively detect and alert when a similar situation may unfold to help prevent a recurrence.

77. Predict ROP prior to and during drilling. Compare predicted ROP to actual ROP to identify suboptimal parameters, dysfunctions, problems, and human error.

78. Purchase drilling mud products, additives, and other drilling items directly from the manufacturers.

79. Replace model assumptions with real data.

80. Strengthen shoulders and other components to provide wear resistance to reduce BHA damage when drilling abrasive formations.

81. Determine supply chain demands and bottlenecks; address them using non-capital-intensive strategies.

82. Identify and mitigate last-minute operational changes that are introducing risk and cost.

83. Use unmanned MWD systems to drill to KOP. [40]

84. Employ high-resolution continuous survey measurements to reduce extreme doglegs, identify and eliminate hidden doglegs, improve geosteering, and reduce costs. [41]

85. Establish a maximum allowed dogleg limitation provision for directional drillers based on actual doglegs compared to directional survey average doglegs.

86. Run expandable liners to add casing points without reducing inner diameter. Allows for multiple wellbore designs and contingencies that reduce total well costs. [42]

87. Release disengaged vendors who provide items without technical or operational field support.

88. Systemize independent checks and balances. For example, drillers tally vs. DD tally, CVD tally vs. truck tally, engineer double checks vs. company man double checks.

89. Use CBL, top of cement, and mud recovery percentage to assess opportunities to improve cement quality, structural integrity, and costs for lost and unaccounted for mud.

90. Identify unnecessary (i.e., invoiced) personnel on location.

91. Drill horseshoe / U-turn wells to achieve 2-mile laterals in 1-mile sections.[43]

92. Reduce the number of human driller inputs with tracking and automation. The goal is to reduce mistakes and variability created by changing personnel (i.e., day / night, on / off, and crew rotations).[44]

93. 30-days after pad ops have finished, systemize an invoice audit conducted by lead consultants as a double-check.

94. Employ computer vision (CV) to visually identify potential issues before they become expensive problems. Computer vision is AI technology that enables computers to process images and video information.[45]

95. Predict changes further ahead of bit with a combination of hardware and software additions and enhancements.

96. Utilize wired drill pipe in select areas to gather increased knowledge that can be correlated to standard systems and utilized without having to run wired pipe continuously.[46]

97. For certain services that lack competitive pricing, form affiliate service providers for improved economics.

98. Locate and support entrepreneurs and new start-up service providers that help reduce costs.

99. Purchase mineral rights ahead of the drill bit.

100. Identify overlooked target formations and development areas with lower rock quality or challenging geology that could offer attractive economics with our maximize free cash flow with cost reduction strategy and tactics.

# 8

## COMPLETION ACTIONS

—

Shale wells typically consume the largest amount of capital during completion operations, often accounting for 60% or more of total well costs. Therefore, completion cost reduction must be the centerpiece of our cost reduction plan. Not only is the most money spent during this period, but it is happening at the fastest pace.

Accordingly, completion design, engineering, and field implementation must be streamlined to reduce complexity, minimize risk, and maximize efficiency. This allows us to reduce costs to the lowest possible amount while delivering optimal economic performance. Our completion strategy is to maximize free cash flow while minimizing operational and financial risk at the asset and corporate levels.

From an investment perspective, when completion operations begin, land and drilling dollars have already been spent and are at risk. As fracturing progresses, more and more capital is invested and, as a result, could be lost should a poor decision occur, a mistake be made, or something go wrong.

From a financial value-at-risk perspective, if we are going to lose a well, it is better to have it happen during surface hole drilling compared to milling plugs post-frac. If operations have gone according to plan, nearly the entire DSU AFE'ed amount is at risk for multiple wells on the pad when post-frac millout, cleanout, and flowback operations commence.

> **Every operational step matters, but there are certain actions where we must be at the top of our game.**

If coil or stick pipe gets stuck, or a well control event occurs during late stage fracturing operations, a considerable amount of money will be lost, brutally impacting cash flow. This potential, combined with other more common risks during completion, must be factored into all cost reduction actions.

## Basin Dynamics and Completion Cost Reduction

Understanding basin dynamics, different surface and subsurface details and challenges within a play is critical to completion cost reduction. What works in one area may not work in another or may introduce unintended or unexpected consequences.

For example, applying an optimized low-cost completion

procedure from the Wolfcamp to the Woodford probably will not work as intended, but it could work, with a little finesse and a few innovative tweaks.

Sometimes, methods transition easily between assets. However, often, intelligent, calculated adjustments need to be made to get the desired result. This is why I continuously repeat the following on location when implementing something new:

> *"Ideas from the office are good, but execution is vital. It is our job in the field to make this work. It might take many new ideas out here on the front lines to get this to work right; and I need your help."*

This is the reality of massive cost reduction when we need to take big actions to get big reductions. When executing a new completion cost reduction idea, there is an advantage to being on the front lines in the middle of the action, to see the challenges as they are happening. Seeing and understanding the risk firsthand is a competitive advantage. If something needs to be stopped or adjusted, or if it does not look or feel right, it is harder to determine from the office.

When trying a new approach, it is important to have the strongest people on location. If everything goes exactly as planned, then it does not matter as much. However, for new tactics or technology, there is a high probability it is not going to work as planned. Therefore, someone who cares about our success must be on location to see what is happening, take action, make adjustments, or stop operations.

Many people say that the oil industry is slow to embrace

new technology. I disagree. My experience is that the oil and gas industry is one of the most open industries to embrace new ways of solving problems and new technology when it reduces risk and cost. The problem is most new technologies dump all the risk onto the operator at an incredible cost, not only for the service but also for the damage.

Some service providers are notorious for skillfully and subtly transferring risk onto a trusting client unaware of the danger. Therefore, when agreeing to try a new technology, take the time to think about it and perform enough due diligence to feel that there is a high probability of success.

The actions suggested in this section are by no means the best or only solution for every company's cost reduction goals. It would be impossible to provide the optimum completion approach or design for every horizontal play in the world. The subsurface is far too complex. However, many cost reduction strategies presented here will translate across basins, although they may need adjustments to work as intended.

As previously stated in the book, never take an "Action," tactic, recommendation, suggestion, or anything else and force it into use in your area of operations. This is dangerous, and you will have problems.

That said, the highest cost process during the completion phase is by far the fracturing "frac" treatment. Since so much cost is concentrated on this one operation, an excellent opportunity exists to achieve considerable cost reduction by attacking frac costs from every possible angle and striking aggressively. Therefore, let's start by focusing our attention on the frac contracts.

# ——— ACTION ———

## Monte Carlo the Frac Contracts

The structure of the frac contract plays a significant role in our cost reduction strategy. Frac design and operational variables, combined with the associated costs dictated by the frac contract, incentivize or disincentivize certain behaviors. Since frac contract structure has changed throughout the years and continues to change, it is beneficial to review a recent history of how these contracts evolved.

Years ago, companies discovered that increasing frac treatments increases production. Service companies pushed operators to pump more proppant because the better the post-frac production, the more the companies would frac and do so with larger job sizes. As a result, frac contracts were structured with high margins attached to the proppant and pumping of the proppant. The more proppant pumped, the more money the service companies made.

As horizontal development progressed with a desire to continue increasing treatment size and stage count, operators wanted to reduce fracturing costs. Therefore, operators went to lower-cost sands and vertically integrated sand supply or unbundled sand from the frac contract. Both actions resulted in frac companies restructuring contracts to a per-stage price, which operators favored because it was predictable. At the time, the preferred structure was a bundled service on a fixed per-stage cost with no NPT penalty, as long as the operator included sand and wireline with the bundled frac service.

With this structure, operators could comfortably estimate fracturing costs and deliver wells at scale with an invoice that equaled the AFE regardless of what actually happened on location. If a problem occurred on location with wireline or the well, the fixed stage price ensured that it would not change the actual cost, making it easier to forecast CAPEX for the operator but not for the service provider.

After a while, frac service companies felt that the fixed stage price with no NPT fees was unfavorable because the utilization rate or stages per day were not predictable due to the many operational problems clients were having.

Per-stage price did not incentivize operators to be efficient because there was little economic impact if a problem occurred. At the same time, operators were unbundling every aspect of the fracturing service, further reducing margins for the service companies. Therefore, if a problem happened, the company most exposed to losing money was the frac company providing the horsepower and associated heavy equipment because it was sitting idle and not generating revenue.

To address this risk, frac companies added and increased the downtime charge and went to a price-per-minute structure during pumping, with an allocation of minutes for swapovers. Many frac service companies added complexities to the frac contract structure to extract the maximum amount of money from the operator regardless of what happens.

As a result, some contract structures are highly unfavorable to the operator if anything occurs other than exactly what was designed and planned, resulting in actual invoices far exceeding the estimated cost to the benefit of the frac service provider.

Since the cost-per-minute structure and downtime charge result in an extreme time-based financial cost to the operator and penalty, if there are problems, the industry moved to maximize the value created per minute while minimizing the risk of downtime. This resulted in a move towards fracturing multiple wells simultaneously. This structure can reduce the risk of NPT while maximizing value per minute, depending on contract details and how the frac equipment is configured to deal with challenges on individual wells.

## Deterministic versus Stochastic Modeling

Since fracturing operations have many moving parts with associated costs, we want to take the frac contracts and all related services (bundled + unbundled) and build a quantitative financial model based on operational scenarios.

When a cost estimate or bid price is presented, the model implied is deterministic—a single value is provided.[1] This is like AFE generation. Both are deterministic models yielding a single result. However, we know it is rare in the real world to have the actual total cost match the bid price. Real-world operations are not deterministic; they are stochastic.[2]

Monte Carlo simulation is a system that estimates the outcome of a stochastic process.[3] The modern Monte Carlo method originated during nuclear weapons operations at Los Alamos, New Mexico. The term "Monte Carlo" was used as a code word taken from the Monte Carlo Casino in Monaco.[4] Monte Carlo simulations were vital in developing the hydrogen bomb[5], and now we will use them to optimize the value generated from hydraulic fracturing operations.

Every contract we operate under should have its own deterministic financial model. These models are often built in Excel to check invoices or normalize bids. Once our models are built, we add stochastic assumptions to all operational and financial variables.

Referencing actual invoices, we can create a model that runs like a real-world frac operation and generates thousands of operational scenarios with corresponding invoices. The distributions, assumptions, truncations, correlations, ranges, sampling, sensitivities, and other Monte Carlo variables should be structured accordingly. This model will help us identify cost reduction opportunities and make decisions on the best course of action to achieve results.

## Complex contracts kill cost reduction opportunities.

What is often found in frac contract Monte Carlo analysis is that the more complex the frac contract pricing structure, the harder it is for the operator to reduce costs. This is intentionally done by the vendors to ensure they make a certain amount of money from the operator no matter what improvements are made from an efficiency or cost reduction perspective.

It is not uncommon for an operator to take many actions to increase efficiencies but not realize a material cost benefit. We may get more stages per day but ultimately end up spending almost the same amount of money, only spending it faster with more uncertainty and risk.

As bad as this may seem, there are scenarios in which we can cut days off the frac but end up spending more money than if we were slower. In those scenarios, there is no benefit to taking many cost reduction actions only to pay more money to vendors. Of course, the vendors will say that we are improving time to first oil, but in reality, there is no financially substantial case to be made unless per-unit invoices are lower, not higher.

Unlike drilling, which is highly correlated to the number of days on location, fracturing operations are linked to the contract much more than to the number of frac days.

Modeling the frac contract with Monte Carlo simulations will help us identify unfavorable scenarios and contractual structures. Based on these projections and actual invoices, we can negotiate changes to the structure of the contract. If the vendor is unwilling to change, or we are locked in a long-term contract, do not worry. We still have many tools available to reduce costs.

From a contractual basis, if the vendor will not change the contract, we need to have them more clearly define certain aspects as they relate to real-world operations.

For example, if we are operating under a pump time minutes contract on the frac equipment, we need to have the vendor clearly state, in writing, when the timer starts.

Does it start when the well is open, when pumping begins, or when a specific rate is achieved? This very small detail, and others like it, will have a significant financial impact on total completion costs. We will address additional opportunities to create value on this topic during upcoming action items.

——————— ACTION ———————
## Simultaneous Multi-Well Fracturing

When operating under a cost-per-minute structure, if we improve our progress per minute, we lower our total costs. Fracturing two or more wells simultaneously (super-zippers, simul- and trimul-frac) allows us to double or triple our progress per minute. With this strategy, we cut pump costs by 10-40% depending on the frac contract, fuel usage, and equipment configuration to execute operations, including super-zipper setups versus simul-frac or trimul-frac configurations with and without remote pad operations.

> **Actions that achieve more per minute liberate FCF.**

Increased horsepower, wireline, and surface equipment are required. We also need 1.1 to 1.5 times a single frac crew to get 2 to 3 frac crews worth of work. If we want independent control of each well, we need additional equipment to ensure design placement and deal with problems during frac.

Determining the optimal configuration from a cost, design, and operations perspective will depend on several factors. If there are concerns regarding screenouts, frac gradient variability, effective diversion, or a desire to have more control and frac design placement certainty, super-zipper (independent frac control) may be more favorable to simul- and trimul frac (grouped frac control). Super-zippers may be more expensive, but not necessarily, depending on how we rig up and the frac contract structure we agree to, especially the cost of NPT.

If the contract cost structure between super-zipper and simul-frac are equivalent, super-zipper will deliver lower costs over time because it is easier to address challenges on individual wells during the frac.

In reality, depending on the subsurface dynamics and the structure of the contracts, in certain situations, super-zippers will have lower costs, and in other situations, simul-frac and trimul-frac will have lower costs. In theory, trimul-frac should provide the lowest cost per minute because we have less equipment per well and accomplish more per minute.

Therefore, trimul-frac is probably the best option from a purely cost perspective. However, we need to account for potential problems. For example, if wireline is unbundled and we have a misrun, is our plan to wait or to frac only one or two wells? There are several factors to determine the best course of action including the type of pressure wave we want to generate.

With super-zipper, we would most likely not wait because wells can be fractured independently and can be started at any time. However, with simul- or trimul-frac we might wait, depending on how many wells are available to frac and the cost impact because we cannot start fracturing one well while we are in the middle of the frac on the other wells. However, advancements in technology could address this issue.

Regarding simultaneous multi-well fracturing designs, modifications come into play since we may not be able to replace the frac rate compared to single well fracs. For example, if we treat single wells at 100 bpm but on simul-fracs we get a total rate of 160 bpm—we are only at 80 bpm per well, assuming an evenly split frac placement.

Therefore, we may desire to modify the design to address the lower rate. To do this, we could reduce the clusters per stage, which increases total stages per well, increasing costs. We must run the calculations to ensure we actually reduce costs enough to make it worth our while. In some instances, simultaneous fracturing may only increase the speed of multi-well completions while marginally reducing the cost. If this is the case, we must work on the structure of the frac contracts to obtain more value for the additional risk.

However, there are other tactics we can utilize. If we reduce the total number of perfs per stage, enhance our diverter strategy, or a combination of both techniques, we could hold our clusters and stage length constant at the lower frac rate.

Of course, this could impact production performance. As a result, not everyone is sold on the value of simultaneous multi-well fracturing, particularly for setups that do not provide individual control. Difficult to fracture formations, the ability to administer diversion, well issues, NPT costs, and added equipment costs are concerns with implementing versions of this technology at scale. Furthermore, increasing frac treatment rates (BPM) per well has demonstrated value. Simultaneous fracturing is going in the opposite direction in many cases.

To optimize our designs while leveraging the full value of simultaneous fracturing, consider utilizing two low-cost technologies to help us reduce frac costs while improving production performance: downhole imaging and disposable fiber optics.

## —————— ACTION ——————
## Downhole Imaging Intelligence

Sight is our most valued sense for processing reality and our most dominant in acquiring intelligence. Unfortunately, we cannot see what is happening in the subsurface environment throughout drilling, completion, and production operations. If we could, the value opportunity would be orders of magnitude regarding risk management, production improvement, and cost reduction.

From a timing and cost perspective, there is a window of opportunity to get eyes on the subsurface to gather intelligence for our cost reduction efforts, and that time is right after the post-frac millout. After we frac the well and millout the plugs, if we can see what is going on in the downhole environment, we can further reduce costs by utilizing information based on changes to the casing from the fracturing process.[6]

By analyzing perforation erosion, we can further optimize perforation strategies, clusters, stage lengths, sand volumes, diversion, and a number of other design variables that have the potential to significantly reduce cost and improve production.[7] Any time we can reduce costs while simultaneously improving production, it is worth considering because of the leveraged impact the dual benefit provides on free cash flow generation.

Like an annual wellness check-up when going to the doctor and doing a bunch of health screenings, we need to do the same thing on the wells, especially when operating at scale and during cost reduction efforts.

When looking to drop a bunch of cost weight, it is prudent to do a thorough evaluation of the internals. Technology, including downhole cameras, ultrasound, calipers, X-rays, and acoustics, can generate images we can utilize to reduce cost.

## Cost Reduction Actions from Downhole Imaging

- Optimize frac design parameters using imaging.

- Quantify frac treatment distribution across clusters.

- Test different designs incorporating cost reduction.

- Assess technologies to maximize reservoir rock erosion at the near wellbore (NWB) area for each cluster.

- Calculate the amount of rock eroded per perf to determine optimal techniques to prevent NWB pitch-offs and tortuosity chokes using strategies to erode the maximum amount of reservoir rock across all clusters.

- Select optimal perforation gun phasing.

- In correlation with a cement bond log, look at perforation erosion relative to good cement, bad cement, and no cement. Identify optimal strategies to incorporate cement bond into perforation plans, frac designs, and diversion strategies.

- Identify flow paths behind casing due to poor or no cement.

- Study the optimal number of perforations per stage and per cluster with an opportunity to reduce costs using different configurations.

- Confirm vendor gun sales propaganda and technical specs.

- Test diversion products for validity and cost reduction.

# Cost Reduction from Imaging continued...

- Determine optimal diverter tech, quantity, number of diverter drops per stage, how and when to drop.
- Validate limited entry designs and optimal BPM ratios.
- Test game-changer perforation strategies and value-creation opportunities (e.g., 8 clusters/stage vs. 80 clusters/stage)
- Enhance comfort level with dissolvable plug technology.
- Identify plugs that move (slippage) during frac treatments.
- Discover plugs that fail during frac treatments, resulting in erosional damage to the casing or double stimulated stages.
- Identify cementing issues contributing to casing problems.
- Identify ideal overflush strategy to reduce risk and cost associated with sand flow during wireline, millout, and flowback while maximizing production using different overflush methods.
- Determine differences between running the ball-in-place and dropping the ball for different frac plugs.
- Identify fracture biases relative to perforation strategies.
- Optimize well spacing, incorporating runaway clusters.
- Determine how geologic variability, being in or out of zone, above or below the target, and around faults impacts frac placement relative to different frac designs.
- Check for casing deformation across wellbore at different dogleg severities, around faults, pup joints, and on joints close to the surface.
- Inspect areas that exhibited potential tight spot signatures during wireline and millout operations.

## Cost Reduction from Imaging continued...

- Inspect depths that took longer to mill plugs for evidence that the problem was ovalized casing, or other casing issue, and not a difficult to mill plug.
- Inspect liner top area, if applicable, to identify potential issues or erosional damage during frac.
- Inspect casing that was drilled through to identify drilling damage and benchmark for future potential to eliminate tieback strings and frac down intermediate strings.
- Look at casing relative to torque-turns data and any questionable connections during casing run (e.g., over-torquing, back-outs)
- Examine wellhead wing valve ports and the area directly below the wellhead (wellhead to casing connection) to look for damage.
- Inspect B-section and other wellhead components for erosion or damage from frac, wireline, and millout operations as a risk management checkup, QA/QC benchmark, and equipment integrity check.
- Correlate real-time readings during fracturing and wireline with subsurface events confirmed with downhole images to program AI systems for Virtual Company Man.
  - For example, if we find evidence of a failed frac plug, runaway clusters, or wellhead damage, we can identify signatures indicative of these events occurring during our completion, and program it into our artificial intelligence systems to alert us in the future.

———————— **ACTION** ————————

# Disposable Fiber Optics

---

The more we know about what is happening in the subsurface, the more opportunities we have to create value. Since the cost of running fiber optic cable, a subsurface monitoring technology, has been dramatically reduced (equivalent to the cost of tracers), we have an opportunity to use it more regularly.

By pumping disposable fiber in place on offset wells prior to our fracs, we can directly monitor the subsurface to help make immediate real-time decisions during frac operations and longer-term big-picture field-wide development decisions to reduce cost, increase production performance, and maximize free cash flow.[8]

With a better understanding of fracture growth, we can optimize faster and cheaper than a trial-and-error approach which eats up a lot of acreage inventory and capital. The old oilfield belief that "We will know how everything works when the last well is plugged" embraces a trial-and-error approach. Indeed, it is the most expensive way to run a business.

If we do not know what is happening and are just guessing, that "last well" will come a lot sooner than everyone thinks. With good technology and good decision-making, we can create more value and keep drilling by getting to the answer as fast as possible.

For example, by using disposable fiber, we can better understand and prevent production underperformance. If the fractures are not behaving as we expect on wells that are underperforming compared to wells that are outperforming, we can identify the variables that could be contributing to this occurrence.

---

**Replacing operational and financial model assumptions with real data enables us to act with confidence during all business endeavors:**

**Strategy, Development, Operations, Acquisitions**

---

With fiber optic subsurface information, we can plan, drill, and frac to get more wells above type curve, improving recoveries. Without this information, very often we are guessing on subsurface fracture dynamics and how they relate to production performance.

The more assumptions we have in our financial and subsurface models, the greater the risk we will make low-quality decisions during all business endeavors. The more information we have on what is happening in the subsurface, the more confident we are in the direction we are moving our company. If we can acquire high-quality subsurface data at a low cost, we are better positioned to achieve our 100X aspirations.

A more complete subsurface picture enables better decisions that create long-term stakeholder value. Spending a little bit of money on thoroughly planned science projects can create significant value and allow us to make cost reduction changes with confidence.

# Cost Reduction Actions using Disposable Fiber

- Optimize well spacing in conjunction with frac design.
- Determine how frac geometry is impacted by frac design.
- Compare frac geometry to lateral placement within target interval flexing perforation design, cement bond, frac rate metrics compared to stage-level geologic parameters.
- Improve geosteering relative to frac placement.
- Connect subsurface frac dynamics to production results.
- Control frac placement in real-time.
- Manage frac hits in real-time by adjusting to deliver minimal impact frac bumps, frac pushes, and frac slaps instead of highly damaging frac kicks, frac slams, and frac smashes.
- Reduce costly screenout situations.
- Test completion designs incorporating cost reduction.
- Identify flow paths behind casing due to poor or no cement.
- Test diversion products in real-time monitoring offset fiber to address methods, volumes, and cost reduction potential.
- Determine how to best address runaway clusters in real-time by using diversion, changes in rate, and fluid volumes.
- How does rock variability, in/out zone, above/below target, around faults impact frac relative to different designs.
- Intentionally frac into natural fractures to see propagation dynamics and correlate risk and opportunity metrics.
- Improve cluster efficiency and cluster designs in real-time by making perforation design changes on subsequent stages.
- Correlate real-time readings during fracturing with subsurface events to program artificial intelligence systems to add guidance delivered from Virtual Company Man.

--------------- ACTION ---------------

## Advanced Casing Program

---

Continuous QA/QC of our casing from the factory to the wellsite should be part of every large-scale shale company's business processes. Additionally, during wireline and millout operations, any tight spots or ovality must be reported daily on the "Operational Issues List."

Tight spots are warning signs. They are trying to tell us something. The wells are always talking to us, but not everybody is listening, interested, or speaks the language.

During any serious conversation, there is a spoken and unspoken element—the text and the subtext.

One of my favorite movies illustrating this dynamic on many levels is The Godfather trilogy. In The Godfather, oranges were used, initially by accident and then intentionally, as part of the subtext—a warning sign of a pending danger. Oranges foreshadow the death of specific characters in the saga,[9] although it's not a perfect correlation, similar to many dynamics in the oilfield.

Tight spots are our oilfield version of The Godfather oranges. We must not ignore them, especially when developing assets at scale with cube or row type strategies.

Drilling dozens wells before fracturing any of them only to realize the casing is bad on all of them is like waking up with a horse's head in your bed—shocking and financially devastating. We must do everything possible to prevent that from happening. We must run an advanced casing program.

When tight spots only cause small delays of a few minutes or seconds, which is not uncommon during a wireline run or on the millout, they are brushed off and ignored. There is no record that they occurred.

The only people who know are a couple of individuals on the front lines. Therefore, requiring these small, seemingly trivial or insignificant events to be reported on the "Operational Issues List" is our method for capturing and putting a spotlight on it. Consequently, if we see patterns or increases in certain areas of the basin, we can react. These could be warning signs that must not be ignored or dismissed.

If we have a tight spot, we must know precisely where it is in our string and look back at the torque-turns including all data from our casing run. Having a third-party post-mill casing inspection helps with the diagnosis. Depending on the situation, after millout, we may run an image log to get a visual.

Regarding casing selection, logistically, it is easier to run the same casing across our entire asset for a given target. However, it might not be the best plan of action if subsurface dynamics change, lateral lengths change, or there are issues in certain areas. Leveraging our non-op data, competitor analysis programs, and reverse engineering program will help advance our casing program without having to experience expensive casing failure situations for ourselves.

Ideally, a 5.5" OD monobore design generally provides the best option for shale development, as most shale technologies evolved around this casing size. However, there are many additional casing variables to consider when optimizing casing design and selection to reduce completion risk and cost.

Depending on simul-frac preferences, cluster designs, lateral lengths, and a few other factors, we can deliver a cost reduction opportunity running 6.0" casing or another larger diameter string to reduce pipe friction pressures and increase rates 25 to 35 bpm. This could provide opportunities to employ extreme limited entry at lower treating pressures, increase number of clusters/stage, eliminate all clusters and perforate every few feet, or get more aggressive with continuous diversion with larger and more frequent diverter drops.

In certain areas or for certain depths (TVD, MD), we may beef up our casing weight per foot, casing connection, or grade. Under certain circumstances, we may need to run a split string if there is an interval with $H_2S$, $CO_2$, tectonic activity, or other subsurface risk. If our operating area is vast, and we are drilling extended laterals, it would be rare to have one casing solution for the entire development. Therefore, taking our casing program to the next level may involve optimizing our string for different areas and situations across our asset.

Furthermore, when receiving large orders of casing, set aside a certain number of joints for more in-depth testing. For example, if we have a casing failure, it can be very difficult, if not impossible, in many cases, to recover the casing from the well. Therefore, to gather more information, we may want to analyze a representative joint from that same mill run.

If there was an issue we did not catch with our independent non-destructive testing, more detailed custom metallurgical, materials, and destructive casing testing can be done. Although it is not as good as recovering the failed joint, it may provide additional insights to help determine the truth.

# Advanced Casing Program Actions

- Transition from cookie-cutter casing programs to situation-based: Depth, $H_2S$, Temp., Press., Frac, Geology, Tectonics.
- Visit and inspect casing manufacturer factories on a regular basis. Ensure there are no issues at the plant.
- Establish regular meetings with casing manufacturers' tech teams and separate meetings with casing distributors.
- Audit casing loading / unloading process during transport.
- Visit and inspect casing storage facilities at pipe yards. It is not uncommon for casing to be stored unfavorably.
- Perform independent 3$^{rd}$ party post-mill inspections.
- Use marking systems to segment joints during inspection.
- Meet thread rep at inspection prior to delivery on location.
- Consider yard-based CVD and truck run ready casing.
- Employ experienced manufacturer thread reps.
- Casing running checklist with limitations on all aspects (e.g., pipe rack stack height, doping, make-up, rotation).
- Document any connection backouts, issues, anomalies.
- During wireline and millout, document tight spots, even if they are minimal and only impede ops for a few seconds.
- Place magnets in tanks during plug millout to identify issues including the difference between a difficult to millout frac plug, ovalized casing, and significantly damaged casing.
- Document and map competitors' casing problems to identify manufacturing defects, subsurface issues, and risks in AOI.
- Set aside joints from casing batch orders for chemical, mechanical, materials, metallurgy, and destructive testing.

---
## ACTION
### Perforating
### Special Interest Group
---

Perforating (perf) design optimization is a high-value, low-cost action we can employ to reduce costs while improving production. Perforations play a critical role in connecting our pay zone to our wellbore and determining how our frac is placed. Perf design also determines total costs in a variety of ways. For example, if we favor 100 FT stage lengths, we may be able to go to 200 FT stages with an optimized perf strategy combined with diversion technology and real-time analytics. This maximizes cash flow by cutting the number of stages in half while potentially improving production.

Advanced perforation strategies can help deliver our FCF goals more quickly by favorably influencing revenue and cost. Even minor adjustments can pack a powerful punch, which is why we must form a special interest group (SIG).

## Perforating Special Interest Group Members

- **Asset Manager:** Lead with cost and production accountability.
- **Asset Geologist:** Incorporate geologic aspects into perf designs.
- **Basin Geologist:** Help select areas and wells for testing.
- **Resource Development:** Assist with well selection for testing.
- **Reservoir Engineer:** Assist in analysis of production results.
- **Gun Reps:** Provide tech recommendations to accomplish goals.
- **Acoustic Engineer:** Assist with interpreting real-time frac data.
- **Diverter Rep:** Incorporate diverter tech with perf strategies.
- **DH Image Tech:** Provide analysis on perf image variables.

## Perforating Optimization

There are many perf strategies to consider testing. Due to the minimal cost and disruption to ongoing operations, testing different perf designs can be incorporated into any operation. With the potential to significantly enhance FCF, consider embracing the philosophy to test perf ideas early and often.

On the next page is a list of strategies to consider. However, our SIG team should produce many novel ideas tailored to our AOI. Consider ranking ideas and identifying the optimal well set to test them on. Often, multiple ideas can be combined into one test or be tested on the same well.

Each multi-well pad is a good opportunity to test some aspect of perforation optimization, even if it is just a simple manufacturing test to see if breakdown pressures can be reduced or fracture treatments respond more favorably to one company's technology over another. These simple tests can help us save significant money in acid costs, pump times, and screenout prevention while improving production.

We can also use high-frequency pressure data to analyze water hammers from small rate drops to determine perf efficiency.[10] Then, adjust in real time. Once we have made changes with production performance validation, we can re-run downhole image technology to compare our current state with past performance. Then correlate our database of images with our surface pressures and use AI to project what is happening with our perfs in real-time. Then, have the Virtual Company Man offer actionable suggestions to increase perf efficiency, increase frac performance, and reduce costs.

# Perforating Strategies to Maximize Free Cash Flow

| Perf Test | Cash Flow Enhancement Potential |
| --- | --- |
| **Spacing vs. Perfs** | Test perf design in combination with well spacing to determine optimal NPV/acre. |
| **Stage Lengths** | Vary clusters per stage with different perf configurations to identify highest NPV. |
| **# Clusters plus Diversion** | Increase # of clusters per stage combined with various diversion methods. (e.g., bull heading, continuous, drops, hammers, combinations). |
| **Extreme Limited Entry (>1,500 psi)** | Reduce total number of perfs per cluster and/or EHD to max out perforation NWB related friction. |
| **Equal Entry Hole Diameter (EHD) Perf Guns** | Test different manufactures equal EHD guns to see lower breakdown and treating pressures to eliminate acid or frac more efficiently. |
| **Angled Holes** | Determine if angled perfs reduce pump times or provide other benefits.[11] |
| **Gun Phase** | Vary phase (e.g. 0°, 60°, 90°, 120°, 0°/180°, 90°/270° etc.) |
| **Geologic-Based Phase Alignment** | Change perf phase based on wellbore position within geologic interval. (i.e., low, middle, high) |
| **Single-Shot Perf Clusters** | One-hole clusters to test improved sand placement, production performance, reduced pressures, reduced pump time. |

# Perforating Strategies continued...

| Perf Test | Cash Flow Enhancement Potential |
|---|---|
| **Tapered Perf Cluster Design** | Change the number of perfs per cluster from toe to heel on each stage to help evenly distribute proppant. |
| **Parallel Design Strategies** | Vary perforation strategies based on lateral location within parallel multi-well units. |
| **Geologic Perforating** | Place perforation clusters based on geologic variables obtained during drilling (MSE etc.) |
| **Maximum Clusters** | Limit-test max guns wireline can deploy on a single run varying spacing and clusters per stage, including single-shot. |
| **Cement Bond vs. Perf Orientation** | Run CBL in lateral. Identify optimal cement bond quality orientation. Then orient perf guns in direction of optimal cement bond. |
| **Cement Systems** | Vary cement properties (additives, foam, density, bond, elasticity) with perfs. |
| **Debris Free** | Identify value from min perf debris guns. |
| **Anti-Gravity Perforating** | Design perforations to address the Earth's gravity force impact on frac placement, perf specs, and production results. |
| **Eliminate All Perf Clusters** | Perforate nearly every foot in the pay zone across each frac stage combined with extreme frac rates plus diversion strategies and limited entry perf designs. |

## Perforating Strategies continued...

| Perf Test | Cash Flow Enhancement Potential |
|---|---|
| **Propellant Enhancements** | Add propellant enhancers to guns to create fractures in each cluster before pumping frac to increase cluster efficiency, reduce pressure, and reduce pump time.[12] |
| **Flex Perf Designs with Propellant** | Vary perf designs with boosters to optimize for reduced frac cost and max production. |
| **Water Hammer** | At the beginning of each stage, get a quick ISIP to generate a water hammer effect to "shockwave" open all clusters. |
| **Burr-Free Perf Guns** | Test different manufactures burr-free guns to identify cost reduction potential. |
| **Max Single Cluster Rate** | Shoot one cluster, then pump frac at max rate to establish runaway "super cluster" signature to incorporate into AI models. |
| **Diversion Technology** | Test diversion (PODS[13], glass, Bio Balls, PLA[14] etc.) combined with continuous far-field diverters. Vary perf designs. |
| **Customize** | Using all field tests combine with lab testing, build custom perf system plus diverter package for area of interest. |
| **Real-Time Strategies** | Establish optimal frac signature based on multiple factors (pressures, offsets, etc.) Then adjust guns, perf configuration, and diversion tactics to achieve that signature. |

# —— ACTION ——
## Unbundle and Bundle

Many people think it always saves money to unbundle fracturing, but it's not true. It depends on the structure of the contracts and what is vertically integrated. Unbundling frac refers to separately sourcing equipment, fuel, chemicals, sand, transport, storage, and wireline. Bundling refers to sourcing more than one of these items through the same provider.

I have seen many cases where unbundling a frac service costs the same or more than a bundled service using a simple deterministic model, not including any downtime, and definitely on a probabilistic basis, when real-world factors are included in cost models. If a bundled service costs the same or is cheaper than unbundling, it does not make sense from a risk perspective to unbundle, especially if wireline is included.

Anytime we enter the subsurface, the risk is elevated. Since wireline is the only operation (hopefully) with equipment entering the subsurface during frac, bundling wireline provides significant value. With a bundled service, if there is a problem with wireline, we are not exposed to NPT, which can be massive in terms of cost reduction. If we can get a frac company to partner with a wireline company to wave each other's downtime, we get the best of both worlds.

Once unbundling became popular, unbundled providers increased prices, and bundled providers decreased prices to be competitive. In many areas, the cost difference is minimal, which makes the bundled service value proposition attractive.

———————————— ACTION ————————————
## Continuous Pumping

On multi-well operations, the time spent switching between wells impacts the ability to maximize value per minute with a full frac spread on location. This non-pumping time increases costs and reduces utilization rates. If we eliminate "swap-over" time, there can be a direct monetary benefit for all entities involved.

There are several technologies that enable instant well-to-well transitions. By automating surface valves using computer systems to open and close valves while pumping, the technology can reduce swap-over times to under one minute. These systems work on standard multi-well operations or simul-frac operations without requiring anyone to enter red zone areas while pumping or switching wells.

It is common to implement continuous pumping or similar technology to enhance operational efficiency, celebrate its success, and then move on, thinking that we are saving money. However, if someone with strong financial intelligence does not take the time to ensure we receive full monetary benefit, there is a good chance that our costs do not go down; they actually go up.

This happens for several reasons. First, just because we become more efficient does not mean we automatically save money. We could just be spending money faster and more money than before because there is a cost to the technology deployed—as complexity goes up, cost and risk go up.

Second, the other vendor contracts, primarily the pump-time contracts, may not appropriately reward us for eliminating swap over times.

For continuous pumping to make economic sense, it must provide a direct cost reduction. Direct cost reductions are reflected directly on the tickets, whereas indirect cost reductions are reflected with increased production on a cost-per-unit of production basis. Continuous pumping does not increase production or EURs; therefore, we must see lower field tickets and invoices to justify the endeavor.

> **Direct cost reduction actions must translate onto the field tickets to make economic sense.**

Every frac contract has an assumption for swap-over time built into the prices. Often, the number of allocated minutes is not stated in writing in the contract, but it is known by the team, such that if we go over "X-minutes," we start getting hit with NPT charges. However, there is often no financial benefit if we transition faster.

Therefore, for all our hard work and investment in the technology to pump continuously or swap quickly, there are no savings on the frac tickets. In fact, our pump-time costs probably go up because we are pumping continuously. Plus, those additional pump-time minutes are often at lower rates than the designed target frac rate, which makes it even worse.

Third, with an unfavorable contract, the additional margin we create by being more efficient is captured by the frac companies, not by the operator. The frac companies make more money per day because we are getting more stages pumped per day. We end up spending more money per day with negligible total savings. Therefore, frac service company cash flow metrics per day increased significantly, while the operator received negligible financial benefit—with increased risk.

To address this, we must ensure we are rewarded by vendors for efficient swap-overs. Many of them will say that since some of our services are based on the number of days on location, if we finish faster with quicker swap-overs, we will see savings. However, if we do the math, the net savings may be negligible or unattractive when adjusted for risk.

If we employ methods that reduce or eliminate swap time, we must request a reduction in the pump-time cost per minute charge or establish an NPT bank that we can deduct from the final invoice. It is more straightforward to get lower costs across the board from each vendor, but it will depend on the specific situation at hand.

Finally, when achieving efficiency with technology that adds complexity, we must confirm the benefits scale in the field and on the invoices. This requires constant diligence on the front lines and in the office when processing tickets.

Vendors know operators will closely inspect the first few tickets when new tech is deployed and then return to what takes precedence. When making technology additions and contractual changes, we must update the "Cost Book" and review the cost savings we expect to see with our field team daily to ensure it translates onto the field tickets and ultimately onto our free cash flow.

—————— **ACTION** ——————
## Breakdown Not Shakedown

The highest cost time on a frac is when the clock starts on frac equipment charges, often called "Pump-Time." Direct contracts structure this cost as a single unit price per minute. Complex contracts have a pump-time line item plus additional equipment and personnel-related costs intertwined with multiple variables. Regardless, as soon as the pump-time clock starts, we get hit with the most expensive charges per minute of any oilfield operation across drilling, completion, and production.

> **Mine each minute for money,**
> **especially pump-time minutes.**

In an ideal world, as soon as pump-time starts, we would be at max rate with sand on bottom. However, in reality, the process of getting the rate up to the designed rate is slow and inefficient, especially when the formation is tight, and we are at max pressure or close to it.

We could easily spend 30 minutes or more getting the rate up to plan. This might not sound like much for one stage, but if it occurs on 50 to 100 stages per well, the cost is massive.

If the total idealized pump time is 100 minutes/stage but we are spending an additional 30 minutes to get the rate up to design, that could increase frac equipment costs 30% to 40%.

This inefficiency often occurs due to the slow process of pumping acid, higher treating pressures until the acid reaches the perforations, and the process of getting the rate up while working through issues with the pumps operating at max pressure. All these factors slow down the operational process to achieve the designed rate.

However, there is a method to solve this problem, and I call it "Breakdown Not Shakedown." When we are charged the full cost per minute, but we are not at the full rate per minute, this situation is what I consider a shakedown perpetrated on the operator, with the term "shakedown" humorously utilized to refer to a crafty scheme in which the operator is overcharged for services.

To avoid this high-cost, low-value time and significantly reduce fracturing costs in the process, we will pump the acid down to the perforations with the wireline pump-down equipment before we transition the well to the high dollar frac pumps. Therefore, we will breakdown the formation and establish injection for "free" before starting the clock on the big money per minute frac equipment.

The process of exactly how this is executed is up to the team. We could set the plug with or without the ball in place, shoot, POOH, drop the ball if it's not already in place, pump the acid down, and then get max rate with the wireline pump down pump trucks for a short period to breakdown the formation and inject.

If we are open to new forms of acid, we could use wireline safe acid, sometimes referred to as eco-acid, modified acid, or hydrochloric acid replacement[15], and pump acid down simultaneously with the plug and guns pump down.

This tactic can save a considerable amount of money we would otherwise incur in pump-time charges, especially on hard to breakdown formations that need acid to establish injection efficiently. This can be critical if we want to run the ball in place, as certain formations prevent that tactic from being practical as injection pressures are too high to get much rate until acid reaches the perfs.

Test eco-acid with our cuttings to ensure there are no issues and run additional tests as needed. We also need to confirm that our wireline vendor does not have a problem with our wireline safe acid and there is not a compatibility issue with the type of wire and equipment our wireline provider is utilizing. Additionally, we want to be careful when doing this with dissolvable frac plugs, which can lose integrity quickly if hit with acid.

On multi-well operations, proactively breaking down the formation with acid while we wait for the frac to finish puts us in a position of power to achieve the designed frac rate within a few minutes because the perforations are already broken down with acid and injection established. We are not spending big money per minute to pump the acid down to break down the formation or burn a bunch of time getting up to the target frac rate to establish injection.

Pumping the acid down with the wireline pump-down equipment can also empower us to eliminate the pad stage and start proppant immediately. With a pre-frac formation breakdown on each stage, as soon as the well is open, we can start sand, further reducing frac minutes and fluid expenses.

---
# ACTION
## No Toe Preps
---

Based on frac contract terms, formation geology, and injection pressures, there can be respectable cost savings from canceling toe preps. The way this strategy works is to rig up the frac iron, pressure test the casing, shift the toe sleeves, and go directly to frac.

For this strategy to work, the toe must be in excellent rock—yet another reason the last 500 FT during drilling is critical. We must be in the target interval because if we cannot inject, it will get expensive to deal with toe issues when the frac spread is on location on standby.

Running two toe sleeves on extended laterals, regardless of whether we eliminate toe preps, ensures we can inject at a reasonable rate. With the toe in good rock and two sleeves, I have not had issues executing this strategy at scale. However, it only makes sense if the financial modeling shows significant cost savings, which is not always the case.

One of the benefits of this strategy is that we can go directly to sand on the toe. Many of the unconventional formations respond favorably to sand, especially on the toe, which can be tricky for various reasons.

Having spent significant time inspecting new casing before and after it is cemented, I have found that casing builds up much internal debris. If we run a frac plug into the well before we pump sand, there is a higher risk of presetting the plug, especially on casing sizes smaller than 5.5" OD.

By pumping sand before running wireline, we effectively sandblast the inner diameter of the casing and push all the debris along miles of pipe into the toe, often referred to as the "trashcan stage" for this reason.

> **Sand before wire is a prudent risk reduction action.**
> **When we reduce risk, we reduce cost.**

Regardless of whether we skip the toe prep, it is good practice to frac the toe with sand before sending a plug and guns into the well for the first time.

The first wireline run is always the most treacherous. Even small metal fragments or minor grime can cause major problems. There is nothing more unsettling on the first wireline run than to get hung up and stuck, especially during cost reduction efforts. It can be very demotivating to the team, particularly if it is happening on multiple wells at scale.

Furthermore, getting hung up on the first stage has been known to cause a chain reaction of events—all negative. Once we are stuck with wireline, we are off standard procedures and exposed to non-routine ops risk, an elevated level of risk. If we cause damage while fishing or milling, we can weaken the casing, wellhead, and stack before starting the first frac stage. Now our wellbore will be stressed repeatedly during fracturing in a weakened state, potentially leading to failure.

"Sand before wire" as it is sometimes referred to in the field, is a solid risk deduction tactic to help avoid these types of high-risk, high-cost entanglements.

---

# ACTION

## Delete Pad Stages

---

Every minute of frac pump time must be under a microscope. We reviewed cost reduction solutions for the breakdown component of fracturing operations; now, we must look at the pad stages and any periods that we are not administering sand.

Consider deleting every line on the pump schedule that does not contain proppant. As soon as we have enough rate to start sand, start it immediately. There needs to be a sense of urgency to start sand. If not, what can happen is that we run long on pad stages because we are distracted with getting everything lined out and end up unintentionally extending the pad and burning significant cash as a result.

> **During frac pump-time, there must be a sense of urgency to be at safe max rate and sand concentration to get the job in the ground efficiently.**

If we feel strongly that pumping slickwater without proppant adds value—don't stop. I am sure there are areas across the globe that have proven it creates a better frac treatment.

From a production performance perspective, I want to pump as much sand as the economics dictate, and I want the sand turned on as fast as possible. The goal is to get the job in the ground quickly and efficiently, then move to the next stage.

Even if we are only at 0.10 ppg sand, I feel better knowing we are getting something down there while on the high dollar pump-time minutes. Plus, the faster the sand gets on the perforations, the lower the treating pressures. Therefore, we can hold max rates with less pressure-related stress on the horsepower, further improving efficiency and reducing pump times.

If we are dropping diversion, we may need spacers before and after the diverter, depending on the product we use. I do not suggest that we remove the required spacers. However, I have successfully dropped diversion while running small concentrations of sand, referred to as "dirty drops." Therefore, we would drop the sand back to 0.50 ppg or lower and then drop the diverter on top of the sand.

When operating at the top of our game, an experienced crew and company man maintain an impressive pace during frac. Treatments can be subdivided into multiple segments for analysis and monitoring. Once we establish optimum metrics for each operational segment, we can work with our teams to elevate performance across our asset base.

An often overlooked dynamic to completion efficiency is rock quality. If we are in-zone, it can make our teams look like superstars. If we are out-of-zone, it can make us look like amateurs. That is how important geosteering is in many unconventional plays. Providing geologic data to our field teams can help reduce costs when making decisions during frac. If we are out of zone, it can help to get sand on perfs quickly, which is another reason to delete pad stages.

---

## ACTION

### Control Pump Time

---

Years ago, when frac contract structure was based on a fixed price-per-stage, frac pumping companies diligently administered the frac within the allotted time.

For example, if each stage was projected to take 90 minutes, the frac service company would do everything possible to administer the frac within that amount of time. They were incentivized to be as efficient as possible because if it went over, they would lose money. However, as contracts evolved, instead of a fixed price per stage, they would assign a certain number of hours, and if the frac went over, there was an additional time charge.

With today's contract structures completely based on the number of pump-time minutes, the operator is totally exposed to any inefficiencies while frac service companies benefit.

For example, if the service provider is inefficient, they dilly-dally after the well is open, they do not lineout rate or sand concentrations, a problem prevents us from maintaining full rate, we lose the hopper, have a blender issue, or any other problems occur, pump-time will go over budgeted time, and the operator will lose money. In contrast, the service provider will make more money than was estimated on the AFE or bid price.

With current unbundled contract structures, in which the frac horsepower provider is only making money from pump-time minutes, the longer the pump-time the more money they make. Frac pumpers are incentivized to be inefficient in many cases. Therefore, operators need to take an active and aggressive role in managing the time aspect of the frac.

Across the oil and gas industry, when it comes to judging stimulation success, there is an elevated awareness towards pumping 100% of the designed sand. We consider it a success if we get all the sand in the ground on any given stage.

It makes sense to focus on the amount of sand as a barometer of success because it is highly correlated to production performance. The more sand we pump, the better the production, for the most part.

However, we rarely focus on how efficient we are in administering the sand. On a fully unbundled frac, if the operator is not diligently monitoring the time component, we are guaranteed to go over the designed minutes.

We will take longer to place the jobs because everyone else on location is incentivized to take their time. An oil company that does not take the lead in managing pump time is operating in an environment in which pump time is uncontrolled. To address these issues and take control of pump time, consider implementing the following actions.

## Prioritize Time

Similar to the emphasis on proppant placement percentages, we must establish a focus on placing the sand within the planned pump time. Therefore, in our procedures, checklists, and meetings, we need to elevate the subject and discuss the topic of pump time.

Highlighting targeted pump time in procedures, putting a

checklist together to address pump time dynamics and how we can manage them, and confirming all the horsepower and equipment are ready to go at full rate to achieve the designed rate, are a few good starting points.

It is not uncommon for a frac company to start a stage knowing that we cannot reach full rate because they do not want to be on downtime. Plus, if we take longer to pump the stage because we can only reach, for example, 70 bpm instead of a designed rate of 100 bpm, they make more money.

## Establish a Pump Time Bank

Frac pumping companies have become experts in managing and documenting all the minutes associated with pump time since that is how they generate cash flow.

However, there are multiple levels of cognitive bias, as discussed in Chapter 4, to be more comprehensive in calculating the time that benefits them, but much less so in calculating the time that counts against them.

To address this issue, consider establishing a "Pump Time Bank" with the Wellsite Leaders as the account holders documenting time-related issues regarding pump time. There are digital technologies that can help with this, but the Wellsite Leaders must take an active and aggressive role on location in documenting and controlling pump time.

Non-productive time (NPT) is well documented in most reports, but not time-related inefficiencies that impact pump time. For example, if the contracted rate is 80 bpm and we can only get 60 bpm because of an equipment related issue, the additional time we spend to get the job in the ground needs to be credited in the Pump Time Bank. Any issues related to the performance or inefficiencies of the frac service need to be captured and credited in the bank.

Another common occurrence is if something happens with the frac equipment, whether it's a leak, we lose sand concentration, or the blender goes down, and we need to shut down. As soon as we are down, we go on NPT. As soon as we start pumping again, we stop the NPT clock and start the pump time clock.

However, it might take 10 to 30 minutes once we start pumping to get back to where we were before the equipment problem. Therefore, that additional time needs to be credited to the Pump Time Bank.

At the end of each stage, we must document with the service provider how many minutes we have accrued the bank. Additionally, we need to document our Pump Time Bank account in the reports so that when it comes time to reconcile the frac tickets, we can confirm that our Pump Time Bank account credits are deducted from the total frac pump time amount. Digital technologies can help backup our time deduction calculations, but they might miss nuances that frac pumpers take advantage of to maximize pump time charges.

## Set a Minimum Acceptable Rate

When frac contracts are negotiated, consider establishing a minimum acceptable rate. For example, if the target rate per well is 100 bpm, and we have a minimum acceptable contract rate of 80 bpm, if we are only able to get 75 bpm, we must shutdown and go on downtime until the issue is resolved.

With a minimum acceptable rate written into the contract, frac companies have an incentive to ensure target rates can be achieved, or they will lose money.

From a design perspective, it is critical that we reach target frac rates quickly before limited entry is lost due to erosion—which occurs within minutes. Therefore, not getting up to the designed rate very fast will impact proppant placement across the clusters and production performance.

Additionally, losing rate during the treatment has been shown to be detrimental to production due to the closing of fractures, which often do not reopen when frac rates are rectified.

Having these concerns integrated into the frac contracts allows us to obtain some method of financial compensation for the economic pain inflicted on us in the form of additional cost and loss of production performance.

## Carry a Backup Gun: The New York Reload

Perforations play a critical role in managing pump time. If we have difficulty getting to max rate due to pressure or other related issues placing the job impacting pump time, we may not have the optimum gun or perforation configuration for the job. Similar to armed self-defense tactics, consider having a backup gun, also known as a "New York Reload," in case of issues with our primary gun.

The term "New York Reload" originated in the 1960s with the NYPD when they carried standard-issue revolvers. When out of ammo during gunfights, it was more efficient to drop the primary gun, draw a second gun, and continue shooting.[16]

A similar situation occurs in the oilfield. Sometimes, due to casing type, cement, hole diameter, geology, or a combination of factors, our primary gun does not optimally fit the current situation. Therefore, we need to have a second gun type available on location that differs from what we are currently shooting.

Maybe a super deep-penetrating charge, low-debris charge, or burr-free charge is what we need. Having a good backup gun that we can quickly draw is a good tactic to manage high-cost adversarial situations during the operation.

## Employ Aggressive Tactical Actions

Even if we have no equipment problems, it is common to exceed pump time due to high treating pressures, preventing us from reaching the planned rate efficiently. Once we get enough sand down there, the pressure usually breaks back, allowing us to increase rate. When this situation occurs, we can greatly exceed planned fluid volumes on the low sand concentration segments of the pump schedule.

If we want to be aggressive, we have a number of tactics to deploy. Once treating pressures start to behave, we can run the sand "hot." In other words, if we are at 1.50 ppg in the schedule, we would bump that up to 1.75 ppg.

If we are at 2.50 ppg in the schedule, we can bump that up to 2.75 ppg or 3.0 ppg. We can gain significant ground doing this and get our time back on track. We still deliver 100% of the planned sand into the formation; we just do it a little differently.

Furthermore, there are situations in which 10 or 20 stages have high pressures and we have difficulty placing the sand according to the design, so we make a bunch of adjustments which typically result in much longer pump times using more fluid to place the same amount of sand. This will put us way over planned pump time for the well as a whole.

When this happens, unless there is a geologic anomaly, after placing half the stages at the midway point in the lateral, due to lower frictional constraints, we can place the jobs at lower pressures. Therefore, we can easily run the sand hot and cut a good portion of pump time and fluid from the operation while placing the same amount of sand.

Understandably, not everyone will agree with this strategy to reduce costs because pumping sand at 3.0 ppg as per the design compared to 3.5 ppg with a more aggressive design might impact production. Consider modeling it out to see if there is a noticeable difference in frac designs and the projected production performance.

Another tactic to reduce costs regarding pump time is to increase the rate. We may have a contracted rate for 100 bpm but based on pressure, the equipment is capable of 110 bpm.

If the frac crew does not have an issue with it and we are rigged up to operate at higher rates, we can save a lot of time and money, as long as there are no additional pump rate fees.

During frac contract negotiation, discussing rates, pressures, and the potential to achieve increased rates based on lower pressures should be addressed and worked into the contract if beneficial. Any rate we can achieve above our standard designed rate is an opportunity to reduce costs.

# ——— ACTION ———

## Cap Chemicals

Whether bundled or unbundled, we must disincentivize the practice known as "Juicing," which is running up the frac chemicals to increase invoices. To address this issue, establish a contractual agreement with chemical vendors that we will not pay invoices over 3% (or an agreed percentage) of the correct chemical volumes. Therefore, if the chemicals are run without attention to detail or intentionally run "hot," we have a contractual agreement indicating that we are not obligated to pay more than 3% above the correct volumes.

This will reduce our chemical costs because vendors are incentivized to be precise and disincentivized to juice us up. To monitor the situation in real time, we must watch the additive charts and confirm chemical usage amounts after each stage.

Every few stages, it is good practice to put eyes on the totes and materials to double-check the chemicals are running correctly. Furthermore, we can take pictures of all chemicals on location, including the product labels and volumes in the totes (if levels are visible). Then, upload the pictures to the Daily Visual Report.

These actions provide another diligence tactic to confirm we have the correct chemicals on location, and they are run appropriately. It is easy to have the wrong materials on location, operator reps forget to double-check the labels, and we end up running the wrong product, damaging the reservoir.

As a strong proponent of performing detailed lab analysis

to replicate subsurface environments when testing chemicals, we must ensure that we administer the chemical package in the field as intended based on lab testing.

Taking the time and effort to perform detailed lab testing to formulate the chemicals correctly from a compatibility, implementation, production performance, and economic perspective pays off in the long run.

For example, an offset operator in the United States selected a frac chemical provider that did not perform the proper due diligence for the target and ended up destroying four $10MM horizontal wells—a $40MM mistake.

---

**Few people deeply care about well performance.**

---

The wells flowed back gummy bears. Weak chemical diligence and field application destroyed millions in value as the wells hardly produced, even after a remedial treatment. In these situations, the operator will blame the chemical vendor. However, as the operator, we must be proactive and ensure the chemical recipe works, the correct chemicals are on location, and they are administered properly.

We cannot leave this solely in the hands of the vendors. Shale wells are usually robust. We can be careless with the chemical recipe and injection into the frac fluid—it will often go unnoticed from a production perspective. But sometimes, it will be devastating. Overtreating can be as bad as using the wrong chemicals. We must take action to prevent both.

—————— **ACTION** ——————

# Work Down FR

Friction reducer (FR) is one of the most expensive additives pumped on the majority of frac treatments in the United States. Since the primary function of FR is to reduce pipe friction, as we come up hole, we pump through less pipe. Therefore, we can reduce FR.

It is common to start a frac at a specified FR concentration, let's say 1.0 gal/M, and leave the FR setpoint at that same concentration for the entire job. We leave substantial money on the table doing this and cause formation damage, because it's overtreating. Consider a variable FR treatment schedule. Additionally, due to the damage FR can inflict on the reservoir, it is often recommended to pump an FR breaker, usually administered based on the concentration of FR.

## FR Cost Reduction Schedule Example

| Frac Stage | FR (gal/M) | FR Breaker (lb/M) |
|:---:|:---:|:---:|
| 1 - 15 | 1.0 | 1.0 |
| 16 - 30 | 0.8 | 0.8 |
| 31 - 45 | 0.6 | 0.6 |
| 45 - 60 | 0.4 | 0.4 |

Furthermore, we can use real-time friction models to reduce FR intra-stage. Not only will this save money on the invoices, but it should also help improve production. Anything we can do that improves production while reducing costs amplifies our FCF value creation capacity.

—————————— ACTION ——————————
## Diversion

Diverters are an excellent tool to create value by offsetting erosion to maintain limited entry, improve cluster efficiency, increase the number of fractures generated, enhance fracture complexity, mitigate runaway (dominant) superclusters, and reduce frac hits. From a free cash flow perspective, diversion has the potential to reduce costs and improve production.

Optimizing frac designs is a complicated process involving various factors across multiple disciplines, including geology, drilling engineering, completion engineering, reservoir engineering, resource development, commodity markets, risk, and economics. To simplify the process, let's go to an extreme level of detail and look at the situation from the perspective of a single perforation or cluster. Our production will live or die based on the amount of oil and gas that travels into our wellbore through the perforation. Therefore, we want to optimally stimulate each perforation to the extent it adds value.

For example, based on years of development and design evolution, let's say our company designs fracs at 2.5 bpm/perf without diversion. If our max achievable rate per well is 80 bpm, each stage has up to ~30 perforations. Therefore, we feel comfortable we are effectively stimulating 30 perforations at 80 bpm. We could have 3 spf with 10 clusters, 2 spf with 15 clusters, or 1 spf with 30 clusters. We believe that our proppant distribution is good.

If we start using diversion, with the plan to transition our current pump schedule from a single sand ramp per stage to three separate sand ramps, segmented by two diversion drops per stage, each with the potential to seal off 30% of the perforations or clusters, based on the change in frac treating pressure after each drop, we could increase the number of perforations or clusters.

> **Completion design optimization is not an exact science. Among all the noise, production results tell the truth.**

We know that diversion seals off a certain number of perfs from taking fluid. We know this because if we drop too much diverter, we will screen off all the perfs and pressure-out. There are several methods to calculate this based on pressure, or we could test it by increasing diverter until we pressure out. Then, fine-tune the amount of diverter we drop. The more cost-effective way to optimize diverter drops is to adjust the amount of product deployed on each drop based on the pressure response from the previous drop or stage.

If we design a diverter program to seal off $1/3^{rd}$ of the perforations or clusters on each drop, at the end of every stage, theoretically, we seal off $2/3^{rd}$ of what we started with.

In reality, this does not happen because erosion occurs immediately when we put sand on the perfs, complicating the dynamics of how things unfold. However, we could conclude that after two diverter drops, we are sealing off 30% of the total perfs. Therefore, we could increase the number of perforations or clusters, stimulating a larger interval.

For example, without diversion, we have 30 perfs per stage arranged across 10 clusters, 3 spf, with 15 ft spacing, and each stage is 150 FT in length. With diversion, we could increase the clusters and/or the perfs to cover 200 ft in length. Therefore, on a 10,000 ft lateral, we reduce the number of stages from 67 to 50 or about 25%.

There are many aspects to consider when doing this, but with two aggressive diverter drops on a small stage length, it is not unrealistic to be able to increase stage length, reduce stage count, reduce frac costs, and improve production performance.

**Every action that has the power to increase production while reducing costs must be investigated.**

Unfortunately, not only do we incur perforation erosion, but we also erode the diverter. Additionally, adding diverter drops increases pump-time and operational complexity. Depending on our simultaneous fracturing rig up, dropping diverter can be challenging. We must calculate the additional cost and compare it to the savings. Therefore, in certain situations, diverter drops may not add value.

Administering continuous diversion is operationally easier if we do not currently have a diverter program. We may also only use diverter to maintain limited entry and not necessarily to seal off perfs. Every situation is different, and we must confirm diversion adds value. The downhole imaging and disposable fiber optics action items, previously discussed, are excellent companion actions to help optimize our designs.

# —— ACTION ——
## Ghost Drop

---

In an aggressive effort to reduce costs, some operators are eliminating diverter products and replacing them with "Ghost Drops." This is when we do everything we would typically do when dropping diversion, except we do not drop anything. The mechanics of dropping diverter involve reducing rate, often down to a very low level, and then ramping back up to full rate.

Reducing rate abruptly may change fracture dynamics, potentially contributing to additional fracture complexity. The water hammer effect may open new fracture paths in clusters that were not previously taking fluid so that when we increase rate back up to max, we end up injecting into new rock.

Some have seen evidence of this during fiber optic testing. While investigating designs with fiber optics, when unplanned shutdowns or rate drops occurred, it was noticed when getting back to fracturing, more clusters were taking fluid.

To obtain the value from water hammers or if pressure is close to safe max frac pressure, I have ghost-dropped stages and have seen a noticeable difference in treatment pressures trends, abrupt changes in pressure, and pressure drops, potentially indicating new rock is opening. We can also use rate drops to run diagnostics on cluster efficiency and then take real-time actions to improve performance and reduce costs. For these reasons, some operators like to do a quick ISIP at the beginning of each stage and do micro-rate drops during the job to reap this tactic's informational and economic benefits.

---
## ACTION
### Just Say No to Acid
---

For certain formations, the potential exists to eliminate hydrochloric acid (HCl). On the first four stages, pumping acid is often necessary to establish injection. After four stages are placed, investigate the possibility to stop pumping acid.

The combination of breaking down the perforations with our "Breakdown Not Shakedown" tactic, implemented before we start frac operations on each stage, having the frac stages mostly in zone, combined with advanced perforating strategies, can allow us to inject without the need for acid. Additionally, in some cases, acid can damage the reservoir and equipment. Therefore, if we eliminate it, we save money, reduce risk, and potentially improve production.

To maximize the savings, obtain raw concentrated acid mixed with additives delivered in truck transports. Then, administer it into the well by cutting the concentrated acid down to the desired HCl percentage with a water-to-acid pump rate ratio before entering the stack.

After four stages, release the acid transports. Not only do we save a considerable amount of money on acid but also on acid tank rentals, acid containment around the tanks, and cleaning costs. This also eliminates the risk of an acid spill.

Although this strategy may not work for everyone, we can still achieve HCl savings if we reduce concentrations. For example, if we currently use 15% HCl and we reduce it to 7.5%, we cut costs by 50%.

—————————— **ACTION** ——————————

## Dissolvable Frac Plugs

Operators are running fully dissolvable frac plugs with operational and economic success. These plugs are more expensive per unit, but the additional cost is justified for several reasons. First, dissolvable plugs have the potential to eliminate the post-frac plug millout, one of the riskiest oilfield operations. Depending on well design and subsurface conditions, this can save $80,000 to over $300,000 per well.

More importantly, by eliminating post-frac plug millout operations, we liberate resource development to further increase lateral lengths as part of our 100X value creation strategy without technical limiting factor concerns regarding millout capabilities on extended laterals.

> **Technologies that eliminate entire operational steps enable exponential stakeholder returns.**

Second, dissolvable plugs significantly reduce the risk of getting stuck during millout. From a financial perspective, the entire well-level capital investment is at risk when milling plugs because almost 90% of the AFE is invested at this point. If we get stuck and cannot retrieve coil, stick pipe, or any part of the work string, the well could be lost in a worst-case scenario.

If we use dissolvable technology but desire to perform a

cleanout run, it still adds value to pay for dissolvables. Usually, if we get stuck during millout, it is on frac plug parts. There are engineering procedures and tactics to minimize the risk when milling composites, but if we can eliminate the risk by milling dissolvables, we are money ahead.

When pulling screens to visually check the contents of the plug catcher during dissolvable plug millouts, there is usually very little to see, as long as the technology has been applied correctly.

Third, it can add value to run dissolvables on the toe stages in case we lose returns or have difficulty reaching TD. As lateral lengths continue to increase to enhance returns and improve per-foot metrics, post-frac millout risk continues to intensify. Dissolvables give the field team options during millout and reduce the psychological pressure to reach total depth even when conditions suggest returning to surface.

Fourth, consider running dissolvables based on observed pressures during frac treatments. The strategy is to start each well off with dissolvables. After each stage, based on pressure, dissolvables continue to be deployed until pressures are favorable for millout with coil or stick pipe. Then, we switch to composites.

Geology and drilling data could also be incorporated. For example, if drilling mud losses were observed at certain measured depths, this dynamic often carries over to completion operations, especially during millout.

Fifth, the first plug run into the well generally carries the most risk. If we get hung up with a dissolvable, the plug can easily be eliminated with acid. If we get hung up with a composite on the first run, we only have a few options. We could erode it away with a plug erosion job, but that can damage the casing. We could mill it out with coil, which could also inflict damage on the casing right before we frac.

As a risk management tactic, I recommend running a dissolvable on the first run for every well. There have been far too many cases in which debris and a micro dogleg or "hidden death-leg" result in a plug presetting on the first run. Then, during millout, the casing becomes damaged, resulting in an expensive quagmire and eventual total loss.

Finally, in recent years, many operators have abandoned utilizing coil tubing for plug millouts due to the risk of getting stuck and other bad coil tubing experiences. Instead, operators are favoring stick pipe. In general, stick pipe provides more options during millout, but it takes longer and costs more.

Running dissolvables may provide a comfort level to transition back to coil. In most cases, coil tubing is the most cost-effective method to use when operating under pressure.

If you have had problems in the past with dissolvable plugs, consider giving the technology another look. The potential to eliminate the millout on part or all of the lateral has significant implications to improve safety, reduce risk, reduce cost, and expand value creation opportunities with longer laterals, moving us closer to our long-term objectives.

If we don't have coil or snubbing on location, we can't get stuck milling plugs (especially on extended laterals) because we are not milling plugs. We are leveraging dissolvable technology to eliminate this labor-intensive, high-risk, high-cost operation.

---
## ACTION
### 4-Plug Plan
---

Running dissolvable frac plugs on 100% of the stages can be expensive, and it is unnecessary if we are going to perform a cleanout run. Therefore, an excellent low-cost, low-risk strategy is to employ what I call a "4-Plug Plan." This plan requires having access to four types of plugs on each well.

Starting from the toe, for the first 20% of the well, our plan is to run dissolvables. With this strategy we significantly reduce risk during the initial plug runs and during millout.

For the remaining 80% of the lateral, the plan is to run a good composite plug. This is our workhorse plug, which we have excellent data on from a running, millout, and downhole imaging perspective. Once we find a good composite, let's stick with it. There is much risk in running new-to-us plugs every other week, and I generally avoid it.

A slim-hole composite or slim-hole dissolvable is the next plug in our plan. If a tight spot develops, we want to have the option to continue with minimal disruption. Having a slim-hole plug within reach enables us to move forward seamlessly. We do not want to be in a situation where we have a tight spot and then start looking for a plug or have to run coil and end up damaging the casing.

A bridge plug is the final plug in our arsenal. We want to have two within reach in case of a leak below the lower master valve. If a leak develops, quickly running this isolation device in a timely manner is often our best option.

# ——— ACTION ———
## License to Shoot

When perforating with high cluster counts, there are several scenarios in which we will not be able to shoot at the exact (to the foot) designed depth for each cluster. Unless we are perforating based on geologic variables, evenly spaced geometric clusters are the most common and easiest to implement, but it is not always possible.

With high cluster counts and a desire to continue increasing them during optimization and cost reduction efforts, I prefer to provide the field team with a "License to Shoot" as needed within the stage interval. For example, if each frac stage length is 200 FT and our perforation design requires 30 clusters per stage, our shooting policy would be to try and shoot each gun at the correct depth, but we are "Weapons-Free" to shoot as necessary within the 200 FT interval.

> **Scenario planning with tactical actions written into the procedures reduces risk and cost.**

When shooting a lot, it is common in some areas for the formation to grab the guns. It would be inefficient to request permission to shoot-loose every time it happens. We want to document it in our "Operational Issues List," but the team must be weapons-free when in the middle of an operation, and a decision needs to be made to prevent getting wireline stuck.

# Perforating Guidelines for High Cluster Stages

- QA/QC the plug, setting tool, and guns. Take images (before/after RIH); upload to Daily Visual Report (DVR).

- When picking up wireline BHA, make sure not to cause damage by kinking the plug to setting tool assembly. Use the proper carrier to roll and lift the BHA up into the air.

- Establish maximum allowable wireline running speeds for pump down and pull out, in vertical and horizontal section.

- Confirm we are at the correct stage depths with the wireline engineer and operators.

- Shoot on the fly at a safe speed for the conditions at hand.

- Pump 2 to 3 bpm while shooting to stay in tension unless more rate is needed to stay in tension. We do this to keep debris away from the wireline bottom-hole assembly and prevent from getting stuck or shooting the wire. This is critical, particularly on high-angle trajectories, undulations, and toe-up wells where the guns will backslide, resulting in an incident, if we are not careful.

- Never shutdown pumping until entire BHA clears top perf.

- Wellsite Leader must stay inside the wireline truck until confirming safe bump-up from the wireline engineer and outside operators.

- Tag the tool trap to ensure wireline BHA is 100% above the frac stack before operating any frac stack valves.

- Confirm nothing was left in the wellbore. Inspect wireline components in detail. Take pictures and upload to the DVR.

# ——— ACTION ———
## Eliminate Screenouts

Screenouts cost significant money in downtime charges and flowback-related expenses, not to mention the additional risk of flowing a live well with a full frac spread on location.

Add in simul- and trimul-frac operations, and screenouts add a level of complexity and cost that is highly undesirable. Therefore, we need to implement a strategy to reduce the probability that we screenout a well.

In most cases, screenouts are avoidable, but the wellsite leader plays a critical role. If this person is not watching the frac treatment plots very carefully and is able to take action quickly, it will be hard not to have problems. Fortunately, we can utilize several processes and technologies to help. See the next page for "Screenout Avoidance Tactics."

> **Elevating our ability to predict and prevent screenouts reduces risk and cost. Robust system incorporating checklists, statistics, rules, alerts, and AI are weapons to consider employing.**

The right people and systems ensure we do not have screenout issues. However, there is one additional action we can take to help our field team make good decisions, and that is to provide detailed stage-based geologic information quantifying geologic risk, which we will discuss next.

## Screenout Avoidance Actions

- On the first 4 stages, if there is a greater than 50% probability of a screenout, sweep sand or go to flush. Do not take excessive risk because the well will not be able to flowback on its own, and we will need coil tubing.

- Get comfortable deploying sweeps. If pressure is rising rapidly, cut sand and sweep. Once pressure lines out, start sand again.

- Add screenout detection algorithms based on historical data to the "Virtual Company Man" automated alert system.

- Incorporate screenout projections into the "Knowledge Bank with Notification System" to provide additional intelligence during frac operations.

- As a backup system or if we do not have automated screenout avoidance alerts, consider employing simple "IF-THEN" algorithms. For example, "IF at full rate and pressure increases 2,000 psi, THEN cut sand and sweep."

- Track screenout details. If there is a wellsite leader with a track record of excessive screenouts, it's probably best to release that person. Screenout are far too risky and costly to tolerate inattentiveness—it's a safety issue.

- When a screenout does occur, have a carefully crafted checklist and/or procedure to handle the situation safely. Flowing back wells on a multi-well pad with a frac spread on location is a high-risk operation. Different consultants will do things differently and it must be addressed.

# —————— ACTION ——————
## Quantify Geologic Risk

Risk management provides the foundation for our cost reduction efforts, especially during completion. However, in our industry, it is customary for the completion field team to have minimal information on the geology of the wells we frac.

Without geologic intelligence, it is difficult to assess a variety of operational dynamics, including issues during toe-prep, formation breakdown problems, pressure anomalies, issues placing proppant, ability to avoid screenouts, challenges pumping the guns down, risks when perforating, millout challenges, and tight spots.

For example, if we have an abrupt pressure drop during a frac stage, without any other information, we might assume the frac plug failed or there is a casing problem. However, if we know there is a fault across the stage, it makes sense that the pressure drop could be a function of the fault. This information is vital when making critical decisions during the frac, on the plug pump down, and during millout.

To increase our field teams' subsurface intelligence, quantify geologic risk, and make it easy to use in an operational setting during real-world situations, consolidate our rock knowledge into at least two quantitative geologic variables:

1) **Percentage of the stage in zone (% in zone).**

2) **Faults across the stage (# of faults).**

With these two variables, we can incorporate geologic risk and intelligence into our decision-making process on the front lines. Therefore, on a stage-by-stage basis, we will list the percentage of the stage in zone and the number of faults present. We could also add a note with additional geologic-related information that adds value for operational execution.

For example, if we lost circulation during drilling or got hung up at a specific depth, providing that intelligence to the completions team elevates awareness, helping us make better decisions and avoid costly mistakes.

Geologic intelligence also allows us to incorporate formation risk into our thinking and mental models when reacting to anomalies during completion. This will help avoid screenouts, and place 100% of the sand. When troubleshooting, it will help our field team reach the correct conclusions.

During completion operations, making the right decisions to avoid costly situations often happens within a time frame of seconds. For example, if we take a few seconds too long to react to a potential screenout or issue with wireline, we will incur an expensive event impacting FCF. Arming our team with geologic intelligence enables us to make the correct split-second decisions that happen too quickly to text or call someone to discuss.

For example, if pressure is acting squirrely on a stage that we know is out of zone or has a geologic anomaly, we can incorporate that data into our decision-making process and field actions to place 100% of the sand. Having zero geologic data puts us at a disadvantage. Plus, we need the geologic data to feed the Virtual Company Man to make the algorithms work.

# —— ACTION ——
## Hobson's Choice Sand

Hobson's Choice is an expression used to describe an illusion that multiple choices are available when only one exists. The term is named after Thomas Hobson, a livery stable owner in England 400 years ago. Hobson had many horses, giving the appearance of choice, but he required customers to take the horse closest to the stable door.[17]

From a cost perspective, sand purchasing can be a Hobson's Choice. There are many options, but it is the sand closest to location that is often the only viable choice. It is not Mr. Hobson that is forcing us to take the horse closest to the door but our friend Mr. Cost, who reminds us that trucking costs are often equal to or more than the cost of the sand, strongly suggesting that we buy the sand closest to the well.

Of course, we don't always have to buy a horse (dry sand); we could buy a mule (mixed sand) at a discount. Purchasing wet sands, unwashed sands, local sands, off-the-ground sands, river sands, and bulk purchasing sand at a deep discount are cost reduction actions to consider. The "highest quality" proppant does not guarantee improved production. There is a case to be made that lower-quality sands provide superior results. Furthermore, it might make sense to vertically integrate sand mining. However, plans change, and the future is hard to predict. A desirable mine today could be undesirable tomorrow. Diligence the sand, its location, and the economics, as this is a decision with long-term impact.

---------------- ACTION ----------------
## Zero Cost Sand

---

Each oilfield location has pros and cons regarding operations and cost reduction opportunities. Wherever we operate in the world, we must look for opportunities relative to our current geographic location. When thinking about costs, what can we find about our specific physical position on the Earth that will give us a competitive advantage?

During full-scale shale development, when planning road construction, consider incorporating the surrounding resources we could use to reduce cost into the decisions on where to build the roads to the wellsites. Easy access to the items we need to grow our business will help reduce costs.

---

**Every area on Earth provides a unique cost reduction opportunity.**

---

For example, in certain parts of the world, where the development area is covered in sand or has sand close by, we can use the process of building the roads and location to stockpile sand that we will use for the frac.

Once the exact location for the well is selected, we inspect the area for sand and lab test our findings. When we clear the road and pad, we take the surface sand and push it into a pile on the side of the pad. After we finish building the drilling location, we use the same construction machinery to gather mounds of sand right next to the location.

From a legal perspective, using surface sand should be incorporated into our SUAs during the negotiation process if we do not control the surface outright.

Before the frac, we run the sand through a simple screen to remove debris and sieve it directly on location at minimal cost. Then, we conveyor it into large piles for use on the frac. This strategy eliminates sand material costs, sand trucking costs, and sand storage costs. We also improve safety by eliminating equipment, road risks, and associated hazards.

If we do not have sand directly on the pad, we can use a similar strategy: trucking wet sand onto location and administering it into a sand pile for use on the frac.[18]

This eliminates costs associated with storing the sand, demurrage costs, silos, dealing with boxes, complex forklift problems, pneumatic sand transfer truck issues, and the associated logistical nightmares that can inflict high-cost non-productive time (NPT) on our economics.

In certain parts of the United States, if we can reduce pad stability issues and concerns with a sand pile strategy for storage and usage, especially during unfavorable weather periods, it should be considered.

Eliminating sand silos, sand boxes, forklifts, extra foundation rock, pneumatic frac sand tanker trailers, and other associated complex sand storage and handling equipment by employing the "Sand Pile" strategy can reduce operational complexity, cost, and risk.[19] Any action that reduces costs while improving safety is worth our consideration.

---

# ACTION

## Go Electric

---

Depending on the area of operations, diesel fuel for fracturing can account for 10% to 20% of total frac expense. Utilizing natural gas turbines to generate electricity to power frac equipment can significantly reduce fuel expenses.

When performing the calculations comparing natural gas to diesel, we must involve our financial team to determine the actual price for our field gas. Selling natural gas is far more complicated than selling oil due to the many associated fees and deductions. This can make using natural gas as a substitute for diesel far more attractive because, depending on commodity markets and fees, natural gas can be the closest thing we have to free energy.

In addition to lower fuel expenses, utilizing an electric frac fleet can reduce the number of pumps required. This reduces the footprint and number of people required.

Location size is built to accommodate the fracturing operation, which is up to 3X the footprint of a drilling rig, as it requires the most people and equipment. If we can reduce the size of the location because the fracturing footprint is smaller, we can reduce construction costs, surface damages expense, resurfacing, and redress costs.

Finally, from an ESG perspective, eliminating diesel eliminates the continuous fuel hauling requirement. From an emissions, traffic, and noise perspective, this should enhance our metrics, providing further support for the transition.

# ———— ACTION ————

## Laws of Frac Stack

Problems with the wellhead and stack are time-consuming, costly, and potentially catastrophic. During frac and millout operations, an issue could cost anywhere from a few thousand for a time delay to tens of millions for a well control issue below the lower master on a high-pressure, multi-well pad.

> **Rules reduce risk.**
>
> **Lower risk equals lower cost.**

Establishing a set of rules or guidelines regarding the wellhead and stack will enable us to reduce risk and cost. Since we depend on this equipment to provide the physical barrier to maintain well control, we must exercise prudent risk management during optimization. Our strategy demands that we approach cost reduction as professional risk managers. Cost reduction efforts provide an opportunity to reexamine all aspects of the wellhead and frac stack.

The optimal approach is asset-specific and not one-size-fits-all. Everyone has personal preferences and strong opinions. Optimizing the wellhead and stack should be based on past challenges within the specific area of operations, which is an ever-evolving process. The best approach today will not be the best approach tomorrow. With that in mind, let's review ten guidelines to consider.

## 1) Single Vendor for Wellhead and Stack Components

Selecting different vendors for the wellhead, frac stack, and production tree components introduces additional operational risk. It is common for the drilling team to prefer one vendor and the completion team to prefer another.

Therefore, the casing head and tubing head are from "Vendor A" and the frac stack is from "Vendor B." When we do this, it is difficult to establish responsibility, ownership, quality assurance, and quality control of the entire well integrity apparatus, especially the connection between the tubing head and the lower master valve, one of the most critical connections on location for well control. This adds complexity, risk, and additional costs to our operation.

Also, if the tubing head wing valves or any other wellhead components get damaged during stack installation or operations, it is not the stack vendor's primary concern. With the recent increase in failures below the frac stack, it is prudent to have a single point of responsibility for the entire assembly.

## 2) Wellsite Leaders Must Oversee Installation

From casing head to frac stack installation, vendors have step-by-step procedures to properly install their equipment on our well. Establish a requirement that Wellsite Leaders must obtain a written copy of these procedures and follow along to ensure each step is performed correctly.

Many technicians have minimal experience since this area of the industry is in a constant state of instability, with workers coming and going. All it takes is for one inexperienced tech to make a small mistake, especially on a B-section installation, and we will have a costly failure.

The risk and cost are elevated on multi-well operations due to the domino effect. This is also an issue when it comes to ensuring all wing valves are installed at an angle and not pointed directly at offset wellheads, possibly triggering a chain reaction event if one were to fail or a mistake be made during drilling, completion, or production operations.

To reduce risk and cost, there must be strong oversight. Therefore, the Wellsite Leader must be present with a detailed installation procedure or checklist and have a copy of the vendor procedure in their hands while physically standing next to the technician during installation, watching what is happening. It is not acceptable to allow vendors to install this equipment without oversight, which is often the case.

## 3) Strengthen and Stabilize the Wellhead

Once the casing head is installed, confirm with the manufacturer where the support level in the cellar needs to be. If we use a spudder rig, we can clean up the area and then fill the cellar with concrete up to the base plate if that is the requirement. Then, we can allow the material to fully set to provide a strong foundation.

If we are doing this with a primary rig, it is difficult to get to the cellar with a ready-mix truck and we are often back to drilling before the concrete has fully set. It is a common recommendation by casing head manufacturers to fill the cellar with concrete up to and just over the bottom of the base plate to provide stability and support.

During drilling, there is a significant amount of weight transmitted on the casing head from the BOPs and additional strings. Additionally, there is a lot of vibration and impact that gets transmitted to the casing head. The concrete foundation helps stabilize the head and upper joints of casing to minimize the impact that these dynamics have on jostling the components causing damaging.

## 4) Clean Surrounding Area

Before installing the casing head, tubing head, and stack, it adds value to have the area in and around the wellhead clean. The casing head, tubing head, and stack installation is very intricate. We want the technicians to focus on their work and not worry about getting covered in OBM sludge.

Usually, after the primary rig moves out of the way, the area around the wellhead is covered in a thick sludge of mud and grease, making it unpleasant to work in. We want this entire area cleaned so that anyone can walk around in white shoes without worry. This way, if someone needs to lie down in the dirt to install something or inspect something, they will not get covered in muck.

We also want the mousehole plugged so no one falls into it working around the wellhead. If there is a shallow cellar area, we fill it with dirt or put a nice cover that we can see through and access. The goal is to make the area around the wellhead organized and clean so we can think with a clear mind and catch problems during installation and during the frac.

## 5) Protect and Secure

Once the casing heads are installed, we need to protect them from collisions. I prefer to surround the wellhead area with large concrete barrier blocks, often called "Bin Blocks," or something comparable to provide a barrier.

These barrier blocks are low-cost and easy to carry on a trailer, so getting them to location is very cheap. Surround the area around the wellheads with bin blocks at least 20-30 FT away on all sides. This way, if a vehicle hits them, the blocks have room to move forward without hitting the wellhead. Compliment the blocks with large traffic cones with a weighted base to prevent them from blowing over in the wind and reflective collars, making them easy to see when headlights are shined on them at night. I also place solar-powered yellow safety lights on the cones that blink at night in areas with fog or weather issues. Finally, install a camera at the wellhead to monitor all the wells.

Some of my team thought all of this was excessive, until I showed them videos of trucks hitting the concrete barriers. Wellheads act like magnets for vehicles.

Just recently, a friend had a situation in which a Company Man backed his F-250 into the cellar and landed on top of the tubing head wing valves, bending the entire assembly and casing—the whole thing, casing and wellhead, kinked over. The incident occurred before frac—an expensive quagmire.

This would be very hard to do with my setup because we have multiple physical alerting systems: concrete bin blocks, cones, reflectors, and flashing lights.

If the wellhead is hard to see or if someone is not paying attention, it is not unheard of for a person to hit the wellhead with their vehicle and not report it. No one wants to be responsible for a costly mistake or get in trouble. Plus, many people do not think it is an issue if a strong piece of equipment, like the wellhead, gets hit with a vehicle. The driver is more concerned about damage to their vehicle, not the wellhead. However, wellhead equipment is not engineered to withstand side impacts or something falling on top of it. It is designed to contain pressure.

## 6) Eliminate Multi-well Confusion

On multi-well pads, it is easy to get confused when dealing with many wells during frac and wireline operations. To reduce the risk of opening or closing the wrong well consider using several tactics to eliminate costly mistakes.

First, place each accumulator or control system at a safe distance directly in front of its corresponding frac stack—not off to the side or in a weird position. Put them directly in front at a safe distance so it is intuitive as to which control system operates each stack.

Second, paint each frac stack and corresponding accumulator the same color. Third, paint a number or put a sign with a number next to each stack and accumulator.

Finally, place a sign on the accumulator and stack that says "Wireline" for wireline operations and "Frac" for fracturing operations. These four actions should eliminate the risk of accidentally shutting in the wrong well.

Four different methods are suggested because everyone is different, and we do not know which method will resonate with each person. Some people respond favorably to colors, while others identify with numbers. Having what is effectively free redundancy is a prudent risk and cost reduction strategy.

## 7) One Touch Rule

To further reduce risk and cost, consider only letting one person operate the hydraulic valves for frac and one person for wireline, or have login access with digital systems.

My preference is that only Wellsite Leaders open and close hydraulic stack valves. Never should someone tell someone else to open or close a hydraulic valve. Only the Wellsite Leader in charge of wireline and frac be the person to open and close the valves. Not another consultant or field hand helper.

If there is one consultant in the wireline truck and one consultant in the frac van, with a third consultant operating the valves, that is how an accident can happen. We have too many cooks in the kitchen.

The lead frac consultant operates the frac stacks and the lead wireline consultant operates the wireline stacks.

Therefore, after the frac is finished, the lead consultant stands up and exists the van to close the frac well, and it is the same for wireline operations.

It will be very difficult for someone to accidentally close the wrong well if we follow this rule because the consultant that is fracturing is not going to get up in the middle of the frac stage and close the well—that's nuts.

Nor is the consultant in the wireline truck overseeing perforating going to get up in the middle of the operation and shut in the well on wireline.

However, if we do not follow this rule and have another consultant or valve man opening and closing valves then it will happen, especially when we are operating at scale.

## 8) Confirm Bump Up Before Operating Valves

If the lead consultants are the only personnel closing valves, it is hard to make a mistake. However, there is a situation during wireline that has contributed to very expensive mistakes: accidentally closing the frac stack when the wireline tools are not 100% in the lubricator.

Consider implementing the following protocols: The lead wireline consultant must stay in the wireline truck until the tools have bumped up in the lubricator. The lead consultant must confirm bump up.

Then, have the wireline engineer tag the tool trap (as long as it is safe to do so) to 100% confirm that we are completely in the lubricator and that the bottom of the wireline assembly is not across the upper stack valves. Then, and only then, can the consultant leave the wireline truck to close the stack valves as per our "One Touch Rule."

## 9) Inspect After Every Stage

After each stage we must inspect the wellhead and frac stack for any potential issues. Are there any drip leaks, especially below the lower master? Is there pressure on the backside based on our analog gauges? We should have each backside digitally monitored, but due to the consequences of an event, we should also have an analog dial gauge at the wellhead that we can double check to make sure everything is matching up.

Recently, there have been major events where the casing or cement did not contain pressure, and the backside pressure increased unnoticed until a well control event occurred at surface. For whatever reason the digital gauges failed, and equipment was not rigged up to safely release backside pressure at proper rates. These accidents can go into the tens of millions in cost, so we need redundancies.

After every stage, look at the backside analog gauges and flowback tanks rigged up to the backside pop-offs to see if there is pressure. Performing these double checks on each stage requires significant physical movement. Some consultants are not interested or capable. If a completion consultant is only comfortable running operations sitting in a chair in the frac van, our risk and cost will be elevated.

Also ensure there is no movement of the stack. We should be rigged up in such a manner that we do not transmit vibration from the pumps to the frac stack, but sometimes it still happens. If vibration is noticed, we need to reconfigure our frac iron rig-up to reduce horsepower vibration being transmitted to the stack and wellhead. Watch the stack for vibration during pumping. It is good practice to have the frac van positioned so we can see the frac stacks from the van windows. Additionally, consider checking the stacks with a digital and spirit level throughout frac operations to see if there is any movement.

## 10) Understand Risks Below Lower Master Valve

Any issue on the wellhead below the lower master valve on the frac stack is a major event. Some operators do toe preps through the tubing head wing valves, flow back through them, pump methanol or chemicals through them, or produce the well via tubing head wing valves.

From a risk perspective, consider not connecting anything to these valves, if possible, until after all intervention is finished. There have been cases where the wellhead had connections on the wing valves while the stack experienced vibration during frac and millout. The flanged connection on the spool to wing valve failed and a major blowout ensued.

The risk of flowing through tubing head wing valves is if there is a failure at the ninety or on the inner valve; there is no way to stop it without taking significant risk if the well will flow on its own. If we are flowing back sand and the tubing head wing valve gets cut out at the ninety there will be a well control event with no valves below to close and stop the situation from expanding.

Remember, we are fracturing past the wings and milling out past them, so they are exposed to sand and pressure. As previously discussed, when we do our downhole image work, we want to take the opportunity to inspect the internal status of our wellhead post-operations to see how frac, wireline, and millout impacts our well integrity components to see if there are any issues or unknown risks occurring that need to be addressed. With this information, we can set more robust rules regarding tubing head wing valve operations.

# New Frac Stack Tech Brings New Opportunities

New frac stack technology has the potential to reduce swap time and reduce risk. However, it can introduce new potential problems that bring new risks, many of which are hidden.

For example, conventional tech requires people to place hands on the accumulator and use force (human muscle) to activate. We cannot accidentally activate the system.

We can close or open the wrong valve, but not by accident. On some new systems we can accidentally activate the valves with an unintentional press of a button, by brushing up against the human machine interface (HMI), accidentally placing a chicken sandwich on top of a mouse button, or any one of several other unbelievable but true events, costing millions.

Introducing electronics to control valves during frac and wireline can add risk during weather events (e.g., lightning, rain, ice, floods, winds, and temperature drops) that can make electronics unpredictable or unreliable. A streamlined conventional system should be able to execute swapovers in under 10 minutes. New systems can do it instantaneously, providing new opportunities to further advance our operation.

> **New technology can enable lower costs**
> **when engineered correctly.**

As frac stack technology evolves, we should perform trial runs with different systems to test the value. Implementing new systems will require new procedures and checklists to reduce risk and leverage the full value of the advancements.

———————————— ACTION ————————————

## Eliminate Use of Frac Stack Valves

Any operation that increases risk and cost should be on the chopping block for elimination if it is technically and economically possible. Opening and closing frac stack valves should be on our list for elimination consideration. Cycling stack valves during frac and wireline adds time and complexity. As we increase the number of wells we simul-frac, the risks and costs associated with stack operations increase to the point that a problem with the valves or a mistake operating them will have a massive impact on our FCF goals.

With advancements in frac stack technology, we can evolve our execution and the "Laws of Frac Stack" to eliminate the usage of the frac stack valves during fracturing and wireline operations. There are technologies that we can place above the frac head that eliminate the need to operate the frac stack. Similar to spaceship and space station docking systems, we can use an integrated multi-chamber valve and quick-connect latch system to access the wellbore for frac and wireline.[20]

The wireline lubricator attaches to the stack with a remote-controlled quick-connect latch. After each stage the multi-chamber system equalizes pressure between the wellbore, chambers, and the lubricator using an automated process, allowing wireline to enter the subsurface. This system removes the need to open and close stack valves and eliminates humans having to manually equalize pressure, thereby reducing the risk and time associated with swapovers.[21]

# Next-Level Frac Stack Actions

- Eliminate opening and closing frac stack valves.
  - Reduce or eliminate costly valve repair and DBRs.
  - Reduce common valve failures and associated NPT.
  - Remove grease from entering the perfs, causing damage. 6.0 lbs of grease is used to open and close a single valve.
    - Less grease, debris, and metals from cycling valves and valve damage pushed into perfs may help reduce breakdown pressures or help eliminate acid.
  - Automate well to lubricator pressure equalization.
- Remote plug ball drops when not running ball in place.
- Eliminate cut wireline events with process, checklists, tech.
- Avoid catastrophic shut-ins during frac using process, checklists, redundancy, and technology.
- Shrink the frac stack and remove certain components.
- Purchase lower master valves to further reduce costs.
- Frac the full 24 hours per day.
- Add sensors across key components. Digitize the wellhead.
- Automate instantaneous well-to-well transitions.
- Automate lubricator to well connection detection.
- Autofill the lubricator to prevent burnt wireline events.
- Automate greasing on all valves.
- Install video cameras with computer vision at the wellhead.
- Utilize AI in conjunction with images of the stack to identify anomalies before they become costly problems.
- In a well control emergency, a remote system with no requirements to function stack valves should be faster, safer, and easier to isolate pressure with a bridge plug.

———————— ACTION ————————
# Millout
## Special Interest Group

During standard well-level operations, plug millouts can carry the highest financial risk of any operational phase because 90% of our capital is invested at this point.

When milling post-frac plugs, undesirable situations are unforgiving, and there is little room for error. Imagine sitting at the tables in Las Vegas with 90% of your chips at risk and you are on your final hand. If you make a mistake, you lose everything.

During millout, if something significant occurs, we can lose our entire investment, and that "something" is often the result of weak process, equipment failure, human mistakes, or low-quality execution.

Historically, millouts received minimal engineering guidance and little management oversight—often just a few written lines in the completion procedures. As a result, many millout trainwrecks occurred, often with well control events or stuck pipe and total asset loss.

Some operators, to this day, give minimal attention to millout operations. For example, a friend of mine recently lost an $8.5MM well due to a small mistake during millout flowback. The team reviewed how to safely conduct ops on the plug catcher in detail. However, the information was not clearly conveyed from day to night flowback hands, and, as a result, the night flowback hand flushed the choke full-open to clean the plug catcher, pulling debris into the well. They stuck coil and lost the well—the coil and BHA were unable to be recovered after spending an additional $2.0MM fishing. A detailed, easy-to-read checklist given to each flowback hand would have prevented this.

Millouts are not as glamorous as drilling or frac, but they're super critical and I strongly encourage an extreme amount of engineering and management oversight on every detail of this serious operation.

A problem during millout will have a huge impact on our economics. Therefore, assign our best people and strongest technologies. For these reasons we will form a special interest group (SIG) to ensure our process is bulletproof. Below is a list of people to consider for our Millout SIG.

## Millout Special Interest Group Members

- **Ops Engineer:** Lead with cost and risk accountability
- **Wellsite Consultants:** Lead initiative on location
- **Snubbing Engineer:** Optimize configuration, operations
- **Coil Engineer:** Optimize coil configuration, operations
- **Plug Engineers:** Reduce millout risks, improve processes
- **Fluid Reps:** Improve equipment, fluids, sweeps, metrics
- **Bit Engineers:** Improve bit selection, options, execution
- **Tool Reps:** Provide solutions, BHA design, improvements
- **Flowback Reps:** Review processes, discuss suggestions
- **AI Engineer:** Process historic data, find insights, build AI
- **Digital Tech:** Build models to support the effort
- **String Provider:** Enhance inspections, minimize damage

## Millout Strategies

Well pressures, lateral lengths, directional trajectories, and casing dynamics play a critical role in determining the optimal strategy and equipment to millout the well. Most strategies and equipment selection are highly dependent on pressure and depth (TVD and MD).

First, we must know if there were any mud loss intervals during drilling or geologic anomalies. Second, run calculations to determine if we can easily get to TD with coil and stick pipe. If there were lost circulation issues during drilling or our programs indicate a potential problem reaching TD, we need to deploy our most dependable dissolvable plug technology that we have tested and confirmed with downhole imaging.

As previously discussed, the initial 20% of the frac stages can utilize dissolvables as a risk management strategy on any given well, but depending on other factors, we may need to increase that percentage. In certain situations, it may be the best or the only prudent strategy to run 100% dissolvable frac plugs and eliminate the millout because the risk is too high. If the formation is pressure depleted or there have been offset issues in the past, it is not worth losing the well during a high-risk millout operation.

If eliminating the millout is the selected strategy, the next question we need to ask is if we are going to do a cleanout run to clear the wellbore of sand. In some cases, even a cleanout can be too risky. Therefore, we may need to adjust our overflush to try and minimize the amount of sand flowback or run a resin coated sand to prevent the wellbore from becoming choked off from dune buildup.

To help gauge the level of risk, we must establish a database with ISIPs after each frac stage compared to opening pressures for the next stage. If there is a significant pressure delta, there is a high probability that we will have problems during the millout. If we see this happening, we must be prepared to continue running dissolvables.

This is a situation in which we must react to what is happening and deviate from the standard frac plan. In fact, we must address this scenario in our frac and wireline procedures, so our team is prepared to execute during operations.

If opening pressures before we start fracturing on each stage are close to zero pressure or there is a temporary vacuum, which we should be able to detect if we are paying attention, the millout is going to be dicey, and we need to continue with dissolvables until opening pressures stabilize.

Fortunately, in many areas, pressure depletion is not an issue, and we have strong pressures. Therefore, operators feel most comfortable running composite plugs that must be milled.

Accordingly, our millout processes and procedures need to be at the highest level to address the risk and minimize the chance of an undesirable situation. Working with our millout SIG we should consider taking our current processes and procedures to the next level based on feedback from the team.

A good written millout procedure is much longer than one might think. In fact, it can be longer than the frac procedure, depending on the amount of detail and scenarios we want to include to reduce risk and ensure success.

I find that in terms of experience, few wellsite leaders are experts in millouts. Most of them have much more experience on the frac side of completion, which is part of the problem.

To address this, I prefer to have explanations included in the procedures explaining why certain things are in our procedures so that the wellsite leaders understand the complete picture. Decisions made during millout are very different than decisions made during frac. In many cases, a bad decision during millout can be irreversible, and that must be explained in the procedures.

The costliest millout situations that I have seen over the years are typically due to a wellsite leader making a bad decision in a difficult situation. With no help from engineering, it ultimately results in a string of stick pipe or coil being left in the well permanently, and the well resulting in an almost total loss. Many of these situations have been coil tubing related, which has been the case across the industry.

As a result, the industry has trended away from coil and towards stick pipe with expanded use of stand-alone snubbing units, especially on the extended lateral length wells. However, from a cost perspective for wells that have pressure, coil tubing is still the lowest-cost most efficient option, if it can be executed safely.

## Millout BHA

Configuring the optimal BHA will reduce our risk and cost during millout operations. We want to engage our SIG for feedback and different opinions. I prefer to run a bit that is 97% of casing ID size or as big as possible to reduce the possibility of coring the plug, reduce plug part size to increase debris transport, prevent sticking, reduce stalls, minimize BHA failures, and ensure the plug is milled to a fine material consistently across all stages.

Motor and agitator sizing and selection are also crucial and should be engineered on a well-by-well basis. We also need to confirm that the BHA is not too stiff. There must be a certain level of flex. If the BHA is too rigid, it will snap, which has happened in many cases.

Furthermore, there are a number of specialty components to consider to improve performance and reduce risk. It is prudent and informative to meet with any entity that has technology for plug millout BHAs, because they may be able to help address area-specific challenges. Even if the tech is not a good fit, many of these vendors have robust databases on millouts with valuable information that they will share.

## Frac Plug Selection

Certain frac plugs are superior to other frac plugs. I do not recommend running an untested or unknown plug because it is $200/plug cheaper or even free, which is a red flag. Selecting the best frac plug for zonal isolation and millout is critical and can mean the difference between economic success and failure. If the plugs slide during fracturing or fail, the frac treatment will fail. If the frac plugs are difficult to millout the cost is exponentially more compared to the cost of the plugs.

Certain vendors have databases on plug performance. Find these vendors or develop our own database. However, the problem with building our own database is that it can financially break us.

For example, do we really want to find out if a certain frac plug does not work by running it on our wells? That will cost tens of millions of dollars. It is far cheaper and smarter to learn from others.

Not long ago, I was operating in a new area and was working on the millout procedures, so I invited a vendor with a millout database to come by the office. He showed the top ten worst composite plugs for millouts based on getting stuck with coil tubing. I knew half of them were problematic, but the other half were a surprise.

An offset operator, which we had a non-op working interest in, selected a composite plug that was at the top of the "Worst Frac Plug List," and they got stuck during the coil millout.

It took 30 days and almost $1.5MM to rectify the situation. Why did they run that plug? Because they got a "very good deal" on it. I doubt they knew it had one of the worst track records in terms of getting stuck during millout.

We do not want to be one of those people who do not know. It is our responsibility to know. However, we also do not want to be one of those people who think they know everything.

## Fluid System

Our mixing plant, chemicals, and sweeps are critical to control risk. It is common to start with fresh water and forget about biocide, clay control, or some other chemical we use on the frac. Therefore, we may accidentally contaminate or damage our formation with freshwater shock if we hit the perfs with untreated freshwater. Before frac leaves location, premix a couple frac tanks with frac chemicals to ensure our millout fluids match our frac fluid.

Regarding fluid systems, slickwater is the lowest cost and may be optimal depending on our situation. Minimizing or eliminating gel sweeps not only reduces cost but can also improve hole cleaning when comparing high-viscosity laminar flow with slickwater turbulent flow.

One goal could be to eliminate gel sweeps and short trips. Instead, we clean the well with annular velocities and high Reynolds number (Re) fluid. We do hydraulic short trips without significantly moving the BHA and circulate with increasing rates as high as pressure allows to maximize annual velocities.[22]

## Millout Tactical Actions

Working with our SIG, we must level up our tactical actions to reduce risk and ensure success. We must get our thoughts in writing so they can be sense-checked and discussed with the collective wisdom of the group. This high-value knowledge also needs to be uploaded into our Virtual Company Man. We do not want someone out there operating using their own methods with our money, which is often the case.

There are many cowboys in this business, except they are not cowboying around with their money; it's with our money and that's what I have problem with. These people want to do what they want to do, but they don't want to take responsibility. This is why we need to consolidate our millout tactical actions into the procedures and Virtual Company Man.

We do not want to be in a rush to mill plugs, which many wellsite leaders always want to do to break millout records and be the hero. Don't be fooled, that is how we get stuck.

Once engaged on the plug, try to mill the entire plug without picking up. How plugs are milled is correlated to stuck events. If we mill plugs too fast, large particles are generated and more likely to hang up the string.

We must inspect plug parts to confirm our process is working correctly. We need pictures of the plug dumps to be uploaded to our cloud-based files for analysis.

If plug debris are not reasonably sized, it could indicate BHA issues or aggressive milling tactics. Lack of debris on surface may indicate ineffective debris transport. We must stop and circulate before continuing.

When milling plugs, we should have a mill time of at least 10 minutes per plug. If we go through a plug before 10 minutes, then we wait the full 10 minutes before proceeding to the next plug. After every plug, we do a freshwater sweep. After every 5 plugs, we circulate a dyed freshwater spacer. After every 10 plugs we do a weight check and dyed freshwater bottoms up.

Once we are at max depth, we pick up 100 FT and pump a freshwater sweep, friction reducer sweep, and freshwater sweep bottoms up. Perform two cleanup cycles minimum.

Once our returns are clean, we begin the pullout process. This is the most critical step. If we are going to get stuck, there is a high probability that it will happen as we are pulling off bottom. There is a debris pile behind us and if we pull out too fast, we will unknowingly pull the BHA into the pile of debris and get stuck.

We must pull at 15 FPM for the first 1500 FT. Not moving the string more than 15 FPM is critical because if we move faster, there is a high probability we pull the BHA into the debris pile.[23]

If we have a list of incidents, we can look to see when hang-ups or stuck events occurred. There is a high probability they occurred when tripping out at a fast rate. Most people are very happy to get all the plugs milled and want to get out of the hole as fast as possible, but this is when we must exercise caution and do the opposite of what we want to do. We can increase speed if there are no issues after 1500 FT off bottom.

Just before reaching the heel, we must circulate a bottoms up at a high rate. We can increase pullout rates once we are in the vertical if there are no issues.

If at any point we get hung-up, stop, drop, and circulate.[24] Do not pull into a tight spot. All issues must be captured in our "Operational Issues List" each day. If there are any hang up events, even just for a few seconds, it must be added to the list.

This critical information is often hidden from the office. We need this data to measure risk and improve our processes and AI systems. A 5-second hang-up is a warning sign. We need to analyze and incorporate it into our models.

Every area is different, and every well is different. These are just a few millout tactical actions to discuss amongst the team. Millouts are a high-risk activity. Each action in our procedures should be fully vetted by every member of our SIG before including them in our operation.

--- **ACTION** ---

## Engineered Flowback

---

Flowing back at max rate has additional direct and indirect costs compared to flowing the same well set at less aggressive rates. To hit max IPs, we must overbuild the facility, incur additional damage to our equipment, and deal with excessive solids production—all at elevated cost.

Most importantly, maxing out initial production can damage our frac properties and reservoir, leading to steeper declines and reduced EURs. Therefore, when establishing flowback strategies to deliver max IPs we must incorporate the additional cost and potential long-term deliverability damage into our decision-making process.

After months of work to get each well online, gratification is realized with IP data that helps everyone breathe a sigh of relief that we have an excellent DSU, an economic area for future development, and, ultimately, a viable long-term project. Strong IPs also give us a data point to showcase in management meetings and investor presentations. All are essential elements and critical dynamics in achieving our short-term objectives and long-term 100X goal.

However, if we want massive IPs, we must overbuild the facility. Facility design and construction teams do not want to be the bottleneck to achieving max IPs, and it is safer to over-design the facility than to under-design it. Plus, we never truly know what max production rate will be for oil, gas, or water. Therefore, it is always safer to go bigger, just in case. This comes at a cost that often gets questioned after several months of production when the facility is clearly too big relative to the current production due to the steep decline.

Additionally, when flowing back wells at max rates, we can cause significant damage to our facility equipment. This can also result in downtime if we must shut in the wells to fix something. However, the additional cost to overbuild facilities or fix surface equipment damage is relatively minor compared to damaging frac properties and reservoir deliverability by flowing back too aggressively.

As an industry, we have been psychologically conditioned to deliver big IPs without questioning or thinking about what we are doing. Oil and gas minerals have been in the ground for millions of years. A lot can happen when we move something that has not moved for a long time. Ripping a well open to max out the rate will cause damage that may not be apparent during initial production but reveals itself in the production decline over six months to a year.

Could these massive production declines be mitigated while also reducing costs? I believe so—with a more sophisticated approach. To address these issues and reduce potential damage to our assets, consider employing an engineered flowback. There are many cost, safety, and production benefits to a data-driven slow-back approach, managing flowback by making slow choke adjustments and analyzing production dynamics, with an elevated level of attention to the production decline. The goal is to enhance value by mitigating the decline to maximize the economics and value of our assets.

A few actions to consider when managing flowback using an engineered approach include:

- Identify historical situations in which flowback operations may have damaged long-term production performance.

- Establish a maximum allowable total fluid rate range based on lateral length, reservoir characteristics, solids production, and historical performance.

- Set a max allowable $\Delta P$ "Managed Pressure" flowback (e.g., 0 psi/day to 50 psi/day) based on lateral length, reservoir characteristics, and historical performance data.

- Engineer a solids production-based approach to minimize sand and solids flowback and damage.

- Consider a combination of all methods.

When employing an engineered flowback, it is not uncommon to restrict choke changes to no more than $1/64^{th}$ to $2/64^{th}$ every two to five days or more. This will depend on our pressure, rate, analysis of historical flowbacks, and choke changes relative to long-term production performance.

With an engineered flowback, we have the potential to provide the highly desirable "Cash Flow Trifecta," reducing upfront cost, enhancing long-term production, and improving operational safety.

Therefore, we do not target max IPs. We take actions to deliver flat production for an extended period to reduce costs, improve safety, prevent reservoir damage, and improve long-term production—increasing EURs, economics, and value.

# Cash Flow Benefits of Engineered Flowback

| Item | Cash Flow Benefits |
|---|---|
| **Preserve Conductivity** | Prevent frac sand and near wellbore damage from aggressive drawdown |
| **Fines Migration** | Prevent fines migration damage contributing to steeper declines |
| **Proppant Crushing** | Reduce crushing and sand flowback with less aggressive drawdown |
| **Fracture Network Pinch-Offs** | Reduce occurrence of frac networks ceasing to contribute to production |
| **Reduce Flowback Complexity** | Utilize less equipment, eliminate flowback tanks, release equipment and flowback hands sooner |
| **Downsize Tanks and Containment** | Install smaller / fewer tanks with a smaller secondary containment |
| **Downsize Vessels, Flares, Lines** | Install smaller separators, treaters, flares, and combustors |
| **Sand Production** | Reduce sand dune buildups choking production, and lateral cleanouts |
| **Damaged Equipment** | Minimize choke, valves, vessels, and flare damage |
| **Lift Equipment** | Change lift method or install lower cost lift equipment and size |
| **Tubing Size** | Utilize smaller diameter tubing |
| **Spills and Emissions** | Reduce spill risk and lower facility gas emissions from stack valves, thief hatches, and pop-offs |

---
## ACTION
### Timed Flowback Release
---

It is easy to spend significant money during flowback and receive little value in return. If streamlined, flowback equipment usually has minimal cost. What carries high costs are the day rate accruals for flowback personnel and associated combo trailers.

To reduce costs, consider establishing a fixed period for flowback personnel to be on location. For example, structure a plan to release flowback personnel within two to six weeks.

There have been multiple situations, usually during busy periods, in which the team "forgot" about flowback personnel being on the clock and left them out there for three to six months on a single multi-well pad.

Every flowback's dream is a "nice long flowback," and they often do everything possible to make it a reality. We need to make sure it is not a reality by having a written flowback procedure, daily task checklist, and plan to release them in a set number of days.

Just like millouts, flowback operations do not get the attention they deserve. Flowback hands are often the least experienced personnel on location, operating with almost zero oversight. Many do not have a production background and do not understand the risk involved in flowing back multiple high-pressure wells. Operators may be better off in certain situations without any flowback personnel on location.

Recently, several significant accidents during flowback

have occurred because flowback personnel unintentionally caused them due to mistakes—operators would have been better off without flowback on location in those cases. Some operators have eliminated flowback hands from their operations. I am not suggesting it, but companies are doing it.

Consider developing a flowback procedure detailing all aspects of the flowback and what we want to accomplish each day. Additionally, we must involve the night hands and give them tasks, or they will fall asleep. Night flowback hands often carry out personal business or another job during the day and then sleep through their shift on location. I know this because we have cameras on our locations and catch them sleeping in their trucks. They need to understand that they must add value, or they will be released.

Just recently, a large offset operator was flowing back a single well and had two flowback hands on days and two on nights. That's four hands on location. The night hands both fell asleep, and the oil tanks overflowed into the secondary containment, filling it up to the point the automated dike wall float shut the well in.

We need to ensure our flowback hands do not do anything dangerous. Most do not have a well control certificate or training on flowing wells. It pays to take the time to find high-quality hands with experience that we can keep on the team.

Finally, we must have our pumpers and foremen on location from day one of flowback. This way we are prepared to release flowback more efficiently. With a flowback procedure, task checklist, and our pumpers engaged, we can transition flowback off location in a timely manner.

## Free Cash Flow Trifecta

A trifecta is a bet that wins if we select the first three finishers in a race. The term has broadened in recent years to include a situation where three things are achieved.

In terms of maximizing shareholder value in the oil and gas industry, a Free Cash Flow Trifecta is an action that reduces cost, improves production, and reduces risk. A cash flow action that accomplishes all three can deliver massive value, helping us achieve our long-term goal of delivering 100X returns.

Finding cash flow trifectas takes work. We can generate a trifecta with an intelligent acquisition, during drilling—by increasing percentage in zone, during production—through flow optimization, and definitely during completion—by embracing technical advancements.

Frac design optimization can provide one of the best ways to deliver free cash flow trifectas if we can identify and unlock hidden value. For example, shale recovery factors are low (10% to 20%), so the potential to improve EURs and decline curves is very much attainable.[25] Achieving this while reducing cost requires a higher level of sophistication.

> **Anyone with money can drill and frac,**
> **but not anyone can reduce costs,**
> **while increasing production,**
> **and reducing risk.**

Spending money to increase production is easy. Doing it at a lower cost and risk than our competitors is not. Anyone can put more sand in the ground or drill a longer lateral, but can we do it while reducing cost and risk? That is what separates the newscaster executive from the next-level innovator.

If we have excellent rock, going bigger on everything is a proven strategy to create shareholder value—the rock will respond favorably. Core areas with thick targets make technical teams look like geniuses.

However, as we exhaust core inventory, things become more complicated. These non-core areas have challenging geology, unfavorable depths, difficult drilling conditions, lower pressures, and other undesirable characteristics, driving up cost, reducing production performance, and increasing risk.

As a result, we must deliver Free Cash Flow Trifectas to transform non-core acreage into core acreage proxies from an economic perspective. Leveraging new technologies and advanced thinking can help make it happen. But we must be open to all possible solutions and believe the answer to our specific problems exists, or we will create it.

When trialing new tech, one of the most common mistakes operators make is to step back and allow the vendor to take the lead. When a vendor presents a new technology that we have never tried, it's hard to provide oversight because we do not know what to do to ensure it works. Therefore, we rely on the vendor, and the new tech often fails to deliver. Then we conclude it doesn't work. To prevent this chain of events, take the opposite approach and get so deeply involved in the new tech that we know it better than our vendors. With that in mind, in the spirit of achieving 100X returns, let's review "100 More Completion Cost Reduction Actions" in no particular order.

ENGINEERING

IS

MAKING THE UNECONOMIC

ECONOMIC

# 100 More Completion Cost Reduction Actions

1. Optimize our development strategy to reduce total number of wells per DSU required to maximize NPV/acre/target.

2. Achieve corporate oil and gas production targets with fewer wells drilled versus plan, using enhanced frac designs and development strategies to increase recovery factors (RF), reducing total development costs per unit of production.

3. Institute a frac pause. If cost challenges are significant or the wells are not yielding optimal production performance, stop fracturing to give the team time to reassess and reorganize. With a fresh plan of action, restart fracturing operations.

4. Identify faults from geologic and drilling data. In areas where faults act as pipelines to water or increase the risk for fault-related shear damage to casing, avoid placing clusters 100 FT to 300 FT on either side of the fault. In certain areas, this strategy reduces casing failure risk, improves oil and gas production, reduces water production, and reduces costs.

5. Fully automate well-to-well swapovers during frac.[26]

6. Build a microphysics model incorporating geology, trajectory, frac design, and artificial lift methods to reduce costs while maximizing production performance.

7. Increase frac horsepower to increase rate, reduce pump time or expand stage length, reducing total number of stages.

8. Time the negotiation and signing of long-term service contracts during the slow periods of the year to get better pricing and make better deals.

9. Establish a water midstream entity.

10. Develop a custom-sized casing string to reduce pipe friction, increase frac rate, and reduce frac cost.

11. Stop using completion packers to reduce complexity and eliminate costly issues when unsetting them.

12. Purchase (direct own) lower master valves (LMV) for frac stacks to exercise closer inspection (QA/QC) and diligence of this critical well control component. Owning the LMV eliminates rental charges over extended periods if flowing back wells up the casing, post millout.

13. Eliminate frac HHP idling with auto shutdown systems.[27]

14. Run factory-assembled perforating systems delivered to location ready to run. This reduces risk, misfires, wiring, assembly time, and BHA build errors onsite.[28]

15. Reuse flowback sand.

16. Eliminate freshwater use with produced water recycling.

17. Reduce winter operations. Fracturing during the coldest periods of the year has increased costs. Additionally, data indicates a higher incident risk during winter months.

18. Establish a detailed "Reject-If Checklist" for specific field tickets to systematically fix incorrect invoices on location.

19. Eliminate short trips during millouts.

20. Low profile, quick connect, rotatable mandrel hangers, lock ring wellhead systems to reduce risk and costs.[29]

21. Pump constant concentration proppant designs instead of gradual sand step-ups or sand ramp designs.[30]

22. Transition from conventional wireline cables to greaseless wireline cables to reduce NPT, reduce friction, and increase running speed.[31]

23. Reduce costs and increase effective perforated lateral length by replacing conventional casing shoe tracks with consolidated shoe technology.[32]

24. Utilize knuckle boom crane trucks to eliminate the need to have a separate crane to handle items.

25. Hire millout tool reps as wellsite leaders during milling operations, removing tool rep charges from millout tickets.

26. Use portable conveyor sand systems to transport sand as close as possible to location before using trucks.[33]

27. Use propellant-enhanced perforation technology to create fractures in each perf cluster before pumping to address NWB issues, high frac treatment pressures, cluster efficiency, reduce frac pump times, and help evenly distribute proppant across all clusters.[34]

28. Digitize the frac stack.[35]

29. Task wellsite leaders with screenout flowback handling operations, removing flowback personnel from location until post-frac plug millout.

30. Establish a reciprocal downtime grace period.

31. Based on real-time completion data, vary stage, cluster designs, and spacing in real-time to improve production and reduce cost.

32. Identify all opportunities to direct source materials.

33. Require wellsite consultants and vendors to provide financial relief if they violate company "zero-tolerance rules." Provide the choice to work off the violation, covering the cost, or be released.

34. Terminate high-risk, low-reward frac stages. For example, if we are out of zone with high erratic pressure close to max pressure, we end frac stage early to avoid further pain.

35. Install vibration sensors on frac equipment and frac stacks to reduce risk, NPT, identify HHP issues, and achieve max fracturing rates.

36. Run plug erosion jobs to remove preset composite frac plugs quickly, avoiding downtime and intervention risk.

37. Use downhole cameras to determine the optimal amount of near-wellbore rock to be eroded to maximize production and prevent pitch offs.

38. Minimize FR loading based on water and lab testing.

39. Use unmanned dry FR delivery systems.[36]

40. Reduce or eliminate fracturing chemicals by reusing early time flowback waters. Filter out solids, and frac without adding new chemicals by testing frac flowback waters to confirm original frac chemical properties are still present in sufficient quantities to reuse without adding more.

41. Fully automate frac treatment operations.[37]

42. Use tracers to confirm entire lateral is contributing to production when eliminating post-frac millout and cleanout operation using dissolvable frac plugs.[38]

43. Run short-interval survey tools from TD to surface to get a detailed look at DLS relative to MWD data to identify hidden high doglegs, risks, and opportunities to reduce occurrence and impact costs.

44. Identify link between hidden doglegs and plug damage during wireline pump down that leads to casing erosion during frac, or poor frac treatment placement causing below type-curve performance.

45. Address nonlinear casing loading (aka point loading) that could damage casing before, during, or after fracturing, contributing to tight spots impacting frac plug placement, effective cluster treatments, and post frac millouts.

46. Transfer prime-up fluid back to active frac tanks to save on disposal costs and water usage.

47. Establish strategic partnerships with key vendors for reduced costs.

48. Run single use (disposable) frac plug setting tools to eliminate assembly and parts that can be left in hole with conventional frac plug setting tools. Eliminates tool redress preventing human error on location.[39]

49. Run micro-proppants, 200-mesh or smaller, to improve proppant distribution, allowing for increased clusters per stage and reduced fracturing pressures.[40]

50. Manage frac fleet service maintenance intervals to align with mobilization periods between pad sites to minimize maintenance or pump issues during fracturing that increase operator costs.

51. Reduce days between rig release (RR) and frac start date. More static time before fracturing operations has been correlated with costly completion related issues.

52. Rent advanced modern forklifts and manlifts. Reject older low-tech forklifts and manlifts that slow down operations.

53. Orient perforations in the direction of optimal cement bond. Perforating where there is no cement or poor-quality cement can result in damage and suboptimal production.

54. Perforate using drilling data (e.g., MSE, depletion, in-out zone, gas shows, heavies C2+) and avoid completing low-value intervals, reducing frac costs.

55. Use automated HHP swapping technology to hot-swap frac pumps to help achieve 24-hour pump times.[41]

56. Engineer nitrogen injection rates to reduce waste and over-usage during post-frac plug millouts.

57. Take lower cost leftover sands (e.g. 40/70, 30/50 mesh etc.) from the mine and start frac with the larger mesh sand first. Then pump the majority of the job with 100 Mesh. Switch back to larger mesh at end of the job.

58. Standardize and synchronize frac and wireline telemetry data to find inefficiencies.

59. Optimize overflush strategy to minimize sand flowback costs and risks during wireline, millout, and flowback. Incorporate overflush strategy to simultaneously enhance production performance.

60. Use microparticles to reduce pressure and pump time.[42]

61. On multi-well pads for parallel wells, administer a smaller frac treatment on the middle wells that receive frac energy and get stimulated from the offset treatments. Thereby reducing total DSU fracture treatment costs.

62. Install permanent overhead lights to reduce or eliminate light plant rental units during drilling, frac, millout, flowback, and facility operations.

63. Use industrial wastewater from other manufacturing industries, tested and approved for frac water feedstock.

64. Use high-viscosity friction reducer (HVFR) systems to reduce water volumes and pump times.[43]

65. Automate wireline pump down to reduce risk, time, cost.[44]

66. Build an app to track frac stack lower master valve history.

67. Analyze past high-cost and undesirable completion events (e.g., stuck wireline, stuck pipe during millout, casing failure, failed wellhead, pressure events). Set automated alarms to proactively detect and alert when a similar situation may unfold to help prevent a recurrence.

68. Work with offset operators to share flowback water.

69. Unbundle frac chemical products, services, and testing. Perform independent testing to reduce bias. Source direct. Establish full control of frac chemical program.

70. For each vendor service, establish a set of required support documents and line-item detail for field ticket submission.

71. On frac, establish ability to safely increase rate as pressure drops. A small increase in rate will help reduce pump-time.

72. Systematically track competitor problems and hold weekly reviews with our team. Critical problems with casing, wellheads, frac stacks, millout BHAs, coil, and stick pipe happen in waves, and we do not want to get wiped out.

73. Request frac equipment provider use single button push stop-start systems to reduce frac fuel costs.[45]

74. Administer fuel additives to reduce fuel consumption.[46]

75. Systemize an action plan for frac hit preparation, real-time monitoring, and countermeasures on offset legacy vertical and horizontal wells to prevent a costly offset incident from unfolding during each frac stage.

76. For rental equipment pricing structure, request vendors not to start charging day rates until frac begins and pause charges during mobilization periods between wells.

77. Transition from an operational approach to an engineered approach for processes that demonstrate cost volatility.

78. Consider fully autonomous flowbacks and sand handling, removing personnel and costs.[47]

79. Work with vendors to develop operator-specific frac chemical blends to reduce cost and increase production.

80. Flow back directly to production facility by using pipeline gas to power facility vessels. This reduces flowback costs, tank costs, water hauling costs, gas venting, and flaring.

81. Identify lost value between disciplines. Since completion ops are between drilling and production, it is often the best place to identify inefficiencies that fall through the cracks.

82. Rather than pay for rental equipment, calculate the cost to purchase rental equipment that can be operated in-house or by contractors. A few examples include purchasing water transfer equipment, light plants, generators, flowback equipment, and communication systems.

83. Streamline contractor day rate structure from day rate plus mileage plus per diem, to a single flat rate per day.

84. Identify the optimum friction reducer type and loading to reduce fracturing pressure and increase the percentage of uniform sand placement across all clusters.

85. Purchase and install permanent tornado shelters. Owning is often cheaper than renting because we must install and remove rented units; then transport them to the next pad.

86. For large-scale fracturing operations, consider paying for certain equipment to stay on standby in case of a problem. For example, if we have high screenout risk, waiting for a coil unit to become available is very expensive. In some areas of the United States, it could take a week to get a coil unit. We are money ahead if we pay for a dedicated coil unit to stay on standby. This way if we have a screenout and need coil, we will incur minimal frac NPT charges.

87. Contract well control specialists and have them audit our completion operation on a periodic basis. It is not uncommon to overlook well control aspects of our frac, millout, and flowback that can be adjusted and strengthened to reduce risk and cost.

88. Systematically test casing, stack, and wellhead metallurgy.

89. In certain situations, designing the same spacing between frac stages as we do between perf clusters is not optimal. If stages communicate into previous stages due to poor cement, geology, or other propagation issues, designing stages further apart from each other can improve production, with more effective stimulation, and reduce costs because we will have fewer frac stages to execute. For example, if we space clusters 10 FT apart, we could space stages 40 FT apart. For a 10,000 FT lateral with 200 FT stages, we go from 50 stages down to 42 to 44 stages and potentially get better production performance, if we suffer from the issues mentioned above.

90. Install surface casing heads with dual wing valves rigged up to iron systems that facilitate rapid pressure and rate release and suppression if casing ruptures during frac.

91. Utilize excess containment liner materials discarded from production facility installation to supplement containment needs on completion operations.

92. Systemize fracturing the heel segment of the well to reduce risk, cost, and improve production. Depending on geology, doglegs, cement quality, offset trajectories, and how high up in the curve we perforate, fracturing the curve can do more harm than good. Damage risk is elevated for the frac well and the offsets. Additionally, if we don't perf high in the curve, depending on our artificial lift systems, we can create more value placing AFL deeper. Perforating high in the curve can prevent that from being an option. We can eliminate at least one frac stage if we cancel the curve frac.

**93.** Automate Stop Work into Virtual Company Man systems. For example, if backside pressure is detected beyond a specific pressure for any wells on the active multi-well pad, remote pad, or on offset legacy vertical wells, within a certain AOI, set automated alerts to Stop Work immediately, demanding personnel shutdown the frac. Automating Stop Work Authority (SWA) for a variety of scenarios like this helps enforce our policies more effectively and in a timely manner to prevent an incident, particularly in situations in which every second counts.

**94.** On parallel multi-well completions, identify the optimal pattern to generate a pressure wave maximizing stimulated reservoir volume (SRV), allowing us to reduce frac size and/or frac rate achieving the same results at a lower cost.

**95.** Phase out vendors who are unable to deliver actual invoices that match the expected bid price.

**96.** Depending on our "Downhole Imaging Intelligence," perforate a certain number of initial clusters based on the max rate we can get with our wireline pump-down equipment. After we perforate the initial clusters, kick on the wireline pump down pumps and breakdown the initial clusters. Then continue perforating the remaining clusters. With this method, we can increase the clusters effectively treated, reducing stages, and reducing costs while obtaining improved production performance.

**97.** Incorporate Virtual Company Man system into voice activated head gear and smart glasses, providing team with instant intelligence and advisory interactions.

98. Reword fracturing contracts to start pump-time clock when a specific fracturing rate is reached, instead of when the frac stack operating valve is opened. Additionally, state that full contract horsepower must be available and effective before starting the pump-time clock. Small contractual changes like these can help reduce costs by preventing frac crews from opening the well and hitting us with charges before they are 100% ready to go. Plus, if we are paying for a certain amount of HHP per minute, unless we have access to that HHP, we should not pay for it and not incur charges until it is ready to go.

99. On millouts with coil or stick pipe, after all plugs are milled, if the situation is adversarial, consider releasing the BHA at the toe to obtain a slick string for easier TOOH. Additionally, consider BHA systems which allow for release of the motor and then trip to planned production tubing setting depth and land tubing. Then install the wellhead, pressure test the tree, and open or remove flapper valves to initiate production.[48]

100. Based on analysis from imaging, fiber optics, and well performance, reduce perforations per cluster, including down to single-perf oriented clusters. Spread out perfs. Assemble the maximum amount of HHP to generate extremely high frac pump rates. Initiate frac using micro proppants in attempt to treat all clusters before losing limited entry due to erosion. Use diagnostics to determine perf efficiency. Then, administer diversion and further increase the frac rate accordingly.

# 9

## PRODUCTION ACTIONS

—

Production cost reduction is a game of pennies and a game of billions. When examining our economics on a cost-per-unit-of-production basis, one of our goals is to collect pennies that add up to billions of dollars over time.

For example, if we reduce our operating expenses by one penny per mcf ($0.01/mcf) and we produce 1.0 bcfd, that adds up to $3.65 million per year in new cash flow we can place in our shale value creation machine. Now, that's a penny worth picking up! The old saying, "Find a penny, pick it up, all day long, you'll have good luck,"[1] reminds us to be on the lookout for tiny treasures that bring good fortune.

However, to get all those pennies, we do not want to elevate our risk exposure into the upper right-hand corner of the risk matrix, as discussed in Chapter One. In financial circles, there is a legendary investment strategy that illustrates this type of risk exposure dynamic. Long-Term Capital Management (LTCM) was a hedge fund founded in 1994 whose principals were well-known and respected, considered to be financial geniuses by Wall Street standards.[2]

Their core strategy was convergence trading, which involved buying undervalued securities while shorting overvalued ones, betting that they would converge to fair market value. Initially LTCM was very successful due to their ideas, models, innovation, and most importantly, an underlying stable global market environment.[3]

However, the 1997 Asian financial crisis and the 1998 Russian financial crisis uncovered weaknesses in their operating model. High leverage, minimal risk management, no safety systems, and an over-reliance on historic data resulted in the spectacular implosion of LTCM.[4]

Its collapse almost took down the entire global financial system before the U.S. government arranged a bailout by a group of Wall Street banks.[5] LTCM's strategy has been compared to "picking up pennies in front of a steamroller"[6] since LTCM was taking huge amounts of risk to collect relatively small rewards. With this type of strategy and risk exposure, if anything goes wrong, you get crushed.

Unlike LTCM, our business model is to operate with strong risk management and safety systems, especially for production cost reduction efforts, which have a different risk profile than drilling or completions.

During production, most of the time, we do not have a constant human presence physically on location. This is a very different type of risk exposure compared to drilling or completions, which have an intense workforce presence on location, with a significant focus on a relatively small group of wells. Our production cost reduction approach must consider this dynamic during our FCF efforts not only to deliver cost reduction per unit of production, but also risk reduction at the enterprise level to minimize our long-term exposure.

There are many strategies to reduce production costs. The approach presented in this book represents the best interests of the long-term owner operator, for the company and people that will live with and produce the assets for decades.

Tactics for the short-term owner, operations outsourcer, or flipper company are not presented here. The actions we discuss may not resonate with those entities or individuals interested in strategies that support shorter-term buy-and-sell business models or a hands-off, outsourced operating approach.

## The Mixed Martial Arts of Cost Reduction

Production cost reduction requires hand-to-hand combat. Keeping wells online, especially in challenging weather conditions, while reducing costs across a wide variety of lift systems, vintages, and facility setups requires a hands-on understanding of each well's uniqueness.

This is a full-contact sport requiring us to get up close and personal with the wells. Like mixed martial arts (MMA), we must use all combat techniques available, including grappling, striking, ground fighting, boxing, jiu-jitsu, kickboxing, karate, judo, and wrestling, from a cost reduction perspective.

However, we cannot place the entire burden of our cost reduction goals on our field team. Leveraging a variety of actions that will be discussed, we must put our team in the best possible position to succeed, without having to constantly tussle with the wells.

Nevertheless, sometimes we have no choice but to get into a street fight and brawl to get the costs down and keep the wells online—addressing problems and challenges on a well-by-well, muscle-to-iron, gritty, relentless, teeth-clenching manner to achieve quantifiable results.

---

**Wells are like people; each one is unique.**

---

There are so many similarities between the human body and hydrocarbon wells that an organization was formed to bring innovators together from the medical industry and the oil industry called Pumps & Pipes. The organization's goal is to provide a platform to share knowledge between medicine, energy, and aerospace, since all three industries deal with pumps and pipes, and have similar challenges.[7]

Just like solving problems with the human body, it is not recommended to perform the same expensive, invasive, and risky surgery or workover on every patient regardless of their condition.

We would also not give every patient the same medication or chemicals. That could cause more problems than it solves. Therefore, we want to approach OPEX reduction with the same way of thinking and mindset. Every patient is different, and every well is different. This individualized cost reduction approach takes more thought, discussion, and work but pays off handsomely in the long run.

—————————— ACTION ——————————

## Eliminate Wellsite Facilities

From an operating cost and risk exposure perspective, the most aggressive and effective action we can take is to completely eliminate the wellsite facility.

With modern technology, including advanced multiphase flowmeters (MPFM), we could measure three-phase production from each well and then pipe our products to a central gathering facility or directly to a local refinery via a multiphase pipeline.

From a regulatory perspective, this strategy is permitted in certain parts of the world and should be employed to significantly reduce CAPEX, OPEX, and risk.

With modern MPFMs, combined with a robust calibration and maintenance program, including adjustments to the meters as production declines, we should be able to leverage this technology to create significant value for our company. This strategy eliminates all the problems and risks associated with operating remote production facilities. Additionally, from an environmental perspective, we drastically remove spill risk and emissions from the storage tanks, often the primary source of greenhouse gases (GHG) at the wellsite.

Unfortunately, depending on flow rates and regulations, MPFMs may not be able to always be used as fiscal meters for custody transfer, and are more often used as allocation meters.[8]

However, this may change in the future, or we may consider engaging with regulators in certain areas or cases to request an exception to the rules as we become more comfortable with the technology.

Furthermore, combining MPFM with AI advancements incorporating virtual and physical check meters helps make this technology increasingly accurate and accepted.

---

**Simplifying and streamlining our wellsite facilities, all the way up to the point of eliminating them, eliminates costs and risks.**

---

Even though we may not be able to eliminate all field facilities and pipe our products to a local refinery, we can optimize our field processing setup by removing vessels and tanks. For large-scale developments, in most cases we should remove all the vessels and tanks from the wellsite and create a central separation facility and central tank battery.

If we want a secondary check, we could install one vessel at the wellsite and measure oil, water, and gas independently and then flow to our central processing station.

If oil or water pipeline access is limited or we want to keep our facility at the wellsite, we could reduce or remove vessels and tanks. Instead of each well having a separator, treater, or other vessels, we could employ commingled configurations. All vessels require maintenance, especially heater treaters. The more vessels we remove, the lower our OPEX will be.

In addition to increased costs, every vessel adds risks in a variety of ways. Most wellsite facility production-related incidents involve problems related to the processing vessels or tanks. Streamlining our facilities by removing vessels removes risk in addition to reducing CAPEX and OPEX.

If an oil pipeline is within reach, we can consider eliminating most or all the oil tanks with a "tankless" approach. If we go tankless, we eliminate oil trucking, oil tanks, oil theft, people on top of the oil tanks, opening the hatches, vapor risk, emissions, and a variety of piping, connections, and valves. With a tankless system, we use a lease automatic custody transfer (LACT) unit to sell oil to the pipeline.[9]

The setup of a tankless facility can be accomplished with a variety of designs that fit the assets production profile, oil properties, and oil pipeline requirements.

For example, we could separate the water and gas. Then, flow oil to a comingled vessel directly to the LACT. There are new systems that eliminate all vessels and handle everything with a single treater to separate oil, water, and gas, yielding stabilized oil for sale in conjunction with a tankless setup.[10]

If we cannot eliminate the entire wellsite facility using modern technology, the next best thing we can do is centralize as much as possible. Then, streamline our design removing as many vessels and tanks as our determination allows. Less equipment means less to repair, less to maintain, and less risk—all of which help reduce OPEX.

During all operations, including drilling, completion, and production, for every item, step, and process, ask the question:

*Can we completely, 100% eliminate this?*

Eliminating unnecessary things is better than optimizing them.

———————————— ACTION ————————————

# Build Facilities that
# Enable Low Operating Costs

Years ago, I worked alongside a highly respected and well-known facility engineer, with decades of experience, building wellsite production facilities for public and private oil and natural gas companies across the United States.

He was semi-retired and visiting my office on a day we were reviewing facility operation tactics with our field team.

We had pictures up on a screen during the meeting and were walking the team through a process we thought was best to safely relight heater treaters, flush vessels, and restart gas lift compressors. I turned to my former colleague and facilities engineering expert and asked him what he thought. To his credit, he was honest and said,

*"I design facilities; I don't operate them."*

At that moment, a revelation hit me like a ton of bricks. This is the reason why many wellsite facilities are built the way they are—in the most user-unfriendly manner possible.

Not because it is intentional but because they are designed and built by people who do not operate them, do not know how to operate them, have never tried to operate them, or have no interest in it.

This dynamic is common across all industries and contributes to higher costs, inefficiencies, elevated risks, and unnecessary problems. It is unquestioned in our society for engineers to be responsible for designing, building, and optimizing things that they do not know how to operate themselves.

Therefore, they may not know how to best design things in a user-friendly manner, from the perspective of the daily operator, especially from a troubleshooting, repair, and maintenance standpoint. This is often the case for facilities.

> **It is more difficult to optimize technology that you do not know how to operate.**

Rigs, frac equipment, production vessels, and wellsite facilities are a few examples of items designed by engineers and others that may not know how to operate them. Similar to aerospace engineers who build airplanes, but who do not know how to fly them, or software engineers who build programs to control machines that they have never operated themselves.

It is not necessary to be a pilot to design a plane, but it helps—especially if we want to deliver revolutionary cost reductions that no one else has been able to achieve.

Wellsite facility designs are often the root cause of higher production costs, elevated risks, and daily problems. To address this issue, we must work with our field teams and vendors to design and build facilities that are easy to operate, maintain, troubleshoot, and repair without requiring costly vendors when there is an issue. Elevating our level of direct involvement and hands-on knowledge is the secret.

# Low OPEX Facility Design Actions

- Form a SIG with the field team to improve current facility designs from the daily user's perspective to lower OPEX.

- Elevate the level of direct involvement and hands-on knowledge by extensively increasing the amount of time on the front lines, observing firsthand how costs are incurred.

- Simplify the design. Everything should make intuitive sense, look organized, and be easy to operate.

- Customize vessels. When ordering vessels, request features for low-cost ops. Fire tubes, coatings, ports, and accessways, must be optimized to operate at a low cost for a long time.

- Label every component with paint markers to support ease of use. The production team always thank me for doing this because it makes it easy to mentally map things out, especially when troubleshooting.

- Install components at an easy to maintain, fix, and replace position. For example, if a treater valve is installed a little too high off the ground, or in a tight position, it will be difficult for a pumper to fix. Then we will need roustabouts. A job that should only cost $30 can end up costing $1,000.

- Build it modular or hot-swappable. Install isolation valves and bypasses on vessels and key components. If there is an issue, we can easily remove a component and stay online.

- Incorporate features to eliminate LOE. For example, downcomers to eliminate chemicals, transfer pumps to recirc bottoms, gun barrels eliminating processing costs, electric line power for lift systems eliminating fuel costs.

- Build facilities that can stay online when drilling new wells.

---
## ACTION ---

# Midstream Control

---

It is common for upstream-focused companies to only receive partial value for their midstream assets by financial markets. Therefore, leadership consensus is often to monetize midstream assets using a variety of divestment strategies. Through this process, upstream entities lose a certain level of control over midstream operations, which can unintentionally impact upstream costs, sometimes significantly.

When midstream assets are sold, it is common for the best people to leave, priorities to change, and focus to shift away from production, especially when midstream companies become further consolidated. Within a few years, assets once controlled by people we knew and processes we trusted are now managed entirely differently, sometimes adversarially. What was once an asset can become a liability, driving up production costs and risks.

From a cost reduction perspective, it is better and easier to reduce OPEX if we own the midstream assets on which we depend. If this is not possible or we have to sell, it is best to retain a controlling percentage or structure the deal so that we maintain as much operational field control as possible.

If we do not control the gathering systems, we must increase engagement with midstream partners, sometimes to an extreme level, to prevent costly situations from damaging us including shut-ins, high line pressures, meter errors, accidents, and many other situations, all of which drive up OPEX.

# Midstream Actions to Reduce OPEX

- Maximize communication to maximize uptime. Have each superintendent, foreman, and pumper maintain regular daily contact with their respective midstream counterparts.
  - o Integrate planning between midstream and upstream
- Debottleneck midstream systems
  - o Compose a detailed map with line pressures, pipe pressure ratings, valves, meters, compressor stations, compressor details, pig launchers, and chem injection.
  - o Identify areas to install bypasses. A simple bypass around a component can be the difference between staying online and being down for months due to mechanical failures.
  - o Add multiple interconnects between gathering systems. Strategically placed interconnects can reduce sales line pressures—reducing costs while increasing production.
- Identify and fix hidden chokes
  - o Find areas that need pigging, closed valves, forgotten partially closed valves, stuck valves, debris, compressor issues, and other hidden chokes increasing pressure.
  - o Replace valves, regulators, meters, bypasses, and other midstream components with a history of problems that could be contributing to higher line pressures.
- Resolve contractual midstream dynamics impacting OPEX.
  - o Line pressures higher than contract specified pressures
  - o Uncoordinated or surprise shut-ins
  - o Gas spec requirements. For example, if contract is 4 ppm $H_2S$ at the wellhead, but we can renegotiate to 20 ppm or a blended 4 ppm downstream, we can reduce OPEX

---
# ACTION
## Facility Modification
---

During efforts to get the most out of our cost reduction campaign, it is prudent for leadership, including directors and engineers, to visit field facilities regularly. The goal is to get boots on the ground looking for cost reduction opportunities that are hard to see from the office.

> **New eyes bring new ideas
> to maximize cash flow.**

Analyzing our facilities with fresh eyes can provide a new perspective. With complex facilities, it is easy to overlook good cost reduction opportunities that could be scalable across our assets.

Before visiting our facilities, let's pull all invoices from the past year and perform a full statistical analysis. This way we are fully informed regarding where our money is going.

Next, have our vendors meet us on location to discuss what we can do to eliminate costs going forward. Meeting on location is much better than meeting in the office because we can look at all the components, look at our invoices, and discuss cost reduction opportunities while looking at historic work that we have performed.

Facility operations provide the rare opportunity in our industry to see everything on the surface. Additionally, we can confirm data quality on the invoices, work quality, and a variety of other statistical metrics to help reduce costs and increase uptime.

For example, high surface pressures generally lead to increased costs and reduced flow rates. This combination is a double whammy to our cost per unit, presenting an opportunity if we can lower pressures. As previously discussed, actions to reduce midstream pressures help but there are many actions on the wellsite facility and artificial lift systems we can take.

The first step is to confirm we are reading pressure correctly. Unfortunately, it is common for our digital and analog gauges to be incorrect. Bad data quality from bad gauges occurs on all assets, including brand new wells. Most people are surprised to find gauge issues on new wells, but in my experience, it is common.

Pressure gauges are highly sensitive and easily damaged, but not to the point that they stop working, and that is the problem. They convey wrong information, but they keep working. Gauges on new wells often become damaged during installation. Roustabouts or technicians install the gauges and then something needs to be changed and we must hammer on the lines, inflicting shock damage to the highly sensitive analog and digital gauges.

Flowback is the other issue. When a well is new and powerful, it is easy to exceed the operating range of a gauge, incur a water hammer, micro-spike, or pressure surge on different parts of the facility, causing gauge damage so that they do not read correctly after that event. It only takes a few seconds for the damage to occur.

During troubleshooting and optimization efforts, evaluating a facility using data that is thought to be correct, but is actually incorrect, results in flawed analysis, confusion, and overlooked cost reduction opportunities, especially if we are overly relying on digital data via SCADA systems. Visiting our wellsite locations in person helps address these issues and ensure data quality.

One might think vendors are unwilling to help us modify our facility to reduce their cash flow, but if we have done our job, as discussed in Chapter One: People First, regarding working with true partners, our vendors will help.

A lot of facility-related work is performed by private contractors with owners who are highly involved in day-to-day operations. Therefore, we want to have vendor ownership, vendor ops leadership, or someone different from who has been doing the work, meet us on location to discuss OPEX reduction actions.

It is also good to have vendor competitors come to location and give us a second opinion. If someone is taking advantage of us, this is a good time to identify it.

 The more time we spend discussing cost reduction with vendors while standing in front of our facilities, and thinking about new ways to reduce facility costs, the more opportunities we are going to find.

If we have thousands of wells, it is challenging to visit all of them in a timely manner. Therefore, once we have a game plan on a group of wells, we can craft written procedures and have the team replicate our actions across the field, high-grading the wells based on our plan of attack.

# Wellsite Facility Modification Actions

1. Pull all invoices. Perform a detailed statistical analysis.

2. Arrange meetings with current vendors at field facilities.

3. Meet with new vendors on location for new ideas.

4. Confirm all gauges (digital & analog) are reading correctly.

5. Calculate actual hydrate-forming pressure and temperature to identify where hydrates are forming. Make adjustments to eliminate hydrate formation from choking production.

6. Inspect every item from the wellhead to the sales meter for cost reduction, modification, substitution, or removal.

7. Add gauges to specific components to identify problems.
   - e.g., Dump lines, storage tanks ($oz/in^2$), upstream and downstream of valves, regulators, checks, and flowlines

8. Confirm wellhead wheel valves are fully open. Since it is difficult to know if they are stuck, get with service provider to confirm rotation count required to be fully open. Stencil the number of turns on each valve.

9. Identify 90° elbows, narrow fittings, and other hidden chokes we can remove.
   - I have had situations where removing certain items and eliminating lines increased production by +25%.

10. Reconfigure chemical injection points, methods, and accessories that improve the application of the chemicals.
    - Get with chemical vendors to optimize where and how we are administering our chemicals to reduce required treatment—pumps, atomizers, nozzles, setpoints etc.

11. Remove vessels if unnecessary as production evolves.

12. Shrink tank batteries on older wells to reduce risk and cost.

13. Add facility features to eliminate invoices.
    - For example, for each invoice, what modifications can we make to eliminate the need to do this work or require the vendor that generated this invoice.

14. Connect low-cost automation to reduce OPEX.
    - For example: chemical sensors, temperature sensors, sunlight sensors, smart chargers, timers, notification lights, motion detectors, trackers, and cameras

15. Remove or reconfigure components that repeatedly break.

16. Downsize or remove rental equipment if possible.
    - Consider purchasing rental equipment on the secondary market during oil and natural gas market downturns.

17. Transact in the used equipment market.
    - Purchase at a deep discount from vendors and operators.
    - Make equipment trades with competitors.

18. Purchase hand tools and equipment our pumpers can carry to reduce the need to incur service charges.
    - Power tools, transfer pumps, specialty equipment, test equipment, long arms, hoses, fuses, cables, cleaners etc.

19. Optimize our preventative maintenance (PM) schedules.

20. Document facility issues causing problems and increasing OPEX in our "Operational Issues List." The more people know what is going on, the better our chances of finding a solution to further reduce our costs.

## —————— ACTION ——————
## No Charge List

Most purchased surface equipment comes with a warranty. When equipment deals are made, it is common for warranty details to be obscure, forgotten, or not passed down to the field team. Therefore, when something goes wrong or breaks, the equipment is repaired by the vendor, a field ticket is presented, signed, and paid for something that was under warranty or included with the service. This happens all the time, and it is not the vendor's responsibility to track which items are covered. It is the responsibility of the operator.

The process for creating invoices is systematic. Whenever work is performed, an invoice is generated. However, the process for operators signing invoices is very loose. Vendors know this. It is a problem in our industry which results in billions of dollars in payments that should not be made.

Since production has a lot of surface equipment under warranty and rental equipment that includes field maintenance as part of the service, we must address this cost reduction opportunity. To make sure our team is not signing invoices that should be no charge items, create a "No Charge List" and include it in our "Cost Book." Items under warranty, rental services that include technician callouts, and all other items that should not be charged for will be consolidated in this list to make it easy for our team and remind us of the structure of our agreements. The list also serves as an internal reference for negotiating new orders and contracts.

# ——— ACTION ———

## Ones Make Millions

In our business, there are one-dollar items or actions we can employ that save thousands of dollars per well, which, if scalable, can create millions of dollars in value. These small items are part of the hidden magic in staying online, preventing expensive issues, and getting our OPEX lower than our competitors. When we identify these opportunities, we must share them among the team to see if they can scale across our assets and impact enterprise-level economics.

Early in the book, we discussed how a pumper found an opportunity to save $50 a month on a well that was scalable. That small savings grew to +$100,000,000 over time. Every member of our team has the potential to create this level of value. Throughout the day, it is common for our team, especially front line production personnel, to employ these one-dollar tactics which generate much value, but they are not letting anyone know about it.

This occurs for several reasons. First, when we are in the daily grind, it is easy to overlook the potential value of something small that we are doing that works. Second, many people do not realize the power of scalability. Third, we may not encourage the free flow of information. If there is too much stress in a company or we have a hostile workplace, it is not conducive to sharing. Finally, people do what they are incentivized to do. We must encourage and reward members that share knowledge.

---

# ACTION

## Weatherproof

---

The annual cost of U.S. weather damage is +$100 billion,[11] with the oil and gas industry as one of the most exposed sectors due to its vastness and direct exposure to the elements.

Society depends on our industry to deliver energy when the weather shuts everyone else down. Therefore, we have a duty to stay online regardless of what happens. Every area has different weather challenges. Additionally, microclimates within specific DSUs must be identified and addressed.

For example, assets located in valleys, in areas exposed to high winds, in areas with constant dust and sand, or in areas that do not get good sunlight—all must be handled differently compared to assets just one section over. Eliminate the corporate mindset that weather events are an acceptable excuse for down production or costly events.

---

### Money doesn't care about the weather.

---

The shale machine needs money regardless of the weather. Therefore, we must take proper actions in advance to mitigate the weather's impact and stay online to support our company and society. Unfortunately, some team members are not weather-aware, not interested, or detached from the reality of real-world field operations and build cookie-cutter facilities regardless of the microclimates. Then, they fail to take action when adverse weather advances on our position.

Recently, a manager from the Permian Basin was placed over an asset in the Anadarko Basin for a large independent, offset to a former AOI. This person built multiple highly advanced facilities, with all the bells and whistles, based on designs from West Texas, for Western Oklahoma, costing $1.5 million each—which should have only cost $500,000.

But that is not the worst part. As soon as adverse weather hit, these million-dollar facilities could not stay online. Any little weather event during the winter months would put these facilities down, resulting in lost production and additional costs. The facilities were overbuilt and not designed properly for the weather.

To avoid this type of situation, we must instill a weather-aware corporate culture. First, incorporate weather into all meetings, especially the morning meeting.

Second, require all operations personnel to receive daily weather alerts from local sources. Most towns have a free daily weather service we can subscribe to from one of the local news stations. This is often the most accurate source of short and long-term local weather forecasts.

Third, require the use of weather apps for smartphones. Good weather apps are very cheap. Have the team purchase the weather apps of their choosing to be proactive about the weather to keep our wells online.

Fourth, restrict vacations during specific times of the year. In most operating areas, there are easy weather months and hard weather months. Within production divisions, it is generally frowned upon for production team members to take vacations during the hard weather months.

This is usually during the colder months of the year. During these months, it should be understood that we require all-hands-on-deck to be in the field when the weather hits to ensure our wells stay online and prevent incidents.

If we are in the middle of a deep freeze, should a well go down, it might be down for weeks or months because it is easy for everything to freeze up once we stop moving fluids. If we are prepared, any issues that occur can be addressed promptly.

If we need to wrap something up, add heat trace on a component, launch extra methanol, or set a small generator, we can do it quickly and move to the next well. When it comes to being proactive about weather, consider the first U.S. President's thinking in 1799 when George Washington wrote,

*"Offensive operations, often times, is the surest if not only means of defense."* [12]

Fifth, there are many actions we must take before the weather hits. In late summer, for most areas, we should perform preventative maintenance, clean vessels, pull firetubes, and add winterization.

Sixth, conduct a weather readiness audit and have someone inspect each facility for weather and microclimate weaknesses that we can address while the weather is good.

During some of the coldest events, I have been able to stay online when everyone around me has gone down. This was possible by being proactive and preparing each facility and lift system before the weather became adversarial.

# ——— ACTION ———
## Label Everything

Establishing a robust labeling program can deliver a solid reduction in OPEX. Labeling every pipe, valve, component, stainless line, and connection allows our production team to address problems easier and faster, making our operation safer and reducing costs. In addition to labeling all the iron, we must label all the wiring for our digital systems.

As we leverage the value of digital tech, our operations become dependent on these systems, which can be hard to troubleshoot. Therefore, when problems occur, we are forced to call technicians, which is an expensive and timely process.

If everything is streamlined during installation and labeled in detail, fixing our digital systems can be accomplished by our pumpers at no cost. If we must call vendors to location, with everything nicely labeled, the troubleshooting process is much less stressful, faster, and safer, further reducing costs. Dealing with systems that are not properly labeled can result in having a tech on location for days on one facility trying to figure out how the system works before doing anything to fix it.

Furthermore, when people quit, often in a blaze of glory, it is not uncommon to find out that they uniquely installed or wired items that no one else understands. Some vendors intentionally do this because they make most of their margin on service fees. In fact, during bidding, they price the hardware and installation at cost and depend on the callouts, maintenance, and repair fees for the profit.

---
## ACTION
### Minimize Tethered Economies
---

Tethering a product or service is a business strategy in which the vendor keeps the operator dependent on them to run its business.[13] If we put ourselves in a situation in which we scale our business with tethered services, it will be difficult and expensive to reduce costs, particularly as we reach economies of scale and try to leverage our size.

Tesla, Apple, Amazon, and many other tech companies employ various tethered strategies, which is one of the reasons they have high valuation multiples.[14] To make sure we are not on the receiving end as we scale, identify vendors, services, and products that might be using this strategy on us.

Digital oilfield services, vessels, artificial lift equipment, chemical systems, BHA items, rental equipment, SCADA systems, software, and ERP systems are a few businesses that commonly employ tethered strategies.

If the value is there, using tethered products makes sense. However, once a seller realizes they have leverage, they often increase price, relentlessly, or service us to death. This can happen while their quality declines or fails to advance.

If we go with a tethered system, we must ensure we have an exit strategy, if and when the decision is made to cut ties, because nothing lasts forever. With AI, the value opportunity is too great to remain stuck with a vendor that is not advancing. Therefore, in many situations we need to switch to providers that are best leveraging AI or transition in-house.

——————————— **ACTION** ———————————

# Eliminate Flares

Flares are expensive to purchase, install, maintain, and repair. If we eliminate our flares, we can save money, improve safety, and benefit the environment. Depending on our production and where we are in the world, it is common to have low-pressure Quad-O combustors on our batteries. I am not suggesting we remove the combustors unless we have a system that contains the VOCs on the tanks, and it is not a regulatory requirement in the country of operation.

What I am suggesting, in certain situations, is we configure our facility to safely remove high-pressure flares. Depending on our production situation and well pressures, if we have debottlenecked our midstream systems, such that we have bypasses, and multiple flow options, it should be rare to have a situation in which we cannot sell gas. If we upgrade our inlet separators to high-pressure vessels and have redundant safety shutdown systems, in the rare case that our midstream pressure approaches max line pressure, forcing us to shut-in, our safety system will isolate the well automatically.

There is an industry push to eliminate flaring. Removing our flares helps address this issue because we cannot flare gas if we do not have flares. Enhancing our alarm system and increasing our oversight should occur in conjunction. With the flares removed on legacy facilities, we can repurpose them on new wells and convert some into mobile flare units to employ across the field as needed.

# Actions to Eliminate Flares and Flaring Costs

- Run bypasses around compressor stations and other midstream items to stay online at higher pressures when there are problems or midstream preventive maintenance.
- Install higher pressure separators to produce when we switch from low pressure to high pressure takeaway via bypasses.
- Stop installing flares on wellsite and centralized facilities. If we do not have flares, we cannot flare gas and are forced to be more innovative to stay online when midstream is down.
- Use what would be flared gas to mine cryptocurrency.
- Apply flare gas to evaporate produced water.
- Lab test produced waters for valuable minerals and set up a produced water (e.g., lithium, iodine, etc.) mining operation.
- Divert gas to repressurize deleted zones injecting into parent wells prior to child frac operations (aka "Boosted Frac").
- Run "Smash-Turtle" ops discussed in gas lift action item.
- Setup closed-loop gas capture systems.
- Modify safety shutdown systems.
- Build natural gas microgrids for electric generation stations.
- Establish basin centralized underground natural gas storage facilities in depleted zones, salt caverns, or aquifers.
- Establish virtual natural gas pipeline receiving terminals.
- Initiate an Acid Gas Injection (AGI) program to eliminate flaring and take advantage of 45Q tax credits.
- Partner with industrial manufactures and offer flared natural gas for equity in locally built plants to manufacture steel, iron, or for direct discounts (e.g., OCTG, chemicals, etc.).

# ——— ACTION ———

## Closed-Loop Gas Capture

If there are gas takeaway interruptions that cannot be bypassed and we do not want to flare gas, consider a closed-loop gas capture system (CLGC).[15]

When we cannot sell gas, instead of flaring, we inject the gas into a well of our choosing by diverting production into it down the tubing and backside. Once we can sell gas, we produce the stored gas putting it back into production mode.

Setting up CLGC may require metering, piping, and permitting, depending on our preferred configuration, ownership situation, and custody transfer points. Although there are additional costs, if we eliminate flaring, it should more than pay for itself and help the environment.

Another opportunity, which is much easier, is to optimize shut-ins. Midstream interruptions are often temporary, only lasting a few hours. If we have gas lift wells or a similar setup with buyback meters, when high line pressure events occur requiring us to stop selling gas, we could set up our system to shut the choke or safety but keep injecting using buyback gas.

This will pull line pressures down while we load the backside, allowing other wells to stay online and keep all our compressors running. Once line pressures come back down, we re-open the wells. With this setup, during a temporary midstream shut-in, we eliminate flaring, keep everything online, and potentially boost production because we rock the valves or push more gas through the GLVs, clearing debris.

---
# ACTION
## Tool Up
---

Having the right tools reduces costs, increases production, and improves safety—the highly desirable "Cash Flow Trifecta." Ensuring our team is properly tooled up is an easy way to capture it. When a problem occurs, it is common to call someone with the right tools to fix it. If we have the tools, we do not need to call anyone or be down until a vendor shows up.

Anytime we have a situation in which having the right tool would have eliminated costs or downtime, we must make note of it. Of course, it is impossible to carry every tool, but we can carry tools to address 95% of problems.

Furthermore, if we are paying for a vendor to be on location and see they have low-quality tools or do not have the best tools to do the job efficiently, make note of it. There are many advanced and digitally integrated tools that make work more efficient. Vendors commonly do not carry these newer tools because they cost money and reduce their billable hours—powerful incentives to stay low-tech.

People want to carry great tools. It's safer and makes the job much easier. Therefore, let's survey our team to determine what tools they want. People may hesitate to say they want a certain tool because of the cost, optics, or maybe others make do. For years, people made do using Excel for many things, but we can deliver more value with better software. If a pumper wants the highest quality 60" heavy-duty pipe wrench or latest technical gadget to make work easier—buy it immediately.

# ACTION
## Pump-by-Exception

Having our pumpers run the same standard route every day is not the best strategy to maximize value. From a cost, risk, and production standpoint, we reduce costs and add value if we pump-by-exception.

A pump-by-exception or exception-based surveillance strategy is an operating model that uses SCADA to proactively focus our team on actions that add the most value.

There are many ways to execute this. Our pumpers could modify their route based on wells that are down, work with team members to prioritize wells, or leverage AI to help guide decisions on which wells to visit. If we have a system that knows the location of our wells and the location of our pumpers, we can route our team to high priority situations.

Depending on the risk profile for each asset, there are probably locations we do not have to visit daily. Therefore, our team can give more attention to high-value assets, situations, and wells that have problems. By doing this, we save even more money because our team has the time to fix issues themselves, as opposed to calling a vendor to do the work.

Furthermore, by automating data collection and entry, our team can spend more time doing preventative maintenance, further reducing risk. Any time we can eliminate low-value tasks including drive time, opening gates, ticket collection, meter reading, and data entry we can spend more time on high-value actions that increase free cash flow.

---

# ACTION

## Do It Yourself

---

If we call vendors whenever we have a problem, it will be challenging to run a low-cost operation. The do-it-yourself (DIY) mentality is just like anything else in life: our first reaction to a problem cannot always be to get someone else to fix it. Doing so only facilitates a transfer of wealth, especially for our marginal producers.

> **Resourceful people reduce costs.**

Personnel selection is critical for a DIY strategy. We must look for people who are resourceful and know how to keep wells online without throwing money at them. We could hire a person with good experience but not interested in DIY. They will call out for every little thing, not catch issues before they become expensive situations, or lack the skills and interest in performing preventative maintenance (PM).

If the default process is to call someone to fix everything, we are less motivated to perform PM. The pumper, for example, does not care about the failure because they do not have a vested interest. They just call it out and it gets fixed.

However, if our pumpers are fixing the problems, they come to location and look at their work to see how it's doing to confirm it's solid. There is a sense of pride and ownership in the work, the assets, and our team.

# —————— ACTION ——————
## Automate Hauling

Every location can automate crude oil hauling, produced water hauling, and run ticket entry, even if telemetry is not installed. If telemetry is available, speak with oil and water haulers to let them know we plan to automate the callout process via automated text and emails. For each tank, set up the alarms to notify our haulers when a load is ready. To ensure nothing gets overlooked, configure the system to continually send alarms to each contact (i.e., every hour) until the load is picked up.

On high rate or new wells, go on auto-haul by having oil and water haulers visit each facility continuously or a minimum number of times per day. With our facilities on auto-haul and haulers receiving continuous alerts, we free our field team of this task, adding hours back to their day to help with new ideas to reduce OPEX.

For locations that do not have SCADA, which are typically lower-rate producers, set up the production recording system to send notifications to our haulers. For example, once our pumpers visit locations that do not have SCADA, tank levels are entered into a database of some sort.

When that occurs, we can set it up to send text and email alerts to our haulers. This is the poor man's version of SCADA, but it works and is necessary for marginally economic wells that cannot be burdened with telemetry and SCADA costs for fear of losing PPQ status.

Even with the most basic system, we should be able to automate this process. Let's say we are using Excel (which I do not recommend) on a group of legacy wells. At the end of the day, our pumper sends the Excel sheet into the office. Once it is in the office, we automate the Excel sheet to send out the alerts by text and email. Bulk emails and SMS text messages can be sent directly from Excel with minimal coding.

With our callouts automated, we free up additional time for our team and significantly reduce or eliminate costs attributed to "no load" situations, which can happen when a hauler shows up on location but does not have a full load to haul and must come back later in the day. Once we have the callouts automated, the next step is to automate the recording of the run tickets.

Dealing with physical tickets is time-consuming and stressful, especially for tank batteries moving large volumes of oil and water. Physical tickets are often lost, unreadable, blown away in the wind, or forgotten to be left in the mailbox on location by the haulers.

When this happens, everyone scrambles to figure it out and fix it, consuming more time. The worst situations occur when an oil ticket is missing because everyone gets concerned about the wells production status, whether a spill occurred or is occurring, or whether an oil theft occurred.

These issues not only burn significant field personnel time but also management time, because missing oil tickets are a big deal and are usually reported up the chain very quickly, inflicting stress on everyone.

To solve this problem, integrate the hauler's data into our

systems. Most oil and water haulers enter load details into mobile digital technologies. Therefore, we connect their systems to our system, removing the pumpers of this burden and all the problems associated with physical tickets.

If we want physical tickets as a check, we continue with our haulers leaving physical tickets in the mailboxes and then reference the digitally captured tickets with the physical tickets to make sure nothing is missed.

Even if our haulers do not have mobile digital systems to capture load data, we can use smartphones. All drivers should have smartphones. Therefore, require them to take a picture of the tickets and send it to us. Then, we use software to read the images and integrate the data into our systems.

> **Smartphones are a tool we can leverage to reduce costs with zero investment in physical equipment.**

Smartphones eliminate the need to invest in additional infrastructure that must sit on location exposed to the elements, increasing our OPEX and adding electronic baggage to our facilities. In addition to automating hauling, smartphones can be utilized in various ways to reduce costs across drilling and completion operations and should be a prerequisite to working in the oil and gas industry for most positions—if for no other reason than to enhance safety on remote locations.

Working with our team to identify new opportunities to reduce costs using smartphones will support our efforts to maximize free cash flow with cost reduction.

————————— **ACTION** —————————

## Install Video and Audio Capability

Sight is our most valued asset. It is something no human would intentionally choose to live without. However, most companies live without sight on 99% of oilfield locations, as there are no cameras to see what is happening.

Historically, setting, maintaining, and streaming real-time video and audio was not possible for many remote locations, or it was cost-prohibitive. However, today, a solar-powered video camera with a variety of features, including audio and night vision, can be installed on most locations for as low as $200. Therefore, cost is no longer a barrier.

Installing SCADA on new locations has become standard practice due to the value of the information. No experienced shale operator would consider not setting a SCADA system on a new drill multi-well location.

The cost of a system, including all sensors and meters, fully installed with labor, is probably between $30,000 and $100,000 depending on the size of the facility, the type of system, and the amount of equipment. The economic case for such a system is easy to make.

Adding one video camera for an additional cost of $200 to $500 for the equipment and installation, enhances traditional SCADA systems and overall business operations by improving uptime with a lower risk and cost basis.

The use cases for cameras on oilfield locations are endless. To test the technology, consider purchasing a few solar-powered 4G LTE cameras with audio, night vision, AI detection, pan, tilt, and zoom. Then get 4-inch X 4-inch X 12-foot or taller posts for each camera. Dig a post hole on the pad and install the wireless cellular cameras.

With full video and audio at a height of 8 to 10 FT and movable anywhere on the pad, we emulate the physical presence of a human on location, virtually accessible 24 hours a day, for minimal additional cost.

Because the cost of each camera is so cheap, once team members see the value, they often install multiple cameras on each location to have the full experience of virtually standing on the pad and walking around.

With video and audio capabilities, including pan, tilt, and zoom, on multiple cameras, it gives the user a virtual experience of inspecting our assets without having to travel to location. This ability opens the door to many opportunities.

> **Audio and visual capabilities installed on the pad provide opportunities to maximize uptime and reduce costs round-the-clock.**

Four cameras on each location— one at the wellhead, one in front of the tanks, one next to the vessels, and one at the front gate—is a thorough configuration. However, depending on what's happening, it is common to move them around, based on personal preference.

For example, if we just cleaned out a heater treater or fixed a compressor and want to monitor it more closely, including listening to what is happening, we put a camera directly in front of it. With the cameras on 12-foot posts, it's easy for one person to move them around within a few minutes.

Having audio is critical. The cameras must have audio capabilities. Hearing what is happening on location adds significant value. There have been multiple situations in which I have had a high-pressure gas leak but there was no indication of it on any of the SCADA gauges or visually on the cameras.

However, it was clearly evident on the audio detection system. Without audio, we would not have caught it quickly and fixed it without incident.

A similar situation happened to an offset operator during an ice storm, but it was not caught quickly. It is believed that the small high-pressure gas leak caused a chain reaction of events which burned down their entire multi-well facility.

In the near future, we will not need to install the cameras on posts and move them around manually. Technology is evolving so that the cameras will be mobile and move on their own or on command. Therefore, we can "walk" the system around each location for a more dynamic experience.

Low-cost entry-level cameras are a good starting point for introducing the technology to the operation and team. In the future, every oilfield location will have a sophisticated visual and audio monitoring system incorporated into drilling, completion, and production operations to provide the benefits of being on the front lines and walking around to provide oversight without the need to physically travel to location.

# OPEX Reduction Actions using Cameras

1. Manage multiple production facility construction projects simultaneously without having to physically travel between locations continuously and have a constant consultant presence on every location.

2. During flowback operations, monitor flowback tank levels without needing to go up on top of the tanks.

3. Evaluate and compare flowback personnel methods and actions, monitoring them 24/7.

4. Release flowback quicker using visual surveillance.

5. Strengthen pump-by-exception cost reduction strategies.

6. Enables 24/7 spill detection and response.

7. Assists virtual pumping wells, eliminating the need to visit location unless there is a problem seen or heard on the cameras.

8. Use visual and audio abilities to immediately put eyes and ears on location in response to concerning SCADA alarms.

9. Monitor locations and proactively address issues before wells go down using SCADA quantitative data combined with visual and audio intelligence.

10. Identify false alarms from SCADA systems to reduce unnecessary travel to location, especially during busy times or weather events.

11. Provide flare & combustor monitoring and issue detection.

12. Crude oil theft deterrent and perpetrator apprehension.

13. Provide equipment theft deterrent and the ability to dispatch law enforcement while criminals are in the act of stealing components.

14. Discourage sabotage or mischief from troublemakers, environmental extremists, disgruntled individuals, locals, released personnel, employees, and neighborhood kids.

15. Cameras are a backup system when SCADA goes down.

16. Help our field team troubleshoot issues as a second set of eyes and ears.

17. Dispatch and direct multiple roustabout crews from a central location without needing to be on location.

18. Monitor chemical tote and tank volumes.

19. Virtually receive and direct physical deliveries without going to the field to spot equipment or manage the physical receival process.

20. Use audio capabilities to identify high-pressure gas leaks that are often not identifiable on SCADA systems.

21. Detect equipment issues using audio anomalies (uncommon sounds from compressors, pumps, chokes, ESPs, and heater treaters) before they fail. Dispatch technicians proactively.

22. Identify people performing unnecessary work by monitoring random contractor visits on location.

23. Reference data documented vendor visits, work, and item drop-offs to audit invoices during approval process before payment.

**24.** Monitor oil and water haulers to determine which drivers are leaving a mess on the pad when they pull loads.

**25.** Observe the local situation and severity of weather events, including windstorms, floods, wildfires, heavy rains, ice storms, tornadoes, and lightning while reacting immediately to observable equipment damage, spills, fires, and explosions.

**26.** Dispatch personnel and vac trucks, to clean up any issues instantly; not waiting until the next day when the impact is discovered, and spills have already made their way off the pad onto private lands or into Waters of the United States.

**27.** Monitor pumpers and other technicians when working on issues alone or on remote locations after hours as a safety precaution. Monitor for unsafe or risky actions.

**28.** Identify emergencies in which we need to remotely shut in the wells.

**29.** Identify loose farm animals on the pad and call farmers before they cause damage or get injured.

**30.** Be alerted to and aware of unannounced surprise inspections from state regulators and BLM agents on our locations.

**31.** Prevent competitors and the public from using our locations as a staging pad, storage location, or parking lot, leading to damage, increased risk, and costs.

**32.** Incorporate computer vision to leverage AI systems to take visual and audio detection and analysis to the next level for additional cost and risk reduction.

# ACTION

## In-House

For every phase of our cost reduction journey, we must perform statistical analysis on our invoices to find opportunities to reduce or eliminate costs.

Regarding OPEX, if we see a constant service-related charge for an operation that will continue or increase in the future, we must investigate whether it adds value to bring that service in-house. Horizontal shale wells will last 50 to 100 years. Therefore, we must look at controlling OPEX from a long-term perspective.

For example, if we do a lot of electrical work and have multiple contract electricians working for us regularly, it might make sense to in-house this position. When we do our invoice analysis, once costs get 2X to 3X the cost of a good annual salary for a position or piece of equipment, it's probably time to consider bringing that position or equipment in-house, if we forecast the current workload to continue or increase.

As oil and gas wells age, repair and maintenance increase, so having good people is critical; finding them is hard. When we have a pool of contractors working for us, it is a good platform to get to know people. In my experience, some of the best hires are from contractors we get to know over time.

Our contractors learn how to do things the way we want by working for us. We are paying for them to get experience. Therefore, if we choose to in-house them, we should position ourselves to be able to do so without contractual restrictions.

---

# ACTION

## Night Ops

---

One of the most straightforward actions we can take to reduce cost per unit of production, is to increase night production. We accomplish this by reducing downtime that occurs between 6 p.m. and 6 a.m.

Production downtime is often measured in days. If a well goes down at night, maybe we get to it in the morning, and there is no indication of downtime hitting our uptime metrics. However, it does impact FCF. To better understand downtime, measure it in minutes. If we do this, we will see more downtime, much of it occurring at night.

If we pump-by-exception, our pumpers can oversee more wells. Therefore, we can transition some pumpers from days to nights, and establish a dedicated nighttime operations team.

**Money never sleeps unless you do when wells go down.**

Some companies do not have a night production team. There are even large companies that have a policy that if a well goes down after 6 p.m., they do not respond until the morning—unbelievable but true. If there are two competitors, one with strong night ops and the other with no night ops, I am certain the company with night ops will outperform on nearly every metric. Obviously, our business is not a 9-to-5 operation, but people forget. It takes a disaster, often occurring at night and discovered in the morning, to remind them.

# Production Night Operations Actions

- Reduce cost per unit by eliminating after-hours downtime minutes with a dedicated night production operations team.
- Sell more oil by handling rejects and working with night oil haulers on BS&W problems and other issues. Getting loads sold efficiently throughout the evening frees up tanks, minimizing capacity issues on high-rate producers.
- Mitigate auto shut-in events impacting production uptime by quickly reacting to occurrences.
- Elevate reaction during extreme weather events, especially winter storms that often inflict the most damage at night.
- Discourage theft and sabotage with a 24/7 field presence.
- Reduce catastrophic risk from multi-well domino effect events due to overpressure incidents, spills, and explosions with a faster reaction speed to alarms.
- Patrol high-pressure wells, areas with sensitive surface offsets, and wells that require extra oversight.
- Address "Operational Issues List" items identified by day pumpers to keep us operating at peak levels.
- Enhance peace of mind, allowing the office team to rest and not worry as much about issues at night.
- Reduce spill risk by catching and stopping issues faster.
- Allow day pumpers and foreman to rest and not have to monitor or react to SCADA alarms at night.
- Communicate issues with our Production Control Room, monitoring SCADA and video camera systems.
- Address stuck valves, stuck dumps, hatch leaks, and vent releases, reducing $CO_2e$ emissions improving ESG metrics.

—————————— **ACTION** ——————————

# Product Knowledge Transfer

Require vendors to provide detailed training for our team on their equipment and service. The goal is to ensure we know the practical details on how to operate and fix any equipment provided. We need this information to reduce costs associated with their products and avoid expensive workovers.

For example, if we rent or purchase electric submersible pumps (ESPs), as part of the agreement, require the supplier to provide detailed training on how to prevent problems, troubleshoot, and repair the equipment. That training must be customized for our company, including detailed checklists, field demonstrations, and microlearning shorts we can video and upload to our Knowledge Bank for access by our team.

One might think this occurs naturally. However, you are incorrect. First, intelligence transfer is rarely required by operators in a formalized structured manner. Operators are good at supplying tech requirements and specs during bidding, but once work is rewarded, most people disengage from the process of optimally running the equipment.

Second, most teams do not set time aside for product training. It's not customary or expected, and most importantly, vendors are not incentivized to transfer knowledge. In fact, it is the opposite. Vendors want customers to be dependent on them so they can continue to extract fees from callouts and other service-related work. This vendor revenue stream is our OPEX, and it must be systematically eliminated.

---
## ACTION
## Batch Work
---

Performing jobs in groups of similar tasks allows us to identify, develop, and deposit efficiencies into our Knowledge Bank. Standardizing batch work processes further reduce costs in a flywheel effect, reducing LOE.

For example, if we are cleaning out heater treaters it will be more cost effective to line up 20 treaters compared to doing the work piecemeal or as we have failures. With batch work, we can also bid it out for a fixed price to further reduce costs.

During the planning process with vendors, we may find additional efficiencies in how best to perform this work should we decide to do it on an hourly basis. Furthermore, as we set the job up in advance, we will identify the people who specialize in this type of work and are familiar with the type of firetubes, gaskets, and nuanced issues on our type of treater so that we do not find ourselves on a remote location looking for a unique component that is no longer available off the shelf.

The economic concept of "Division of Labor," with workers skilled in specific tasks, is a crucial aspect of manufacturing efficiency. Due to the nature of lease operating work, it is difficult to leverage the power of assembly line manufacturing. The remoteness of our factor floor, combined with our reactive nature in performing maintenance, repair, and intervention, make it difficult. However, if we proactively address issues before they 100% fail and group similar tasks in batches, we will be more efficient and reduce our costs.

—————————————— **ACTION** ——————————————

## Chemical Reset

As production volumes and fluid composition change, so should chemical treatments. However, it rarely does. Why? Because it takes time and is not an exact science. Therefore, once issues no longer appear to be a hassle, many don't dare touch it. Even if we are running the chemicals a little heavy, as long as we are online selling oil and gas, we're good.

It is difficult to determine the best chemical concoction for a given issue or the most economical recipe while simultaneously injecting products. Therefore, in certain situations, to fully optimize our chemical program, we may need to turn off all the chemicals for a short period. This action is often referred to as a "Chemical Reset."

If we do a reset, we test to see when our chemicals are out of the system or if there is evidence of a problem. For example, on wells with $H_2S$, depending on the severity, we want to test immediately to be safe, to see how fast it returns, and what the concentration is. I have had situations in which we eliminated $H_2S$, cut the chemicals and it did not return. However, until we cut the chemicals, we did not know that. I have also had situations in which we were paying significant money for chemicals that were causing a problem. Therefore, don't assume that cutting a chemical out will be a negative. Many have side effects or can do more harm than good in specific situations. Safety is always paramount, so if we can eliminate chemicals without issue, we will reduce cost and risk.

# Additional Chemical Cost Reduction Actions

- Consider switching from a vendor-managed program to self-source model, purchasing directly from the factory.
- Unbundle chemical products, services, and testing.
  - Perform independent testing to reduce bias. Hire the best chemical technicians. Establish full control.
- Have multiple companies provide recommendations.
  - Visit vendor labs for capability and competency audits.
  - Performance test vendor chemicals in head-to-head competitions in 3rd party labs.
  - Test vendor chemicals over time to confirm quality and ensure suppliers are not diluting down blends.
- Automate chemical injection.
  - Consider technology that will adjust chemical rates based on production, issue detection, and temperature.
  - Auto track rates. Set over treating alarms: "Friday crank ups." Track deliveries for invoice reconciliation.
- Install low-cost items to eliminate or reduce usage.
  - Gun barrels and recirc pumps to fix tank bottoms.
  - Plumb natural gas into the bottom of tanks to stir oil, eliminating hot oiling, dry ice, and over-the-top lines.
  - Run downcomers to put heat directly on BS&W areas.
  - Atomizers with specific nozzles. Pulsation dampeners.
- Optimize how chemicals are being administered.
  - Batch treatments down the backside, in vessels, or solid chemistry in tanks, line treatments, or during frac.
  - Consider cap strings or poor-boy surface-to-subsurface injection methods to achieve a similar result.

--- **ACTION** ---

## Delete Lift Phases

---

Horizontal wells have multiple phases of lift during their lifetime. A common production phase lifecycle starts with free flowing the well up casing until it dies. Then, transition to tubing until it dies. Next, we go from ESP to gas lift. And in the end, most oil wells end up with a rod pump system.

> ### Horizontal Oil Well Production Phase Lifecycle
> *Flow up Casing → Flow up Tubing → ESP → Gas Lift → Rod Pump*

The cost to switch phases, combined with suboptimal production leading up to the switch, plus the downtime, can be significant. Having large groups of wells go through this process is highly inefficient, affecting many metrics.

To address these issues, some operators skip all the above production phases and go directly to rods and flump produce new wells directly after millout with a large rod pump setup. This is done in areas where offsets are free flowing or on ESP, so the volumes are significant. Of course, this strategy is not for everyone, and I am not recommending it, but it is an option.

Going directly to rods is probably not the most economical strategy due to suboptimal production rates. However, if we can eliminate several production phases while simultaneously enhancing EURs and recovery factors, that is a strategy worth considering. There are actions we can take to get the wells to stay online for longer periods, allowing us to skip over or delete several lift phases, which we will discuss next.

---
## ACTION
## Extend Free Flow
---

If our current assets free flow for 3 to 6 months, but with a few adjustments we can get them to free flow for over 12 months, we will put ourselves in position to delete a lift phase, improve most production-based metrics, potentially increase EURs, and improve our economics.

A lot of damage occurs when production begins. No well is impervious to this fact. Moving reservoir fluids at high rates that have not moved for millions of years, inflicts damage on reservoir permeability and porosity, in addition to damage on the proppant from reservoir solids, proppant crushing, and proppant flowback out of the pack, leading to fractured areas becoming pinched off. All of these factors contribute to steep declines and low recovery factors.

As previously discussed, if we flow the wells in a controlled manner, using a managed pressure approach, there are many benefits, including cost reduction and improving economics by extending free flow.

Instead of flowing the wells up casing, consider installing tubing right after millout, and initiating flowback up the tubing. If we are already initiating flowback up tubing, consider downsizing the inner diameter for a better long-term result.

For example, if we are flowing up 3-1/2 inch or 2-7/8 inch, let's try 2-3/8 tubing. I have had great results in some areas by downsizing tubing, especially when it comes to reducing flow instability and slugging, which can be highly damaging.

# Actions to Extend Free Flow

- Upgrade millout flowback with advanced monitoring, plug catchers, choke systems, and detailed reporting to prevent flow hands and habitual processes from causing damage.

- Use multiphase flowmeters during millout and flowback to quantify sand production, react to sand slugs, prevent stuck events, improve flow control, and increase production.[16]

- Increase EURs by reducing formation damage from fines, and from proppant pack NWB + far field pinch outs, by incorporating accurate solids flow into choke management.

- Switch from flowbacks up casing to up tubing to reduce slug flow hammer reservoir damage and frac mutilation.

- Downsize tubing to smooth out and streamline fluid flow.

- Run tubing deep into the horizontal.

- Run an engineered or managed pressure flowback process.

- Use internally coated tubing to increase production (e.g., laminar flow), reduce downtime, address hidden subsurface chokes from bridges, corrosion, scale, and paraffin.

- After millout, run tubing with gas lift valves. If the well loads up, kick it back online and go back into free flow.

- Install a capillary string (in tubing or casing) and inject foamer or a surface tension modifier to stay in free flow.

- Plumb up a ball valve swappable wellhead to gas lift compressor. As tubing pressure declines to meet sales line pressure, kick on the compressor in wellhead mode. This way, we pull surface pressures way down below sales line and stay in free flow. If the well needs a kick or continuous gas lift, we quickly switch, eliminating downtime.

---
## ACTION
### Improve ESP Economics
---

People always ask me what my favorite form of artificial lift is, which I find hard to answer because each reservoir is different, and each operational situation is different. However, in general, when asked, I usually say "tubing," and everyone giggles because tubing is not considered a traditional form of artificial lift. But I like it because it has no moving subsurface parts to worry about, and we can make the well flow for a long time on tubing with a little engineering, boots-on-the-ground front line tactical savviness, and a few clever techniques.

With that in mind, electric submersible pumps (ESP) are on the opposite end of the spectrum due to their advanced technical complexity and subsurface moving parts.

However, for high-volume production performance, immediate production response to surface control adjustments, and reservoir drawdown ability, nothing beats ESP in most situations. Therefore, in terms of cash flow from production on high-volume oil wells, ESP is my favorite form of lift—I just don't like the cost of it.

During one of the recent commodity price downturns, there was a large private company that had most of its horizontal wells on ESPs, which they were struggling with due to short mean time between failures (MTBF). To compound the problem, they were also struggling financially due to the company's capital structure.

After a while, the difficulty of the situation intensified. As a result of weekly ESP failures, relationships became stressed. The investors and management team were not happy with each other, so the investors terminated most of the management team, which is common in these types of situations, and replaced them with new leadership who were more aligned and had more robust financial experience.

The new team was composed of a variety of people, including highly respected engineers and technicians. After careful analysis, the new team determined they could save significant money and strengthen the economic status of the company by removing almost all of the ESPs when they fail and transitioning the wells to gas lift.

The process took over a year, but eventually, they eliminated all the ESPs. The principals were happy and preparing to hold a celebratory dinner with the management team when several employees presented data on projected cash flows, based on the new decline curves.

> **Any intelligent professional can reduce costs.**
>
> **Maximizing cash flow with cost reduction,**
> **is something else.**

Soon after flush production subsided from the wells sitting static, the new gas lift-based decline curves projected a dramatic underperformance compared to initial expectations and compared to the prior ESP-based decline curves.

The new management team was successful in reducing production costs and workovers, but production was also reduced. As a result, many wells were transitioned back to ESP. Not all reservoirs transition favorably from ESP to gas lift due to several reasons including the added backpressure—remember, with gas lift, we are injecting into the well.

The solution to uneconomic ESP runs is not always to eliminate the ESPs. I have had several situations in which poor MTBF was traced back to component manufacturing defects from the ESP supplier that impacted entire fields. This is a very expensive and painful situation that is not easy to identify because most manufacturers are less than forthcoming when these things occur.

There is no doubt that unless the expense aspect of ESPs is addressed in terms of equipment failures, downtime, and power costs, in certain situations ESPs become an operational and economic nightmare.

To avoid the nightmare, we must be on top of our game. Often, ESPs require a lot of babysitting, which not everyone is interested in or able to do, particularly when scaling production and dealing with thousands of wells.

Getting up to speed on a well-by-well basis using ESPs can be cost prohibitive. It is not practical to sit on every pad babying the pumps. Therefore, we need to leverage the full power of artificial intelligence to the limit of what it can do. Fortunately, the cost of AI has declined to the point that we can implement it at scale.

To improve ESP economics, we will take many actions. When properly done, our ESPs will stay online in a cost-effective manner, until we decide to pull them.

# ESP Cost Reduction Actions

**1.** Set a goal to keep ESPs running for three years or a period to bring production to its final phase of artificial lift.

**2.** Request an equipment purchase structure to have rental costs deducted off purchase price for the first 3 to 8 months to financial protect ourselves from manufacturing defects or other problems.

**3.** Strengthen ESP warranty terms and conditions.

**4.** Incorporate used (previously run) ESP equipment certified by the manufacturer if the discounted price is attractive.

**5.** Configure ESP for as wide a range as possible, focusing on the lower production rates expected in 12 months' time.
- o Max rate is only for a short period. Therefore, it should not be the primary design focus point, but it often is, resulting in an oversized pump within a few months, leading to failure in less than 12 months.

**6.** Run advanced downhole sensors with the maximum amount of features included and installed.

**7.** Upgrade intelligence by maximizing sensor and facility data resolution with high data frequency (i.e., second-level data).

**8.** Establish an inspection and failure analysis program to get to the root cause of failures to prevent them in the future.

**9.** Employ an ESP specialist for installation, startup, pulls, and dismantle inspections and failure analysis (DIFA).
- o This is a separate 3rd party ESP expert consultant in addition to the standard company man and ESP techs.

10. Work with the team on operational procedures.
    o Installation reports should be extremely detailed, including step-by-step documented process inspection, verification, equipment checks, and startup.
    o Include documented images of onsite processes with closeup pictures of the ESP components.
    o If anything looks even remotely off, don't run it.

11. Install components to handle solids, including fallback.

12. Perform fluid testing to optimize a chemical program specifically for ESPs.

13. When designing and operating the ESP to run for an extended period, incorporate the relative increase in gas production as fluid production declines (increasing GLR) that occurs in most formations as reservoir pressure drops below bubble point.
    o Taper the stages with higher capacity pumps on bottom.
    o Use a gas separator, handler, or specialized intake.
    o Reduce pressure drop between stages.
    o Minimize dead zone transitions from diffusers to impellers.

14. Start ESP at lowest possible frequency (Hertz) while monitoring intake pressures and solids production. Take daily samples of solids and estimate amounts to fine tune ESP settings to minimize damage to the frac, formation, and ESP internals.

15. Compare actual production to theoretical potential to identify problems and opportunities.

16. Leverage real-time artificial intelligence, incorporating area specific knowledge, to reduce the amount of oversight required to maintain uptime and avoid failures.
    o Build algorithms on a well-by-well basis to prevent costly failures and workovers, including:
        - Real-time recommendations with AI estimated economic impact
        - Gas lock and slugging detection with auto avoidance response modes (e.g., constant current, constant intake pressure modes, discharge pressure, etc., to stabilize the ESP)
        - Low flow rate alarms from downhole sensors
        - Temperature trend and risk alarms
        - Irreversible failure probability updates
        - Pull ESP alerts
        - Sand erosion, including daily damage reports
        - Scale buildup detection and alarms
        - Tubing leak detection and tubing blockage alarms
        - Unknown anomaly detection and alerts
        - Setpoint recommendations and auto adjusts to keep ESP smooth as liquids decline and gas increases

17. Consider auto control systems that predict the future and take action to prevent failures and optimize power costs.
    o Establish ranges the team is comfortable with to allow the system to operate freely for AI auto control.

18. Institute sand flow and slug flow proactive actions, when detected, to prevent ESP failure and equipment damage.
    o Auto-detect if slugging is below and/or above the ESP.

19. Manage voltage using data, digital twins, and motor curves to reduce power consumption without impacting production. Power optimization will reduce electric costs and increase run life by reducing motor temperatures and stress on all components.
    o Compare actual voltage to optimal voltage.
    o Reduce load or voltage, volt/speed, in small steps at a fixed frequency or speed to reduce power consumption.
      ▪ This tactic reduces power wasted at a certain speed.
    o Reduce hertz while opening choke to reduce power consumption while maintaining production.
      ▪ This tactic reduces power wasted across the choke.

20. Work with utility providers to minimize electrical disturbances. A lot can be done if we have good partners.

21. Consider using generators during unstable power periods.

22. Install lightning protection.

23. On multi-well pads, arrange and protect VSDs and j-boxes for safety, including number and color coordination to prevent confusion, operator mistakes, and electrocution.

24. Capture all ESP issues, no matter how small, in our "Operational Issues List" daily to add into our AI systems.

25. Consider installing a casing compressor to drawdown annular pressure extremely low to allow ESP to max out tubing oil production while minimizing gas lock potential.

26. Develop specific ESP best practices, checklists, quick analysis, and diagnostics for field team to prevent failures.

# ── ACTION ──
## Optimize Gas Lift

From an operational perspective, gas lift is at the top of my list since most issues can be solved from the surface. Plus, if we have solids production, complex trajectories, slugging, high GLRs, or power instability, it's a non-issue with gas lift. Additionally, having a compressor on location provides us with many secondary benefits to help reduce costs and increase production.

Since gas lift MTBF is often the longest of any form of lift, the opportunity to reduce cost is from optimizing production, fixing hidden issues, and reducing compression costs. Even if we have a poor design, with failures downhole, and compressor issues, gas lift still works, providing respectable production. Therefore, it is hard to identify and accept we are inefficient.

To motivate our efforts, let's assume all our wells on gas lift are inefficient for one reason or another and have a potential 10% to 20% uplift in production (common average) if we can find the inefficiencies and fix them. GLVs often perform differently than vendor models suggest. Unloading valves are not 100% open when opening pressure is reached. They throttle in combination with other valves. These realities contribute to instability and slugging, costing us money.

We have many actions to review to reduce gas lift costs, but a good place to start is to run a pressure/temperature survey on slickline to get a better idea of what is happening in the subsurface to fine-tune our models and direct our actions.

# Actions to Reduce Gas Lift Costs

1. Establish an upstream and midstream compression division. Take direct ownership of various segments of compression and incorporate them into our production operations, eliminating rental costs.

2. Purchase compressors during market downturns from other operators or vendors who have them stacked and are motivated to sell at a discount to generate revenue.

3. Run a pressure / temperature survey on slickline to locate injection points and find issues, including bad valves, poor designs, incorrect models, and holes in the tubing.
   o Update models based on real-world data to optimize designs for current and future wells.

4. When on the bottom GLV, drop suction and RPMs to a minimal level and monitor production to see if reduced injection rate maintains or increases production. Consider downsizing the compressor.

5. Configure SCADA polling frequency to the highest level to identify inefficiencies or problems with the compressor or gas lift valves affecting production that might not be evident on the standard 5-minute polling frequency.

6. Run buyback gas through a facility separator before going into the compressor suction to increase light oil production coming from the pipeline and improve compressor injection rate efficiencies.

7. Set compressor scrubbers to dump into oil tanks to increase oil production and reduce rejected loads.

8. Reduce scrubber dump overflush settings to maintain stable gas injection and minimize tank battery vapor pressures, reducing emissions.

9. Compressor issues (e.g., pulsation) will make injection read higher than it actually is. Therefore, we may not be injecting as much gas as we think, causing suboptimal production. Pinch a valve upstream of injection meter run and tinker with the compressors to see effects on injection readings and production performance.

10. Open the pockets to try and increase injection rates.

11. Track preventative maintenance especially on rented units.

12. Connect the compressor instrument panel to SCADA or add sensors for additional insights and cost reduction.

13. Max injection does not always maximize production and can vary by compressor, GLVs, and overall well setup due to various idiosyncrasies. Therefore, test many injection configurations to find the sweet spot based on production.

14. Add a compressor cooler bypass to send hot gas into the subsurface for paraffin elimination, hydrate prevention, and to prevent freezing during cold months to stay online.

15. Use heat generated from the compressor to prevent facility freezing by connecting the radiator system to a heat trace line across the production facility components.

16. Build a chemical launcher safely away from the wellhead to pump batch treatments or shoot budget treatments into the subsurface using the injection line.

17. As production declines and less injection is needed, remove compressors and use one compressor for multiple wells or design the setup with a larger compressor for multiple wells if economics are beneficial relative single well compressor setups.

18. Set up a push-pull system to use excess compressor capacity to reduce wellhead pressures, acting as a dual-purpose gas lift and wellhead compression unit to increase production and deal with high line pressures preventing shut-in events.

19. Test the "Smash-Turtle" method by injecting the maximum amount of gas for one month and then injecting the minimum amount of gas for one month or a period that works for your wells.

20. Inject a foamer or chemical that foams (e.g., biocide) down the backside to further reduce density increasing production while simultaneously cleaning the wellbore and GLVs. Try different foamers via batch treatments and continuous injection to increase production.

21. Consolidate our gas lift knowledge into artificial intelligence systems correlated to our SCADA readings from our facility, compressor, and subsurface to provide recommendations to increase efficiencies, identify problems, and provide solutions in real time.

22. Once we have identified changes that increase efficiency, utilize AI to find other wells across our AOI with similar characteristics that could also respond favorably.

# ———— ACTION ————
## High Pressure Gas Lift

When we reduce subsurface complexity, we reduce cost and risk. High Pressure Gas Lift (HPGL) provides this opportunity by removing subsurface gas lift valves, injecting at high pressure (+1,800 psi) down open-ended tubing, and producing up the casing.[17] This can deliver production volumes equivalent to ESPs without the complexity, reliability issues, and costs. With HPGL, the entire lift system is on the surface. All the complexity is on the surface, making problems a lot easier to fix and far less expensive to deal with.

Additionally, if the formation flows high solids volumes combined with gas slugging, HPGL can handle it without the issues that damage ESPs. Also, if we do not have access to stable 3-phase electric power, we can run the compressors on wellhead gas, which is often preferred due to its reliability.

With HPGL, we can place the tubing deeper into the horizontal to drawdown the reservoir further and blast out any sand dune buildup in the heel that could choke production. Running ESPs into the horizontal can be done effectively but is often uneconomic if the well has high solids production.

HPGL also makes it easy to batch treat the wellbore or inject chemicals without a cap string since we are already injecting. Furthermore, with a simple setup, we can switch between injecting down the tubing or the casing if we want to test different lift strategies. When looking for a low-cost alternative to ESPs, HPGL is a technology to investigate.

---

# ACTION

## Overpumping and Underpumping

---

Once a well transitions to rod lift, it is often ignored as long as it is online. Over time, it is common to build large inventories of horizontal wells on rod lift that are operating inefficiently, especially if we are acquiring large volumes of legacy assets during periods of market consolidation.

Overpumping and underpumping are the two most common dysfunctions on rod pumped wells.[18] Since most production departments are understaffed, there is not enough available time to spend on each rod-lifted well to keep it optimized, especially when ESPs and gas lift wells take priority, due to the higher volumes and OPEX impact.

This dynamic is an excellent opportunity to deploy AI at scale on rod-lifted wells to help identify and automate setpoints to increase production and reduce costs. Continuously optimizing rod-lifted wells 24 hours a day is impossible for a human to do. Since costly damage and failures can occur quickly due to the physics and mechanical intensity of rod lift, automated systems should be considered.

As more horizontals transition to rods, if we are not careful, damage related problems on unoptimized wells can quickly expand from a minor nuisance to a major problem, overwhelming workover crews, and becoming noticeable to financial analysts. Nobody will complain about employing AI too soon on our rod-lifted wells, but they will if we are too late.

——————————— ACTION ———————————
## Strategic Intervention

"Run-to-failure" (RTF) is probably the most common production strategy utilized to minimize OPEX regarding subsurface intervention, artificial lift, maintenance, and facility-related work. RTF strategies are designed to minimize total cost and eliminate non-critical operations. For many situations, particularly on low-volume legacy vertical well rod-pumped assets, RTF is the most economical strategy. Of course, if a safety or environmental situation is imminent, we must perform the work immediately.

That aside, when calculating the cost of RTF compared to proactive strategic intervention, we must consider downtime, particularly on large-scale, high-volume assets. For many high-volume situations, RTF should be avoided.

For example, if we have a 500 BOPD well that our AI is alerting us of potential irreversible damage, when calculating the economics of what to do, we must estimate the downtime to put the job together.

If an ESP goes down on Friday afternoon, and we cannot get to it until next Friday, we will have at least a week of downtime, assuming it's a straightforward ESP pull. However, when certain equipment fails catastrophically during RTF, the replacement work is often much more complicated, taking longer to accomplish (e.g., fishing an ESP in pieces). The risk of this occurring needs to be calculated and factored into our decision-making process.

The cash flow lost from production downtime is lost forever; we never get it back. Therefore, it should be treated in a similar fashion to a direct vendor expense with the appropriate nets and deductions.

500 BOPD for a week is 3,500 bbls or $315,000 at $90 oil, plus the cost of the ESP, which probably has no salvage value since we ran it to failure. Therefore, an RTF strategy will probably cost us +$300,000 on a net basis in this example compared to one day of downtime and minimal ESP costs using a strategic intervention program if we can rerun the ESP on a future well and credit it back. For assets with this magnitude of OPEX impact, I prefer to be proactive and run a strategic intervention program rather than an RTF program.

There are many ways to implement a strategic intervention program. As part of the strategy, consider employing a combination of systems, including preventative maintenance at predetermined intervals, AI-driven predictive maintenance based on continuous computer monitoring, and risk-based maintenance using an integrated analysis, including testing and human-based risk assessments. These three systems form a strong foundation for a strategic intervention program focused on maximizing cash flow with cost reduction.

Depending on availability and how many workover rigs and roustabout crews we are running, multiple wells will need attention simultaneously on a regular basis. Therefore, we will need a system to high-grade and organize our work schedule to maximize free cash flow.

Forming a "Strategic Intervention Team," including members from reservoir engineering, production engineering, and field operations, to plan operations collectively, make decisions on which situations are priorities, and how to schedule the work, will add value and help reduce costs.

To ensure we are never waiting on equipment, consider having a consigned inventory of items ready to go, held by our preferred vendors. When running a large program, most vendors should agree to provide this convenience at no charge until we run the equipment. The most effective strategic intervention schedules are flexible, adjusting to real-time events as they happen throughout the week.

Therefore, we may have an idealized schedule on Monday regarding which wells each crew and pulling unit will hit, but it is subject to change if one of our big producing pads has several wells indicating, through our AI alert systems, that we have a high probability of an imminent failure.

> **A run-to-failure production strategy is no match for artificial intelligence assisted operations.**

If there are two large-scale competitors, one is running its production operations solely using human feedback, and the other is leveraging the full power of AI, it is unfair to compare the two. Personnel turnover and attrition alone will drain the knowledge from the company depending solely on human feedback. Add that to the production mistakes, missed opportunities, and inefficiencies that AI can help address—the value creation opportunity from AI is too significant to ignore.

---
# ACTION
## Hidden Chokes
---

Every well has hidden chokes. We don't think about them, but they are there. When looking to liberate free cash flow and reduce costs, inspect as many locations as possible to find the hidden chokes. The power of finding them is in the ability to scale our discoveries across our assets, exponentially increasing the free cash flow impact for our company.

To help find hidden chokes, put your hands on the components that could be limiting production. You might feel a pressure drop, vibration, or temperature change. Being on location and listening to different components will help identify hidden chokes.

Sometimes, it may only take a few minutes to find the hidden choke; other times, it may take several hours or days to identify an obscure component of the surface facility or artificial lift system, limiting our cash flow. We may have to visit the wellsite multiple times and take pictures and videos to further study in the office.

For example, on one location, we felt like we had a hidden choke, but could not find it. It took several hours over the course of a week on this location to identify what we thought could be limiting production.

The well was on gas lift, and the compressor suction line was connected to the sales line, but it was behind an obscure backpressure regulator. We put the regulator in the fixed full-open position with minimal effect. Then, we measured pressure upstream and downstream of the device while monitoring compressor suction pressure. The minor pressure differences were evident but would not jump out as a problem.

Then, we had the compressor mechanic look at the compressor to see if he could find something, and his assessment was, "It's running good without problems."

But I could hear the compressor stress a little every so often. However, there were minor differences in injection based on the gas injection meter run. The noise of the compressor (which the mechanic said was normal), combined with the pressure variance, made me continue to push to find a hidden choke.

Once I felt confident about the location of the hidden choke, I called for roustabouts, and we replumbed the suction line to be in front of the regulator. There was enough of a restriction and delay when the well was not flowing strong that it reduced the compressor's efficiency, but it was not visible on SCADA due to polling frequency, injection rate variance, and micro-fluctuations. It took a couple of hours to cut the line and make-up the connections in the new configuration.

This minor adjustment added 50 BOPD on a previously 300 BOPD well, for a +16% increase in production. A huge prize for our efforts that could be scaled across multiple locations. The hidden choke was preventing the gas lift compressor from delivering a stable injection to allow the well to produce up to its full potential.

This choke impacted subsurface production, but it was easy to fix because the problem was addressable from the surface. Pure subsurface chokes are more complicated to identify and can be high-dollar to fix, but the prize for removing them is significant, especially when scaled.

> **Every well has hidden chokes on the surface and subsurface.**

Odd noises, pressure drops, and slugging are all potential indications of a hidden choke. Any abrupt production change or well that falls off the decline curve is a sign of a potential hidden choke. Before dropping the decline curves due to suboptimal production performance, it pays to notify the team and have them try and find the hidden chokes.

If we think the choke is in the subsurface, before we decide to pull everything out of the well to try and find it, let's look at options we can deploy from the surface for a low cost.

Scale buildup, paraffin, and other subsurface chokes can often be remedied by pumping chemicals or acids from surface down the backside and tubing. Using slickline or braided line to do analysis, investigations, and subsurface operations is another option.

Many subsurface issues can be fixed "laparoscopically" from surface, as opposed to "open surgery," or pulling everything out of the well during a major workover. Before deciding to perform an invasive operation, which could unintentionally cause additional costly problems and expose us to elevated levels of operational risk, we want to exhaust our options to remove the hidden chokes without mobilizing a workover rig and performing major surgery to the wells.

---
## ACTION
### Enhance Cash Register Systems
---

Selling crude oil and natural gas is one of my favorite things to do in this business. I enjoy working with the oil haulers when pulling nine feet of freshly produced, piping hot crude, working with midstream meter techs when testing custody transfer meters to ensure we get full credit for our natural gas, and working with production accounting teams when auditing our oil and natural gas product sales statements to confirm we are paid correctly—I like to see the cash flow.

It is challenging and serious business to get these minerals up to the surface and into marketable conditions to safely sell them. So, when midstream purchasers shortchange sales volumes—it disappoints me.

---

**Oil and gas sales "errors" are almost
always against the operator.**

---

The meters and tank measurements are our cash register systems, and if they are not correct, we are guaranteed to lose money. Just like invoices, 95% of the mistakes will be against us. If we are not diligent with production accounting and reconciliation, we will not realize the full value of our work.

In my experience, it is rare to have a midstream company accidentally pay for more oil than was sold, but I have had plenty of situations when they severely underpaid for oil and gas that was sold. As hard as it is to believe, it is not uncommon for purchasers to make massive mistakes to their benefit.

Missing multiple loads and forgetting to pay on entire wells happens. Big mistakes are easier to catch. It's the smaller mistakes that are often overlooked.

For example, if a hauler pulls a load, which we have a record of, but the load gets omitted from our run statement, if we are not diligent, we lose full value.

It is not uncommon for these "missing" loads to be random and spread out across thousands of wells. With checks and balance systems, these should be easy to catch, but not always, especially if the driver forgets to leave a ticket on location or our automated systems do not import it on the backend.

Missing tickets is one thing. How about incorrect volumes? Most oil haulers will not admit this, but it is common practice to pull a little "extra" oil on every load.

No driver wants to be short at the LACT. Short loads are how drivers get fired, so the unwritten rule is to cover your cost and pull a half inch on the bottom and on the top. Maybe midstream accepts a 5 bbl to 7 bbl variance due to expansion and other factors, but we should know what those numbers are.

To address this, the best thing to do is to request the LACT "metered oil" volumes when the haulers offload what was purchased. Midstream companies will push back on this request. They don't want to share those numbers. But we need those numbers to see the variance between what was pulled and what was sold.

Good drivers will have numbers that are very close. Slick drivers will always sell more oil than they buy. We want to identify those drivers and ban them or keep a close eye on the volumes. It is easy for drivers to "make" oil between the wellsite and the LACT. To ensure these numbers are not too egregious, we must monitor them.

The other thing to look out for is the BS&W. It is not uncommon for some haulers to always put 1.0% or whatever is the maximum amount allowed by the purchaser.

I have had drivers tell me that they were told by their managers to always put the max regardless of what it is.

This is a lot of lost cash flow if it should be 0.1% BS&W, but we are getting hit with 1.0% BS&W.

To see if they are doing this we may need to hide behind the tanks and then jump out and catch them red-handed. Or get to know the drivers, and they will probably disclose it. Then, we can handle it in the field.

Natural gas is just as bad but often more complicated, especially when dealing with buyback gas, NGLs, and older processing facilities. For example, it is common for buyback meters to read volumes being utilized when in fact, no buyback is flowing. It's just vibration sending a false reading.

Establishing robust processes will help us prevent these issues. To assist our efforts, employing automation and artificial intelligence-supported systems to help us catch problems can add significant value. Underpayment on sales volumes, whether from missing tickets, midstream mistakes, or oil theft, are issues AI can help us identify and address. Rectifying these problems directly translates to increasing free cash flow without making operational changes or exposing our team to new types of operational risk.

---
## ACTION
# Virtual Pumper
---

Installing video and audio capabilities on each location, combined with advanced SCADA, including automated hauling and ticket entry, allows us to "virtually pump" most locations.

By placing low-cost cameras with audio and visual, at the wellhead, tanks, and vessels, we can virtually walk across the pad, inspecting our assets in a similar fashion to when we are physically on location. If there is a concern with a specific item on the pad, we can place a camera directly next to that item for additional oversight.

This further liberates personnel from physically traveling to each location on a daily basis unless we want to. When we do go to the wellsite, we can focus on high-value tasks that reduce risk and increase production. Virtually pumping wells take the pump-by-exception model to the next level.

Some operators have not had success employing pump-by-exception or are not comfortable with it due to the high cost of spills and other incidents that occur but are not caught quickly.

Adding camera technology can help mitigate these types of concerns, but of course, no system is perfect. From a risk and value perspective, the combination of a virtual pumper system and boots on the ground is unmatched.

If we have a 24/7 production surveillance and optimization group, combined with visiting our wells daily, we can help guide our ground team more effectively.

This approach is similar to how the military uses overwatch methods, including drones and other technologies, to enhance situational awareness and guide ground troops during combat missions.

Installing cameras to provide overwatch support to our field production team is a powerful, free cash flow maximizing strategy that should be considered.

For example, with cameras, we can see when a small leak is developing (slight staining or dampness on the pad), which is often not obvious when walking around the pad.

We can also compare images throughout the day to identify when something small is happening, which is often missed during short 15-minute human field visits.

I have avoided a number of undesirable situations using these tactics, combining camera technologies with boots on the ground. For example, with camera technology that includes night vision, which most cameras do, we can use infrared radiation, or thermal energy to identify problems that are not visible to the human eye.

If there is a problem on a production facility, it will often reveal itself with increased heat on various components. For example, it is common to spot a problem by monitoring the combustors on the tank batteries. A balanced operation should have minimal heat on the combustors, but if a valve is stuck open or something is continuously dumping, additional gas will be on the tanks. Standing on location, this may not be visible to the human eye. However, in infrared (IR) mode, if we see the combustors illuminated in white when they are typically dark, it is a good indication of a bigger problem.

# Expanding AI Observation to Optimization

Standard SCADA systems with alarms have become common practice in our industry. However, most only function as passive observation systems. This is scratching the surface of what these systems can do. Using simple algorithms, we can leverage numeric data to provide real-time prediction and optimization notifications.

For example, when a lift system goes down or is not working optimally, there are often a confluence of signatures that occur. If we set our systems to recognize these events, we can provide alerts with probabilities to our team, enabling them to proactively address issues before impacting production or causing undesirable events.

By analyzing instances of past events across our assets, we can develop programable systems and predictions that we incorporate into our AI.

With a virtual pumper strategy, we liberate ourselves from repetitive, time-consuming tasks. This enables us to strategically target situations that increase production and take actions that reduce risk and downtime with the help of our experience-based collective wisdom, programed into our automated notification systems. Additionally, incorporating computer vision to analyze images, video, and audio in real-time, combined with our gauge-based standard systems, will further advance our ability to identify opportunities.

For example, if our audio system detects abnormal sounds from a gas lift compressor combined with a variance in discharge pressure and higher vapor pressure on the tank battery, we could program an alert to send out to the team with a short video clip indicating the probability of a stuck scrubber dump that maybe hung open on the unit—an easy fix to get production back to optimal rates. Without an advanced system, an issue like this may go unnoticed by our team for several days until the unit goes down or fluid blows out of the thief hatches on top of the tanks—events AI can help prevent.

Small cost reduction achievements like this compound and can grow into billions of dollars in value. It is exciting to think about the opportunities this provides. However, the reverse is also true: small costs add up to significant sums of money, which often go unnoticed until it is too late.

> **Lack of attention to cost details, combined with repeated small costs over extended periods, is how large fortunes are lost.**

There is a fine line between economic success and failure. Small costs, individually, are relatively insignificant. However, this is rarely the case in the real world, as small costs accumulate into large sums that determine our future. These costs sneak up on us until our company becomes unviable. This is especially true on the production side of our business, where seemingly insignificant costs accrue. As production declines, small costs become more significant and unsustainable.

In the spirit of achieving 100X returns and maximizing cash flow with cost reduction, here are "100 More Production Cost Reduction Actions" in no particular order.

PRODUCTION

COST REDUCTION

IS

A GAME OF PENNIES

AND

A GAME OF BILLIONS

## 100 More Production Cost Reduction Actions

**1.** Strategically divest high OPEX non-core assets.

**2.** Enhance asset development economics and reduce financial risk by instituting a series of "Key Data Checkpoints," including drilling gas shows and production performance, to drive decisions and development schedules in real-time.

- o Construct multiple drilling schedules, including backup drilling locations that are ready to go, in case drilling schedule changes must be made based on feedback from real-time drilling and production key data checkpoints.

- o Structure the drilling schedule to obtain production data and proxy data from key wells before increased financial exposure through the drill bit.

- o This strategy will proactively reduce OPEX by reducing the risk we bring on poor producing wells, which dramatically increase cost per unit of production.

**3.** Establish systems to maintain production (SIMOPS) while drilling additional wells on already-producing pads.

**4.** Establish AI systems to detect divergences between actual production and theoretical optimum production to identify artificial lift and flow related issues—providing opportunities to increase production.

**5.** Analyze annual workover expenses to determine which rental items to purchase (bring in-house).

**6.** Transition from 100% "stick built" facilities to modular and pseudo-modular constructed facility systems and strategies.

7. Optimize workover rig schedules to incur less road travel time between wells, improving OPEX and field-level production-based performance metrics.

8. Renegotiate midstream and electricity contracts.

9. Catalog already-owned, underutilized parts, vessels, and equipment across all field yards and on production locations.
   o Consolidate information in an organized, easy-to-review format: "Available Parts and Equipment List."
   o Include items that are in use (hooked-up) but can be removed because they are unnecessary due to production decline and changing needs (e.g., VRTs, tanks, chem totes, pumps, valves, and meter runs)
   o Provide a list to all production personnel.
   o When something is needed, personnel can reference our internal inventory list before purchasing the items.

10. Consolidate and structure purchasing of widely used items for bulk discounts.

11. Establish a refracturing program to augment or replace production from our higher-cost new drill program.

12. Eliminate paraffin hot oiling, scraping, solvents, and continuous chemical injection with single-treatment nanofluids customized for our specific crude oil reservoirs and chemistries.[19]

13. Add financial accounting IDs and QR codes to each piece of equipment to identify individual equipment problems, individual cost issues, and cost reduction actions.

14. Use digital $H_2S$ testers to eliminate the cost of test tubes.

15. Develop troubleshooting checklists and step-by-step procedures to address problems for each method of lift.
    o Add to our "Solutions Book" to share with the team.
    o Many problems can be solved with simple hand tools and tactics without an expensive workover.
    o Consolidating the collective wisdom of our team in an easy-to-use format will reduce our workover expenses.

16. Utilize strategic production curtailment tactics to reduce produced water volumes.

17. Universal wellhead systems for all forms of artificial lift, eliminating the need to purchase multiple wellheads when transitioning between lift methods.[20]

18. Pump low-rate, low-cost, acid jobs from surface with water and diverter to remove iron sulfide, scale, salts, and other chokes that are plugging perfs across our horizontals.

19. Use mobile High Pressure Gas Lift (HPGL) units to unload wells after significant frac hits.[21]

20. Slickline deployed paraffin melting tools to remove paraffin to full drift to extend time between treatments.[22]

21. Perform sucker rod failure analysis in a lab settings (not in the field) to determine root causes to address to prevent reoccurring problems.[23]

22. Use a continuous (one-piece) coiled sucker rod to minimize failure in wells with high side loads, doglegs, and other costly issues.[24]

23. Transition to rod lift sooner with a high-volume rod lift unit / design combined with a one-piece coiled sucker rod.

24. Chain off tank battery stairs with "Do Not Enter" signs and require crude oil haulers and water haulers to pull capacity loads only. Establish a rule that no one is allowed on the tanks accept the pumper when necessary to check bottoms, switch tanks etc. This will reduce risk, reduce cost, and reduce rejected crude oil loads.

25. Use drag-reducing agents (DRA) to reduce midstream pressures, increase pipeline capacity, and address bottlenecks. [25]

26. Design midstream systems with DRA incorporated into operating plans to reduce required equipment and construction costs. [26]

27. Standardize items with high-cost volatility.

28. Establish a vessel flushing program to proactively identify internal issues (separators, treaters, etc.) before expensive and dangerous failures occur—so we can take low-cost actions to prevent them.

29. On rod pump wells, based on observed wear patterns, switch from auto tubing rotation to manual tubing rotation. This should extend tubing and rod life. [27]

30. Capillary string well treatments to address specific issues.

31. Increase load line size and request reduced oil and water haul rates since oil and water can be pulled faster.

32. Construct mobile sand separators to move between wells.

33. Specialty stuffing box packings based on production specs for rod pump wells to reduce spills and repacking costs.

**34.** Mount cameras to look into pipelines using existing tap points to see liquid streams in gas thought to be dry gas by conventional measurement. Adjust to recover lost NGLs and improve dry gas measurement accuracy, increasing revenue. Reducing liquid carryover also reduces costs from measurement and equipment damage. [28]

**35.** Establish a consigned inventory program with vendors to reduce or eliminate capital tied up in inventory until items are needed.

**36.** Install external fluid level tank gauges that can be read on the side of the tanks from the ground on older wells, eliminating the need to go on top of the tanks, reducing cost and risk for pumpers, haulers, and others.[29]

**37.** Use external fluid level smart gauges with the ability to see tank levels visually and digitally. If SCADA systems are down, placing a camera in front of the tanks can be used as a low-cost backup system to see tank fluid levels.[30]

**38.** Boron-carbide (B4C) treated steel couplings in rod strings at key depths based on models and field data to reduce tubing wear, increase lift efficiency, decrease loads, and increase production.[31]

**39.** B4C treated plungers, rod pumps, gas lift mandrels, valves, surface equipment, and ESP components to reduce operating costs and increase mean time between failures.[32]

**40.** Consolidate gas meter runs and electricity meters to reduce contractual costs based on per-meter operating structures, maintenance, and calibration.

41. Establish a "Pre-Pull" system to ensure all options to fix issues have been attempted, surface issues have been ruled out, the most likely subsurface failure mechanisms have been identified, we have the replacement parts gathered, and are ready to perform subsurface intervention.

    o Establishing a Pre-Pull systemized checklist process increases workover efficiency, reducing total costs.

42. Establish a list of the optimal product cleaners, lubricants, sealants, and glues for our operation and conditions to fix problems without incurring expensive replacement costs, roustabout work, or workovers.

    o Often, non-oilfield consumer products can be used to achieve excellent results, avoiding expensive ops.

43. Identify vibration-related failures and costs. Then, adjust set points, reconfigure lines, stabilize issues, and remove items to reduce vibration and the associated costs.

44. Rigless ESP systems.[33]

45. Install automated chokes to maximize production by managing drawdown and prevent the shutting in of wells or flaring when issues occur with midstream assets.

46. In remote areas with limited midstream availability, use flares, VRUs, and heat sources (e.g., compressor exhaust) to evaporate produced water, eliminating water hauling and water disposal costs.[34]

47. Eliminate liquid biocides using ozone generated from air.[35]

48. Purchase equipment during favorable pricing periods.

49. Automate sand handling during production.[36]

50. Phaseout standard tubing anchors and replace with slim anchors to reduce gas locking from sand bridging, scale, iron sulfide, paraffin, and other choking agents that tend to form around anchors in rod pumps.[37]

51. Chemical-free scale control (calcite, barite, halite, salt) using radio frequency electromagnetics transmitted into the wellbore and surface equipment.[38]
    o This is a low-cost option to eliminate chemicals, acid wash treatments, expensive subsurface intervention, and the related downtime.

52. Fix annular gas migration, micro-annuli, backside pressure or annular gas flow with wireline conveyed localized casing expander tools.[39]

53. Establish a 24-hr production surveillance and optimization team with a focus on reducing risk and improving metrics.

54. Leverage the power of field knowledge by holding a weekly "Empirical Intelligence Meeting." Engage with our field team to gather data for discussion to incorporate into our "Virtual Pumper" artificial intelligence systems.

55. Eliminate the practice of signing deals with surface or mineral owners that result in reoccurring expenses.

56. Nanocoat components and vessels.[40]

57. Consolidate compressors, using one compressor for multiple wells, and right-size other AFL equipment as production declines.

58. Structure vendor-managed inventories to reduce cost and risks associated with operator inventory responsibilities.

59. Increase production data resolution. With increased telemetry data sampling rates, we will reduce OPEX because we can identify, troubleshoot, and address issues that are hidden between the sampling rates, preventing costly problems, failures, and downtime. For example, if we poll RTUs every 30 minutes, we will not see critical events and details visible with 1-minute data or less.

60. Audit the process of loading, hauling, unloading, and setting of equipment (e.g., sucker rods, ESPs, compressors, pumps) to identify transport habits causing damage impacting run time, and MTBF.

61. Virtual flowmeters (VFM) as production check meters to ensure measurement accuracy.[41]

62. Enhance production by altering NWB wettability.[42]

63. Spoolable thermoplastic pipe for flowline and pipeline networks to address corrosion, permeation, and paraffin issues reducing installation, inspection, and repair costs.[43]

64. Specialty rod lift pumps to prevent gas locking, deal with horizontal well slugging, and place pumps at $90°$.[44]

65. Multi-well controllers to monitor and control all pad wells from a single device, eliminating individual controllers for each well.[45]

66. Rigless well intervention methods and tools to reduce risk and cost.

67. Install solids jetting systems on heater treaters and separators to easily remove solids from the bottom of vessels without opening them up for cleaning.

**68.** Identify optimal economic production strategies other than producing wells 100% of the time. Consider intermittent intraday production cycles and shutting wells in for several days per week. Although this is common for rod pump wells, it has also added value for other types of artificial lift, including gas lift, and ESPs.

**69.** Reduce power costs and improve production by combining off-peak electric power rate discounts with off-peak cyclic and intermittent production strategies.

**70.** Plunger-assisted gas-lift (PAGL) and gas-assisted plunger-lift (GAPL) to extend economic life and reduce OPEX of artificial lift systems.

**71.** Permanent magnet motors (PMM) to drive ESPs and rod pumps, reducing power consumption and improve asset reliability.[46]

**72.** Develop intraday SCADA algorithms since daily averages are not representative of what is actually happening on location.

**73.** In conjunction with lab testing, dilute certain production chemicals with the appropriate base fluid. Depending on several factors, including seasonal temperatures, improved performance can occur at a lower OPEX, especially during colder months when full strength chemicals tend to congeal, clogging atomizers and other chemical equipment. Diluted blends could be administered more effectively, reducing the required amount of chemical needed.

74. Determine the optimal mix of electric powered, solar powered, natural gas- and diesel-powered artificial lift systems to diversify risk during weather events and other power disturbances to avoid being overexposed to one source of power such that an undesirable event could take down our entire company's production cash flow.

75. Route volumes to premium markets for optimal pricing. Manage gas flow to achieve increased NGL recoveries, sending production to optimal processing facilities.

76. Consider self-insurance options and company structures to supplement or replace standard insurance policies.

77. Install retention lines to reduce required chemical treatment volumes.

78. Establish a plan of action to address road icing conditions that prevent pumpers, oil haulers, and water haulers from safely reaching the wells. Purchase large amounts of ice melt in advance of ice storms and spread it accordingly. This has allowed haulers to safely reach my locations, reducing cost per unit of production and minimizing the need for steamers when everything freezes up if wells must be shut-in due to reaching storage tank capacity.

79. Remove invoice approval minimum thresholds to ensure every invoice is reviewed and approved by multiple people in the field and office before payment is made.

80. Systematically clean wellhead / facilities to spot problems. Equipment issues and spills often have small warning signs that are easier to identify when everything is clean.

81. Incorporate vehicle cost, leasing vs. buying, fuel economy, and maintenance when selecting field fleet vehicles. It is common for a company to have a default vehicle that they always select. Consider broadening the variety of vehicles and testing new options to help reduce costs. A few changes to fleet vehicle specs can save money on fuel and increase range, allowing us to be more efficient.

82. Install real-time vehicle tracking. GPS tracking will help us reduce risk and cost. If there is an accident or breakdown, we can quickly find and help our team. We can optimize pumper routes. Real-time vehicle location can be incorporated into our AI to direct our team when pumping-by-exception. We can program automated alerts into the tracking software. Tracking can also help lower our insurance premiums.

83. Use multiphase flowmeters to eliminate test separators, individual well vessels, lines, manifolds, and provide more accurate real-time production data.[47]

84. Install remote control artificial lift systems to prevent workover operations and keep wells online. If an anomaly is detected on SCADA or visually with our camera system, we can prevent subsurface intervention or significant surface damage by shutting down or slowing down our lift systems.

85. Install hydraulic rod pump units to provide options on how we pump our wells and make real-time adjustments including stroke length, acceleration, and dwell times, to prevent workovers and resolve issues remotely.[48]

86. Request manufacturers provide a written warranty on production-related equipment. This is especially critical if we purchase large volumes of equipment, including vessels, ESPs, compressors, tanks, pumping units, and there is a manufacturing defect resulting in failures. The cost to pull the equipment and replace it should be covered.

87. Consider long-stroke pumping units to reduce OPEX, increase production, and transition to rod lift faster.[49]

88. Use permanent and handheld thermal cameras to monitor tank levels, fire tubes, flares, and other surface items.

89. Test electric gas lift systems.[50]

90. Run fiberglass sucker rods to reduce OPEX, reduce pump unit size, reduce gearbox loads, reduce power costs, reduce downtime, provide corrosion resistance, achieve longer stroke length, increase flow area, and increase production.[51]

91. Use water conformance tech to reduce water-related costs.

92. Eliminate rejected crude oil loads and the need to use chemicals by producing into one tank and floating oil over into a second tank that we use as our sales tank.

93. Identify when unbundling is used against us with surprise charges, shrinkflation, additional fees, and trickle charges that are drip invoiced after the primary invoice was paid.

94. Upgrade valve program with custom items and internal components based on conditions to reduce risk and cost.

95. Install downhole gauges to address inefficiencies, reduce cost, and replace model assumptions with actual data.

96. Capillary tubing to increase efficiencies and reduce costs by injecting water, chemicals, or gas at depth to address $H_2S$, paraffin, salt, scale, and corrosion. Or create foams and surface tension modification to increase production.

97. Add custom metrics that pumpers will enter in daily production well-level reports, which allow us to prevent workovers, identify problems, and reduce costs.

98. Set up tank batteries to handle bad bottoms and properly juggle when we need to sell multiple loads of oil from tanks with high BS&W bottoms, not incur any costs, spend time to process, or use chemicals.

99. Enhance systems to capture operator incurred production costs to allocate to non-op WI partners. It is common for operators to not diligently capture non-invoiced costs incurred internally for production ops. This becomes especially critical for minority owner operators who incur costs that non-op partners will not pay unless documented and joint interest billed (JIB). These missing costs can make wells uneconomic for the operator but highly economic for the non-op WI partners.

100. Structure our production "Operational Issues List" review meetings to include representatives from drilling and completions to help identify production issues that can be solved upstream of production during drilling, completions, and wellsite facility construction—thereby eliminating issues, problems, and costs from impacting daily production operations.

# 10

## TRACKING PROGRESS

—

A numeric cost reduction objective and time frame to get there establish the quantifiable foundation for our efforts, focusing our company on a specific outcome. The exact cost reduction metrics selected are a matter of personal preference but should resonate with our team.

Whichever metrics are chosen, they must be easy to compile and track. Some metrics can be resource-intensive and more open to adjustment or manipulation.[1] We want to avoid a situation where staff spend more time compiling data, preparing reports, and sitting in meetings than working on actions that deliver sustainable cost reduction results.

Setting a target and time frame that everyone understands and can track helps us achieve our objectives because people naturally engage in things they comprehend and appreciate.

For example, if we initiate a cost reduction campaign with a target of reducing well costs by $1,000,000 per well in 30 weeks, it is easy to understand, remember, and participate in. Additionally, it is easy to track progress on a weekly basis.

During cost reduction efforts, I have found success by engaging with all team members daily, in the field and the office. Then, following up with an end-of-week update distributed across the company or business unit, showcasing our collective efforts and progress.

A "Weekly Cost Reduction Update" can include our cost reduction progress, how many weeks we have left to reach our goal, and details on the current week's efforts.

Cost reduction is challenging and stressful. Therefore, when we have an accomplishment, it adds value to give credit to all people who were involved and what they did. Putting this in the weekly updates provides team members with well-deserved recognition and appreciation for their hard work. It also motivates others to engage and contribute.

In large organizations, many team members rarely receive company-wide recognition for their contributions. It's not that each team member is not appreciated; it's just that there is not always a venue that commands the full attention to convey individual achievements across the entire organization.

However, one of the benefits and opportunities cost reduction campaigns provide is that they can act as a vehicle for individual and group contributions to be recognized and celebrated at the highest levels within our company.

Cost data and cash flow metrics are critical financial parameters incorporated into valuation models by analysts and investors. Anyone helping the company improve its economic status with cost reduction deserves to be recognized. Cost reduction campaigns provide that opportunity.

We must also act with a sense of urgency. The time component of cost reduction initiatives is critical because if the period is too long, there is less interest in taking action—people procrastinate. If it is too short, we disengage the team by attempting objectives that seem unrealistic or convey panic.

The most effective cost reduction programs in any industry take place over months, not years. If we need massive reductions, it might take multiple cost reduction campaigns over the course of years, but each campaign may only be three to nine months, allowing us to track our progress in greater detail on a weekly basis.

Cost reduction is nonlinear. We will have weeks during our cost reduction efforts when costs do not move, or they go up. Not only do we want to reduce costs, but we also want to reduce risk, and sometimes costs can go up as a result.

Deadlines and specific targets combined with tracking our progress on a regularly scheduled basis help focus our efforts and drive results. Not having deadlines or tracking targets is like playing a competitive sport without structure or keeping score. It may be fun, but it's not taken seriously.

As we track our cost reduction progress, we must confirm that our invoices reflect the reductions we perceive we are achieving. If our reductions are significant, it is not uncommon for vendors to take measures on their side of the equation to retain revenue. The only way to address this is to exercise extreme diligence during our invoice approval process.

# 11

## MOONSHOT ACTIONS

—

Every cost reduction campaign should contain a group of low-cost, low-risk moonshots. The term "moonshot" has multiple meanings but is most associated with the project by the United States to land a human on the Moon.

The actions to deliver what was considered impossible became known as a moonshot. The U.S. spent $25 billion on the Apollo program ($250 billion in 2024 dollars), culminating in the successful Apollo 11 crewed landing on the Moon, with five subsequent successful lunar landings.[1]

Moonshots do not have to be expensive. A significant amount of work can be performed before spending any impactful sums. Before attempting moonshots, let's perform as much of the operation as possible in the office and lab to minimize risk and ensure a high degree of success.

Where many operators get in trouble, based on personal experience, is when a vendor presents a new technology that looks excellent and promises game-changing results. The operator is sold on the new tech and tries it. However, it does not work out as promised, costing a small fortune. After a failure analysis, what is usually found out is that the salespeople were unaware of or did not disclose all the small issues. Before considering new tech, ask the uncomfortable questions multiple times to multiple people.

## New Technology Risk Management Checklist

1) What is the value proposition of this technology?
2) How many other operators have tried this technology?
3) How many continue to use this technology today?
4) Who developed this tech? I want to meet the people who invented this and dealt with development problems.
5) What are the direct and indirect risks with this tech?
6) What are the potential unintended consequences?
7) What problems have you had?
8) What is the cost of these problems?
9) How did you address the unintended costs with the client?
10) How do you quantify success?
11) What is your success rate?
12) What is the estimated cost versus actual cost spread?
13) How much money or time is saved with this technology?
14) How much money did each client save?
15) How are you compensated for selling this tech to me?
16) What incentives are you offering to take the risk?
17) How could this damage the well or reservoir?
18) Will you guarantee that it will work?
19) List the clients who have used the tech successfully and unsuccessfully. We want to speak with them.
20) Can you arrange a job visit to see it in action on location?

Many innovative companies embrace moonshot thinking to deliver massive results that are thought to be impossible. Moonshot actions in the oil and gas industry are more common than you may think. Unfortunately, they often fail and are not highly publicized. However, sometimes they work and change the world, as with unconventional oil and natural gas development—a successful moonshot and one of the most underappreciated accomplishments in recent history.

Not long ago, there were concerns that the world was imminently approaching peak crude oil production.[2] Energy provides the foundation for modern society, so this perceived reality led to many actions that negatively impacted humanity.

Arguably, the military action in Iraq, which cost over $2.0 trillion[3], thousands of American troops[4], and over two hundred thousand Iraqi civilians[5], was partially rationalized and driven by oil supply concerns, the value of oil to modern life, and the fact that oil is vital to the security of the United States.[6]

Unconventional oil and gas production has addressed the concern that we are imminently running out of oil and gas. Still, the cost of drilling, completion, and production must be further reduced to help make operations economic, expand unconventional development, and provide the foundation for affordable, sustainable energy over the next several hundred years as we continue to increase recovery rates above the current 10% to 20%.

Having been involved in multiple moonshot actions in the United States and overseas, the reality is that sometimes they work, but often they don't. The risk of failure should not be a deterrent. It should, however, make us hesitate to implement them if the risk and cost are significant.

Consider looking for moonshots that do not require taking excessive risk. Many things can be tried on sections of an individual well's development. For example, new BHAs or geosteering technology can be tested on a portion of the well instead of drilling a well with all new technology. New completion technology can be tried on a portion of the lateral instead of the entire lateral.

> **Implementing new technology in steps**
> **is a prudent strategy to reduce risk.**

Starting with rigorous testing on paper in the office or lab is a crucial first step. This ensures that the technology is thoroughly vetted and ready for the next phase, which could involve partnering with service providers or other operators and testing in steps. By testing the technology on segments of our operation, we further reduce risk and enhance our confidence in its effectiveness before broader implementation.

Research and development have shifted to service companies in recent years, so leveraging their testing abilities can significantly reduce cost. Service companies have test wells that new technology can be run on before it is deployed in the field. Once this looks promising, partnering with other operators during tech trials further reduces risk. A lot can be learned from technical partnerships. Additionally, leveraging our extreme benchmarking and reverse engineering programs will also assist in implementing low-cost moonshots.

# Moonshot Risk Management Actions

1) Test every conceivable scenario in the office on paper.
2) Construct computer models for new tech.
3) Work with service providers.
4) Test in the lab or on service provider facility test wells.
5) Form tech partnerships with universities and governments.
6) Partner with competitors, private investors, and NOCs.
7) Take non-op WI in competitor new tech test wells.
8) Consider new technology information exchange deals.
9) Employ reverse engineering and benchmarking programs.
10) Test new technology in segments on live operations.

Our goal is to test moonshot-level technologies that have the potential to exponentially increase value without exposing our company to significant financial risk. However, technology and risk are relative. What might be a moonshot for one operator may already be in use by another in a different area, reservoir, or application.

As we increase FCF, we enable opportunities to attempt moonshots that financially stressed competitors cannot. Visionary teams should be motivated by these opportunities to push costs down beyond current levels.

We are competing not only with our peer group but also with other forms of energy and entities attempting to destroy our industry. The more efficient we are, the more free cash flow we generate and the stronger we become. With that, please enjoy a thought-provoking list of "100 Moonshot Actions" in no particular order to help us achieve next-level innovation, moving us closer to our 100X value creation goals.

FREE CASH FLOW

LIBERATES US

TO ATTEMPT

THE

IMPOSSIBLE

# 100 Moonshot Actions

1. Capture $CO_2$ from atmosphere and store it in hydrocarbon reservoirs for dual purpose net negative $CO_2$ EOR.[7]

2. Boosted Frac: Inject natural gas in parent wells and legacy verticals prior to child frac wells until pressure rises and equalizes on child wells via the toe prep stage. After frac and millout, once child wells are all online, slowly flow back injected gas on booster wells (parent and verticals.)[8]

3. Replace the standard rotary drilling method with rock disintegration plasma technology.[9]

4. Chemical Bank: On child wells or new wells that will frac-hit operator-owned offset parent horizontals or legacy verticals; for stages with high frac hit potential, bullhead stage with frac-hit enhancement pad loaded with reservoir altering chemicals, remediation systems, or delayed advanced acid systems to increase production on offset frac-hit wells. Tailor system to stimulate offsets addressing paraffin, fines migration, condensate dropout, water blocks, permeability issues, scaling, gummy bears, sludges, emulsions, precipitates, iron sulfides, and swelling clays.[10]

5. Foamed natural gas fracturing with 60% standard slickwater and 40% injected natural gas using staged booster pumps stepping up to 10,000 psi foamed injection.[11]

6. Drill geothermal wells in central locations and use steam for power generation, winter heat trace, and reinjection into oil producers for thermal EOR including steamfloods, cyclic steam injection, and hot waterflooding.[12]

7. Saltwater Disposal Power Generation (SWD-PG): Use stored energy from saltwater disposal reservoirs to generate electric power. This can be accomplished in several ways, similar to pumped storage hydropower, a type of hydroelectric energy generation.[13]

8. Anthropogenic liquid $CO_2$ waterless fracturing.[14]

9. Particle impact drilling for hard rock ROP issues.[15]

10. Phase out wellsite facilities with multiphase flowmeters and virtual check meters. Use multiphase pipelines to deliver total production directly to upgraded refineries.[16]

11. Eliminate plug and perf with chemical packers pumped down and administered in flush stages combined with pulse-activated perforations integrated into the casing.

12. Full robotic drilling systems to unman rig operations.[17]

13. Process produced water for agricultural use, wildfire suppression, and reforestation.[18]

14. Develop high-energy plasma pulse fracturing technology to eliminate large volumes of water and chemicals.[19]

15. Steer-At-Bit directional drilling tech using Bernoulli's principle to replace mud motors and traditional RSS.[20]

16. Vacuum lift to drop BHP below standard AFL systems.

17. Titan Frac: $N_2$, $CH_4$, $CO_2$ mixed with methanol base fluid.

18. Eliminate risk of getting stuck with acid dissolvable drilling BHAs. If stuck during drilling, pump acid across BHA. Let soak to dissolve acid soluble BHA metal composite components over serval hours. Then trip out.[21]

19. Monetize minerals, including lithium (Li), rare-earth elements (REE), and rare earth metals (REM) from produced water—converting a cost into a revenue stream.[22]

20. Leverage the power of push-button vibration and acoustics to refracture wells at production decline-based scheduled intervals without subsurface intervention.

21. Improve zonal isolation, eliminate cement, increase burst pressures, and reduce pipe friction with large bore next-gen expandable casing.

22. Tractor disposable fiber optics into wells post millout to identify production characteristics on a per cluster basis.[23]

23. Replace frac plugs with sand plugs administered in flush.

24. Expand the use of drilled cuttings for soil farming, cement production, sintered bricks, road construction, lost circulation material and a variety of other products, transforming a cost into a cash flow stream.

25. Flameless treaters, line heaters, GPUs using waste heat.

26. Eliminate production casing and hydraulic fracturing with openhole explosive pulverization completions.

27. Using data obtained from shale development programs, develop lower cost uphole (shale, conventional, hybrid) reservoirs with modified unconventional technology.

28. Process produced water for use as industrial feedstock replacing freshwater usage in drought-affected areas.[24]

29. Eliminate proppant-based fracturing using reservoir rock erosion technology to create fractured voids.

30. Drill toe down with lift systems installed at the toe.

31. Identify rock that binds $CO_2$ molecules. Drill two horizontals connected at the toe in the $CO_2$ binding formation—complete open-hole or with a slotted liner. Then, inject raw atmosphere directly into one side, circulating "Clear Air" without $CO_2$ out the other side back into the atmosphere.

32. Develop AI, computer vision, automated pumpers, robotics, and overwatch drones to monitor ops.

33. Extreme low-pressure high-rate fracturing with advanced friction reduction (chemical + casing material) drastically reducing horsepower requirements.

34. Create AI robotics for real-time inspection of BHA, drill string, and casing during installation, drilling, tripping, reaming, stuck events, and well control situations.

35. Reduce flaring, emissions, trucking, and individual product pipelines with multiphase multiproduct flow measurement advancements. One pipeline for co-mingled oil, water, gas, and sand production.

36. Utilize natural gas, natural gas liquids, or cryogenically processed natural gas from offset wells or pipelines as the frac fluid proppant delivery system, replacing water.[25]

37. Consolidate drilling, completion, and production division corporate structures by leveraging the power of AI.

38. Use produced water to cultivate mangrove farms, which grow in saltwater and sequester 4X more $CO_2$ than trees.[26]

39. Replace proppant fracturing with customized acid blend fracturing based on geologic properties.

40. Eliminate location pad construction using oversized OTR tires and equipment stabilizers. Perform all oilfield operations on raw land and centralized facilities, eliminating road & pad construction cost and maintenance.

41. Process impurities ($CO_2$, $H_2S$, $N_2$) into products. [27]

42. Generate proppant anywhere using mobile rock-to-proppant technology that processes raw deposits and drilled cuttings into fracturing proppant.

43. Replace traditional metals and alloys with composites, enabling lower costs and new design options.[28]

44. Pickup mounted rugged 3D printers that can generate metal and plastic oilfield fittings, rubber gaskets, and other components on demand.[29]

45. Casing integrated frac-through subsurface safety valves.

46. Pursue horizontal geothermal development.

47. Casing plungers set at the toe that travel the full length of the horizontal sweeping liquid, sand, scale, sediment, and other chokes or undulation generated buildups.

48. Rigless runnable braided hose flexible production tubing.

49. Supplement rotary drilling with millimeter wave tech.[30]

50. Digitized tanks and vessels that provide real-time health status, diagnostics, 3-phase flow rates, internal images, and production uplift recommendations.

51. Perf rings that limit erosion to stabilize $\Delta P$ during frac.

52. Hydro-Blast Drilling: Increase ROP using frac horsepower combined with a high-rate jet drilling BHP to hydraulically erode and waterjet fail the rock.

53. Prevent production damage and smooth flow rates with horizontal slug production mitigation technology.[31]

54. Use tank battery (VOCs) combustible vapor to evaporate produced water.[32] Then, collect the solids (salts) and use as diverter during fracturing and acid block jobs.

55. Downhole computer vision automated geosteering.

56. Closed Loop Gas Capture Stimulation. When injecting gas into wells for temporary storage, administer ultralight gas transportable flakes to seal perfs taking gas and distribute injection with ultralight proppant across all perforations.

57. Digitized wellhead and frac stacks with internal seal isolation health monitoring and status updates.

58. Multiphase pumps to drop commingled wellhead pressure across multiple wells and reduce costs associated with producing at higher pressures.[33]

59. Develop quick seal well control technology connected to wellhead wing valves and activated in an emergency.

60. Identify expired human medications, drug rejects, and drug recalls for use in production chemicals.[34]

61. Develop transparent metals or equivalents to visually see into vessels, tanks, pipelines, frac iron, frac stacks, and wellheads to identify problems and opportunities.[35]

62. Build an AI robotic mud logging system.[36]

63. Work with regulators to approve drilling horizontal wells directly on section lines / drilling spacing unit boundary lines / hardlines, to recover stranded hydrocarbons.

64. In tightly spaced DSUs convert a few horizontal producers to injectors and inject water and/or natural gas. After a period, switch injectors to producers and producers to injectors to further increase recovery volumes. [37]

65. Identify waste products from other manufacturing sectors that can replace or reduce the cost of items used for drilling, completion, and production operations. [38]

66. Drill SWDs into faults to intentionally generate earthquakes in areas with high producing well counts. Induce low-magnitude controlled seismicity to increase recovery volumes with an earthquake refracturing program, approved by federal and state regulators.

67. In central development areas, build micro refineries and reinject flue gas (exhaust gas) into reservoirs for EOR. [39]

68. Run subsurface compressors to reduce BHP. [40]

69. Inject or drop microchips, encapsulated sensors, and nanodevices into drilling fluid, frac fluid, and millout fluid to gather subsurface data. Transmit data to the surface or recover the chips during circulation and production. [41]

70. Test produced water for brine mining operations including the production of iodine, magnesium, potash, bromine, boron, zinc, and tungsten. [42]

71. Eliminate frac with fishbone-style drilling sidetracks combined with acid jobs. [43]

72. Repurpose frac ponds, produced water tank farms, and wells on the P&A list, for Geomechanical Pumped Storage. When there is excess low-cost power, use it to pump frac pond water or produced water into a tight formation, supercharging it. When power has a high price, flow the well back through a turbine to generate power and fill the frac pond or tank battery back with water.[44]

73. Pump thermochemicals (Exo-Frac) into the reservoir to create a pressure-pulse that reduces rock strength, reduces the stress cage, and generates microfractures. This reduces hydraulic fracturing breakdown pressures, horsepower requirements, fluid volumes, and pump time. Tests suggest if slickwater fracturing is at 10,000 psi, Exo-Frac would be at 2,500 psi, dramatically reducing HHP costs.[45]

74. Combined wireline conveyed heated acid vapor with plug and perf operations to eliminate the need to pump acid.[46]

75. Drill two horizontal wells that meet at the toe. Leverage configuration to test new AFL technologies.

76. Drill wells North-South and East-West in the same DSU in a cross-trajectory pattern to increase recovery volumes by maximizing frac complexity.

77. Administer magnetite dust (nanometer-sized) into the frac fluid. Once the well is online and producing, establish an electromagnetic charge form of artificial lift to pull oil mixed with magnetite out of the reservoir.[47]

78. Glass Acid Frac: Replace sand with crushed glass and replace fresh water with acid systems.[48]

**79.** Downspace wells with the intention of achieving maximum frac hits. Trimul-frac the outer wells while confirming massive frac hits on the middle wells, using pressure signatures to modify diversion. Perforate the middle wells in stages and breakdown the perfs, but do not directly pump frac treatments on the middle wells. Then bring on all five wells with the goal of paying for frac treatments on three wells but getting five wells worth of frac treatments. Therefore, we get two wells for free.

**80.** Nanocoat proppant and reservoirs to increase recoveries.[49]

**81.** Analyze legacy logs, cuttings, sidewall, and whole core for non-hydrocarbon opportunities, including uranium, helium, cobalt, and storage formations.

**82.** Bioengineer rock-boring and rock-ingesting shipworms (engineered Lithoredo Abatanica) to bore into reservoir rocks, increasing porosity and permeability. There is a species of shipworm that ingests rock as it bores and is considered a rock-boring macro-bioeroder.[50]

**83.** Pump bioengineered rock-boring macro-bioeroders into the wells as part of a production enhancement program. The creatures bore into the rock increasing reservoir quality. Instead of production following a typical decline curve over time, production increases over time as porosity and permeability increase.

**84.** Multilaterals with simultaneous fracturing.

**85.** Acid dissolvable tubing that can be run on deeper joints into the horizontal past perfs and dissolved if stuck.[51]

86. Use a trimul-frac worth of HHP to inject at +200 bpm into one well during fracs. Using FR, larger casing, and increased perforations, frac larger intervals and generate more SRV improving production and reducing cost.

87. Develop a pressure control system that can tolerate significant water hammers. Then, on each stage, increase the fracturing rate to maximum and instantly cut rate to generate a massive water hammer to shatter the formation creating maximum complexity. Continue to ramp the rate up to maximum and instantly kill all the pumps creating water hammers in a cycle until injection pressure is dramatically reduced indicating that we altered the structure of the reservoir. Then pump the proppant.

88. Magnetize the casing and frac fluid to inject proppant via a maglev-type system. After the frac job, reverse the poles to remove the fluid, leaving the proppant behind.[52]

89. Use ISIP water hammers as a proxy for a minifrac to make the decision on a stage-by-stage basis on frac design size or to skip the stage. Correlate water hammer signatures to frac design and production results. Then take a quick ISIP at the beginning of each stage. Based on the signature, make the decision on optimal frac size or to skip.[53]

90. Limited entry maintenance tech. Guns coat perfs with erosion resistant material that also sticks to a pumped diverter to minimize erosion and maintain limited entry.

91. Run a capillary string to TD and inject foamer and natural gas for a poor-boy toe chemical gas lift system.[54]

**92.** Anti-Gravity Completions: Land low in zone. Perforate at zero degrees. Frac at extreme rates +150 bpm. Incorporate NWB and far field diverters. Vary the pump rate with drops and increases. Place on production and slowly drop BHP to near vacuum equivalent.

**93.** Perforate a quarter of the lateral. Pump at an extreme rate +200 bpm based on perf design. Drop aggressive diverters. Using pressure and surface tech determine SRV. Once lateral has been effectively stimulated, end stage. Goal is to reduce a typical 50 stage frac down to 5 to 10 stages.

**94.** Convert rigs, frac, and field to natural gas power. Drop midstream pressure to feed systems. Remove wellsite facility back pressure regulators and increase well production due to lower sales line backpressure.

**95.** Sand slurry pipeline systems using produced water to transport proppant to fracturing operations. [55]

**96.** Rod pump systems that can operate on multiple wells independently with a single unit. [56]

**97.** Automated capillary strings that are run in the horizontal and can move a programed distance each day to administer chemical, water, or acid directly across the perforations.

**98.** Floating proppant to increase effective frac geometry. [57]

**99.** Wireline perforating systems that allow the well to be perforated while pumping proppant.

**100.** Ramp frac rate up and down while pumping proppant and dropping diverter to flex the reservoir rock and create more SRV with a smaller treatment across more clusters.

# 12

## PERSEVERANCE

—

Cost reduction is not easy. It takes dedication, endurance, and hard work. At times, it can be a grind but do not get discouraged. Standard Oil's J.D. Rockefeller set his company's goal to be:

*"the best...at the lowest price."* [1]

He acquired many companies, making them stronger with efficiency and cost reduction. This enabled Standard Oil to develop over 300 new oil-based products, contributing to the progress of humanity and becoming one of the most influential companies in history. [2]

Rockefeller was obsessed with cost reduction. [3] Contrary to popular belief, the expansion of Standard Oil was not Rockefeller's objective—he utilized it as a strategy to reduce costs. [4] Even when Standard grew to a massive company with thousands of employees, Rockefeller meticulously inspected invoices line-by-line, looking for errors to help reduce costs. [5]

According to Rockefeller, the most important human quality to have for success is perseverance. He famously said:

*"I do not think there is any other quality so essential to success of any kind as the quality of perseverance. It overcomes almost everything, even nature."* [6]

Dealing with nature's forces at the ultimate level of creation—Earth, elements, gravity, and pressure—require strength and perseverance. To achieve massive cost reduction, we must be aggressive and willing to work relentlessly.

For example, many experts and industry leaders thought the Haynesville Shale would never be economical due to its depth, temperature (HP/HT), and resulting costs. Perseverance and a mindset to never give up resulted in Haynesville evolving into one of the most valuable energy assets in the world. [7]

This same dynamic occurred in the Permian Basin. Due to difficulties with the initial horizontals, the thinking at the time was that vertical oil well development was the better approach, and shale technology does not work at scale in the Permian. Fortunately, perseverance paid off, as Permian Basin assets developed with shale technology are currently the most coveted oil resources in the United States. [8]

When grinding towards a reduction objective, there will be times when it will be tempting to settle or give up—don't. If we have taken the time to set a realistic stretch cost reduction target and time frame, we will get there. It is more honorable to miss the target and acknowledge it than to abandon our goals. If we fall short, we take a break, analyze what happened, and gear up for another cost reduction campaign, another cost reduction push. Some cost reduction efforts take years. It's not ideal, but might be necessary, depending on the challenges. Regardless, there will be adversity when pushing for industry-leading results. Sometimes, no matter what we do, achieving the cost reduction targets seems elusive or impossible.

Perhaps efforts to reduce costs end up increasing them. The cost reduction effort may be successful initially, but a few months later, costs creep back up to their original levels. Sometimes, a change has unintended consequences, resulting in reduced production performance. No matter what happens, don't give up. Any level of safe cost reduction is a success in the long-term big picture—it's an incremental process.

New ideas and technology emerge daily. Therefore, we can never operate at the most efficient or lowest cost. If we have a selfless commitment to succeed, combined with perseverance, maximizing free cash flow with cost reduction will enable us to reach our value creation goals.

---

**During our journey, we reviewed over 700 actions to maximize cash flow with cost reduction.**

---

If you get anything out of this, please leave a book review.

I would greatly appreciate it.

Thank you for spending your time with me.

I look forward to meeting you one day.

# SOURCES

Regarding the experiences, cases, stories, and examples in this book, any resemblance to actual persons, situations, operations, or companies is purely coincidental.

## Chapter One: People First

1. Charles T. Munger, *Poor Charlie's Almanack: The Wit and Wisdom of Charles T. Munger*, Walsworth Publishing Company (2005); Benjamin Graham, *The Intelligent Investor: The Definitive Book on Value Investing*, Harper Collins (2009); Mary Buffet, *Warren Buffet and the Interpretation of Financial Statements: The Search for the Company with a Durable Competitive Advantage*, Scribner (2008); Thomas Phelps, *100 to 1 in the Stock Market: A Distinguished Security Analyst Tells How to Make More of Your Investment Opportunities*, Echo Point, Lightning Source (1972, 2015); Christopher Mayer, *100 Baggers: Stocks That Return 100-to-1 and How To Find Them*, Laossez-Faire Books (2018)

2. Paul Garvey, *Risk Matrix: An Approach for Identifying Assessing, and Ranking Program Risks*, Air Force Journal of Logistics, Volume XXII (1998)

3. D.L. DeMott, *Human Reliability and the Cost of Doing Business*, NASA, Johnson Space Center, Conference Paper, Science Applications International Corp. Houston, TX (2014); Red Bull, *More Than Sixty Percent of U.S. Workers Admit to Workplace Mistakes Due to Tiredness*, PR Newswire (2015); Catherine Reed, *17 Human Error Statistics*, Firewall Times (2023); Allianz, *Casualties-Behavioral and cultural risk still needs addressing* (2018), *available at* commercial.allianz.com/news-and-insights/expert-risk-articles/casualties-behavioral-and-cultural-risk.html; FAA, *Glider Handbook*, Chapter 13: Human Factors (2022); Jannik Lindner, *Human Error Statistics*, Worldmetrics Report 2024, worldmetrics.org (2024); OSHA, *OSHA Regional Instruction*, U.S. Department of Labor (2021); JJ Keller Safety Management Suite, *Preventing human error in the workplace* (2019), *available at* www.jjkellersafety.com/resources/articles/2019/preventing-human-error-in-the-workplace; NHTSA, Traffic Safety Facts, *A Brief Statistical Summary*, U.S. Department of Transportation (2015); EU-OSHA, *Human Error, European Agency for Safety and Health at Work* (2022), *available at* oshwiki.osha.europa.eu

4. DK, *Simply Artificial Intelligence*, DK (2023); Ronald Kneusel, *How AI Works*, No Starch Press (2024); IBM, *What is artificial intelligence (AI)* (2024), *available at* www.ibm.com/topics/artificial-intelligence; NVIDIA, *What is AI?*, NVIDIA, (2024), *available at* www.nvidia.com/en-us/glossary/ artificial-intelligence

5. Time, *Time Special Edition – Artificial Intelligence: A New Age of Possibilities*, Time

Magazine (2024); Authority Hacker, *149 AI Statistics: The Present & Future of AI at your Fingertips* (2024), *available at* www.authorityhacker.com/ai-statistics/; PWC, *Sizing the Prize*, PWCs Global AI Study (2017)

6. Matthew Hatami, *Oilfield Survival Guide*, Oilfield Books (2017)

7. Meyers & Briggs Foundation, *The 16 MBTI Personality Types* (2024)

8. Psyche, *Schadenfreude: why do we find joy in the pain felt by others?* (2024), *available at* psyche.co/ideas/schadenfreude-why-do-we-find-joy-in-the-pain-felt-by-others

9. Gallup, *State of the Global Workplace: 2023 Report* (2023)

10. *See* n.9, *supra*; *see also* Voucher Cloud, *Survey Reveals Employee Productivity* (2018); Salesforce, *What's Holding Employees Back from Being More Productive?* (2023) *available at* www.sales force.com/news/stories /state-of-work -stats-2023; Asana, *How work about work gets in the way of real work* (2024); Bostontec, *Employee Productivity Statistics* (2023), *available at* www.bostontec.com; Apollo; Rich Russakoff, Mary Goodman, *Employee Theft: Are You Blind to It?*, CBS News (2011), *available at* www.cbsnews.com/news/employee-theft-are-you-blind-to-it; Tessa Kaye, *Crime/Employee Theft*, Introduction to Risk Management Group Project, Temple University (2016);Willis North America, *Executive Risks Alert*, Willis Group (2008); Anne Willkomm, *The Impact of Too Many Meetings*, Drexel University (2023), *available at* drexel.edu/graduatecollege /professional-development/blog/2023/February/The-impact-of-too many meetings; Ben Tobin, *It's not just you: 4 in 5 Americans stressed out from poor office communication*, USA Today (2019); Asana, *Embracing the new age of agility*, Anatomy of Work Global Index 2022 (2022); Megan Milliken and Leo Lindner, *The Future of Energy & Work in the United States*, True Transition (2023); Jim Harter, *Employee Engagement on the Rise in the U.S.*, Gallup (2018), *available at* news.gallup.com /poll/241649/employee-engagement-rise.aspx; Kotter, *Does corporate culture drive financial performance?*, Forbes (2012), *available at* forbes.com/sites/johnkotter /2011/02/10/does-corporate-culture-drive-financial-performance; Newswire, *Eagle Hill's New Workplace Culture Survey*, Eagle Hill Consulting (2019); Harvard Business Review, *The Impact of Employee Engagement on Performance*, HBS Publishing (2013); Scott Keller, *Focus on the Five Percent*, McKinsey & Company, (2018), *available at* www.mckinsey.com /capabilities/people-and-organizational-performance/our-insights/the-organization-blog/focus-on-the-five-percent; SHL, *Find and Retain your HIPO Talent*, SHL.com, *available at* www.shl.com/solutions/talent-management/hipo; Workleap, *Statistics on the importance of employee feedback* (2014), *available at* workleap.com/blog/infographic-employee-feedback; Jim Harter and Kristi Rubenstein, *The 38 Most Engaged Workplaces in the World Put People First*, Gallup (2020), *available at* gallup.com

/workplace/290573/engaged-workplaces-world-put-people-first

11. Britannica Money, *Human Capital definition*, Britannica.com (2024)

12. *See* n.9, *supra.*

13. *Id.*

14. *Id.*

15. *Id.*

16. Voucher Cloud, *Survey Reveals Employee Productivity* (2018), *available at* vouchercloud.com /better-living/office-worker-productivity

## Chapter Two: Shale Economics

1. Jim Schleckser, *Why the Golden Rule of Business is Don't Run Out of Money*, Inc (2017)

2. *See* Chapter One n.1, supra. *also see* Bank of America, *Understanding Free Cash Flow*, BOA (2023), *available at* business.bankofamerica.com/resources/free-cash-flow; Joel Stern, *The EVA Challenge*, Wiley (2001); George Christy, *Free Cash Flow*, Wiley (2009); CFI Team, *FCFE*, CFI (2024), *available at* corporatefinance institute.com/resources/valuation/free-cash-flow-to-equity-fcfe; Tim Vipond, *The Ultimate Cash Flow Guide*, CFI (2024), *available at* corporatefinanceinstitute .com/resources/valuation/cash-flow-guide-ebitda-cf-fcf-fcff; CFA Institute, *Free Cash Flow Valuation*, CFA Program (2024), *available at* cfainstitute.org/en/ membership/professional-development/refresher-readings; Joel Stern, *Theory and Policy of Modern Finance*, Columbia University (2008)

3. *See* n.2, *supra. also see* Roger Mesznik, *Advanced Corporate Finance*, Readings and Cases, Columbia University (2008); Navin Chopra, *Corporate Finance*, Columbia University (2007); Zvi Bodie, *Investments*, 7th Edition, McGraw-Hill Irwin (2008); Richard Brealey, *Principles of Corporate Finance*, McGraw-Hill Irwin (2008)

4. OPEC, *OPEC Share of World Crude Oil Reserves*, OPEC (2023), *available at* opec.org; EIA, *What Drives Crude Oil Prices?*, EIA (2024), *available at* eia.gov

5. OPEC, *Our Mission*, OPEC (2024), *available at* opec.org/opec_web/en/about_us

6. EIA, *What is OPEC+ and how is it different from OPEC?*, Today in Energy (2023), *available at* eia.gov/todayinenergy/detail.php?id=56420

7. Dave Lavinsky, *Pareto Principle: How to Use It To Dramatically Grow Your Business*, Forbes (2020); Richard Koch, *The 80/20 Principle*, 3rd Edition, Hachette (1999)

8. Joel Stern, *Theory and Policy of Modern Finance*, Readings and Cases, Columbia University (2008); Stern Value Management, *Four Decades of Value Creation* (1972), *available at* sternvaluemanagement.com/about-us/our-history

9. Joel Stern, *Theory and Policy of Modern Finance*, Class Discussion, Columbia University, Columbia Business School (2008)

10. *See* n.8, *supra.*

11. *See* n.3, *supra.*

12. Brice Morlot, *Reserve-Based Lending & Insurance*, SCOR (2019); Gillian Tan, *Regulatory Warn Banks on Loans to Oil, Gas Producers*, WSJ (2015), *available at* wsj.com/articles/banks-facecurbs-onoil-gaslending-1435866277;Andrew Scurria and Christopher Matthews, *Banks Cut Shale Driller's Lifeline as Losses Mount*, WSJ (2020); Office of the Comptroller of the Currency, *Oil and Gas Exploration and Production Lending*, Comptroller's Handbook, OCC (2016)

13. Office of the Comptroller of the Currency, *Oil and Gas Exploration and Production Lending*, Comptroller's Handbook, OCC (2016)

14. *See* n.12, *supra.*

15. *Id.*

16. *See* n.11, *supra, also see* Amir Azar, *Reserve Base Lending and the Outlook for Shale Oil and Gas Finance*, Columbia (2017)

17. Caleb Fielder, *Marginal Wells and the Doctrine of Production in Paying Quantities*, Landman Magazine (2011); Patrick Ottinger, *Production in "Paying Quantities" in These Challenging Days: How Muh Financial Stress Can Your Lease Withstand?*, 67 RMMLF-INST 6 , 6-35 (Vol. 67 Rocky Mountain Mineral Law Institute) (2021)

18. Nassim Taleb, *The Black Swan: The Impact of the Highly Improbable*, Random House (2007)

19. Michele Wucker, *The Gray Rhino: How to Recognize and Act on the Obvious Dangers We Ignore*, St. Martin's Griffin (2017)

20. *See* n.17, *supra.*

## Chapter Three: Systemize & Automate

1. Will Parker, *What Went Wrong with Zillow? A real-estate algorithm derailed its big bet*, WSJ (2021), *available at* wsj.com/articles/zillow-offers-real-estate-algorithm-homes-ibuyer-11637159261; Ronny Reyes, *Zillow reaches deal to sell 2,000 of its 18,000 homes after winding up its disastrous house-flipping program*, Daily Mail (2021), *available at* dailymail.co.uk/news/article-10187631/Zillow-reaches-deal-sell-2-000-homes-winding-house-flipping-program.html; Curt Devine, *Boeing relied on single sensor for 737 Max that had been flagged 216 times to FAA*, CNN (2019), *available at* cnn.com/2019/04/30/politics/boeing-sensor-737-max-faa/index; Henrico Dolfing, *The $440MM Software Error at Knight Capital* (2019), *available at* henricodolfing.com/2019/06/project-failure-case-study-knight-capital.html; SEC, *In the Matter of Knight Capital Americas LLC*, SEC (2013), *available at* sec.gov/litigation/admin/2013/34-70694.pdf; Nathaniel Popper, Knight

Capital Says Trading Glitch Cost It $440 Million, NYT (2012), *available at* archive.nytimes.com/dealbook.nytimes.com/2012/08/02/knight-capital-says-trading-mishap-cost-it-440-million; U.S.-Canada Power System Outage Task Force, *Final Report on the August 14, 2003 Blackout in the U.S. and Canada, Causes and Recommendations* (2004), *available at* ferc.gov; ELCON, *The Economic Impacts of the August 2003 Blackout* (2004), *available at* nrc.gov; Reuters, *Spike in deaths blamed on 2003 NY blackout* (2012), *available at* reuters.com/article/us-blackout-newyork/spike-in-deaths-blamed-on-2003-new-york-blackout-idUSTRE80Q 07G20120127; Christopher Kock, *Nike rebounds: How Nike recovered from it s supply chain disaster*, CIO (2004), *available at* cio.com; Andrew Pollack, *Two Teams, Two Measures Equaled One Lost Space*craft, NYT (1999), *available at* nytimes.com/www.nytimes.com/library/national/science; William Bulkeley, *A Cautionary Network Tale: FoxMeyer's High-Tech Gamble*, WSJ (1996), *available at* wsj.com/articles/SB847835969502714000; David Hoffman, *I Had a Funny Feeling in My Gut*, Washington Post (1999), *available at* washingtonpost.com /wp-srv/inatl/longterm/coldwar/shatter021099b.htm; David Wright, *How Could a Failed Computer Chip Lead to Nuclear War?*, Future of Life Institute (2016), *available at* futureoflife.org/nuclear/failed-computer-chip-lead-nuclear-war; NASA, *What was Mariner 1?*, NASA (2024), *available at* science.nasa.gov/mission

2. David Wright, *How Could a Failed Computer Chip Lead to Nuclear War?*, Future of Life Institute (2016), *available at* futureoflife.org/nuclear/failed-computer-chip-lead-nuclear-war; PACOM, *U.S. Indo-Pacific Command documents* (1980), *available at* nsarchive.gwu.edu/nukevault

3. Dan Neil, *Sleeping in Self-Driving Cars? It's No Pipe Dream*, WSJ (2020), *available at* www.wsj.com/articles/sleeping-in-self-driving-cars-its-no-pipe-dream-11585165420

4. Eric Schlosser, *Command and Control: Nuclear Weapons, the Damascus Accident and the Illusion of Safety*, Penguin Books (2014)

5. Curt Devine, *Boeing relied on single sensor for 737 Max that had been flagged 216 times to FAA*, CNN (2019), *available at* cnn.com/2019/04/30/politics/boeing-sensor-737-max-faa/index

6. *See* n.4, *supra.* see also, Matt Stevens, *Causes of False Missile Alerts: The Sun, the Moon, and the 46-Cent Chip*, NYT (2018), *available at* nytimes.com

7. Will Parker, *What Went Wrong with Zillow? A real-estate algorithm derailed its big bet*, WSJ (2021), *available at* wsj.com

8. Bill Highleyman, *The Great 2003 Northeast Blackout and the $6 Billion Software Bug*, the Availability Digest, Sombers Associates (2007)

9. Roger Franklin, *5 Steps to a successful ERP Change Management Plan*, SYSPRO,

(2016), *available at* syspro.com

10. Julia Werdigier, *BP Envisions Leaner Future Under Its New Chief*, NYT (2010), *available at* nytimes.com; David Gelles, *Boeing Fires CEO Dennis Muilenburg*, NYT, (2019), *available at* nytimes.com/2019/12/23/business/Boeing-ceo-muilenburg; U.S. DOJ, *Former Massey Energy CEO Sentenced to a Year in Federal Prison*, DOJ Office of Public Affairs (2016), *available at* justice.gov; Robert Tuttle, *Suncor Replaces CEO Mark Little After Oil Sands Mine Death*, Bloomberg (2022), *available at* bloomberg.com; David Shepardson, *U.S. issues arrest warrant for former VW CEO*, Reuters (2018), *available at* reuters.com; Katherine Blunt, *CEO of PG&E Steps Down Amid California Wildfire Crisis*, WSJ (2019), *available at* wsj.com; Liz Moyer, *Equifax CEO suddenly 'retires' following an epic data breach*, CNBC (2017), *available at* cnbc.com; Bart Meijer, *Philips parts ways with CEO in midst of massive recall*, Reuters (2022), *available at* reuter.com; Rob Copeland, *Silicon Valley Bank Ex-CEO is 'Truly Sorry' but Deflects Blame*, NYT (2023), *available at* nytimes.com; BBC News, *Rio Tinto boss Tom Albanese resigns over $14B write-down*, BBC (2013), *available at* bbc.com/news/world-middle-east-21056203

11. Theodore Rockwell, *The Rickover Effect: How One Man Made a Difference*, iUniverse (2002); Marc Wortman, *Admiral Hyman Rickover: Engineer of Power*, Yale (2022)

12. *See* n.11, *supra*

13. *See* n.11, *supra*, also *see* Rudy Abramson, Rickover, *Creator of U.S. Nuclear Navy, Dies at 86*, Los Angeles Times (1986), *available at* latimes.com/archives/la-xpm-1986-07-09-mn-14301-story.html

14. Paul Cantonwine, *Caught in the Leadership Paradox: Insight from Admiral Rickover*, Nuclear Newswire, *available at* ans.org/news/article-1592/caught-in-the-leadership-paradox; Hyman Rickover, *The Never-Ending Challenge of Engineering: Admiral H.G. Rickover in His Own Words*, American Nuclear Society (2015)

15. PBS Documentaries, *Rickover: The Birth of Nuclear Power*, PBS (2014)

16. *See* n.11 and 13-15, *supra*.

17. Taylor Locke, *Jeff Bezos: This is the 'smartest thing we ever did' at Amazon*, CNBC, (2019), *available at* cnbc.com; Lex Fridman, *Jeff Bezos: Amazon and Blue Origin*, Interview, Lex Fridman Podcast #405; *available at* youtube.com

18. *See* n.17, *supra*.

19. *Id.*

20. OSHA, *Near-Miss Incident Report Form*, OSHA (2024)

21. Howard Duhon, *Bhopal: A Root Cause Analysis of the Deadliest Industrial Accident in History*, JPT (2014), *available at* jpt.spe.org; Allan McDonald, *Truth, Lies, and O-rings*, University Press of Florida (2012); Adam Higginbotham, *Challenger: A True Story of Heroism and Disaster on the Edge of Space*, Simon & Schuster (2024); Matthew

Hatami, *Oilfield Survival Guide*, Oilfield Books (2017); Adam Higginbotham, *Midnight in Chernobyl: The Untold Story of the World's Greatest Nuclear Disaster*, Simon & Schuster (2020); *Chernobyl: The History of a Nuclear Catastrophe*, Basic Books (2020); ABC News, *Toyota to Pay $1.2B for Hiding Deadly 'Unintended Acceleration'*, ABC (2014), *available at* abcnews.go.com; Avery Hartmans, *'Antennagate' just turned 10. Here's how the iPhone 4's antenna issues became one of Apple's biggest scandals of all time*, Business Insider (2020), *available at* businessinsider.com; Yukari Iwatani, *Apple Knew of iPhone Issue*, WSJ (2010), *available at* wsj.com; Michael McCarthy, *The Hidden Hindenburg*, Lyons Press (2022); John Geoghegan, *When Giants Ruled the Sky: The Brief Reign and Tragic Demise of the American Rigid Airship*, The History Press (2022); Airships, *Hydrogen Airship Disasters* (2024) *available at* airships.net/hydrogen-airship-accidents; NASA, *Columbia Accident Investigation Board*, NASA (2003); Adam Keiper, *A New Vision for NASA*, The New Atlantis, (2003), *available at* thenewatlantis.com/ publications/a-new-vision-for-nasa; Rachel Gordon, *The contribution of human factors to accidents in the offshore oil industry*, Elsevier (1996); The Hon. Lord Cullen, *The Public Inquiry into the Piper Alpha Disaster*, Department of Energy. Volume 1 and 2, H.M.S.O. (1990); NASA Safety Center (NSC), *System Failure Case Study Details- The Case for Safety, The North Sea Piper Alpha Disaster* (2013), Case Study and Presentation available at nsc.nasa.gov; Stephen McGinty, *Fire in the Night: The Piper Alpha Disaster*, Pan Macmillan (2017); Christina Rogers, *Takata Learned Early of Air-Bag Problems*, WSJ (2014), *available at* wsj.com

22. Charlie Pittock, *Chernobyl: 'Critical' design flaw 'too sensitive to be made widely known'*, Express (2022), *available at* express.co.uk/news/world/1603877/chernobyl-news-ukraine-russia-war-kgb-files-soviet-union-1986-disaster-spt

23. Avery Hartmans, *'Antennagate' just turned 10. Here's how the iPhone 4's antenna issues became one of Apple's biggest scandals of all time*, Business Insider (2020)

24. Christina Rogers, *Takata Learned Early of Air-Bag Problems*, WSJ (2014)

25. Douglas Martin, *Warren Anderson, 92, Dies, Faced India Plant Disaster*, NYT (2014); CBS, *Wife: Ex-Exec 'Haunted' by Bhopal Gas Leak*, CBS News (2009) *available at* cbsnews.com/news/wife-ex-exec-haunted-by-bhopal-gas-leak/

26. *See* n.25, *supra.*

27. Warren Buffet, *Letter to the Shareholders of Berkshire Hathaway*, Letters (2014)

28. Prateek Joshi, *Artificial Intelligence with Python*, Packt Publishing (2020); Ronald Kneusel, *How AI Works*, No Starch Press (2024); Laurence Moroney, *AI and Machine Learning for Coders*, Oreilly Media (2020); James Phoenix, *Prompt Engineering for Generative AI*, Oreilly Media (2024)

29. Fred Barnard, *One Look is Worth a Thousand Words*, Printer's Ink (1921)

30. Indeed Editorial Team, *What is Institutional Knowledge?*, Indeed (2022)

31. IEF, *Upstream Oil and Gas Investment Outlook*, IEF and S&P (2024)

32. Matthew Hatami, *Oilfield Survival Guide*, Oilfield Books (2017)

33. Jamie Condliffe, *Dueling Neural Networks*, MIT Technology Review (2020)

34. *See* n.33, *supra*.

## Chapter Four: Risk Management

1. Matthew Hatami, *Oilfield Survival Guide*, Oilfield Books (2017)

2. Warren Buffet interview, *available at* youtube.com/watch?v=8OcegOGAGIs

3. *See* n.2, *supra*.

4. Bettina Casad, *confirmation bias*, Encyclopedia Britannica (2024), *available at* britannica.com/science/confirmation-bias

5. Brian Lund, *Common trading biases and how to overcome them*, Encyclopedia Britannica (2023), *at* britannica.com/money/behavioral-biases-in-finance

6. Editors of Encyclopedia Britannica, *delusion*, *Encyclopedia Britannica* (2024)

7. Databricks, *Automation Bias*, *at* databricks.com/glossary/automation-bias

8. Merriam-Webster, *Herd mentality*, *available at* merriam-webster.com/dictionary

9. Merriam-Webster, *Groupthink*, *available at* merriam-webster.com/dictionary

10. Eric-Jan Wagenmakers, *Origin of the Texas Sharpshooter*, Bayesian Spectacles, (2018) *available at* bayesianspectacles.org/origin-of-the-texas-sharpshooter/

11. *See* n.10, *supra*.

12. *Id.*

13. Sean Kenehan, *Not Invented Here Syndrome explained*, Learnosity, *available at* learnosity.com/edtech-blog/not-invented-here-syndrome-explained

14. Ivy Wigmore, *backfire effect*, TechTarget, *available at* techtarget.com

15. Cristel Russell, *Why We Don't Like to be Told What to Do*, Psychology Today (2021), available at psychologytoday.com/us/blog/the-savvy-consumer

16. Gallup, *State of the Global Workplace: 2023 Report* (2023)

17. Cambridge Dictionary, *tribalism*, *available at* dictionary.cambridge.org

18. Kevin Cook, Kitty Genovese: The Murder, the Bystanders, the Crime that Changed America, W.W. Norton & Company (2014)

19. *See* n.6, *supra*.

20. Deb Amlen, *We Do Not See Things as They Are*, NYT (2017), *at* nytimes.com

21. Tavarish Roberts, *I Knew It All Along...Didn't I?*, at psychologicalscience.org

22. Caeleigh MacNeil, *How the sunk cost fallacy influences our decisions*, Asana (2024)

23. UX Hints, Law of the Hammer in UX Designs, at uxhints.com

24. *See* n.5, *supra*.

25. William Samuelson, *Status Quo Bias in Decision Making*, Journal of Risk and Uncertainty, Springer (1988)

26. The Knowledge, The Expert is Always Right? (2022) *available at* theknowledge.io

27. Clay Halton, *Outcome Bias: What is Means, How it Works* (2022), *available at* investopedia.com/terms/o/outcome-bias.asp

28. Wayne Rosenkrans, *Normalization of Deviance*, Flight Safety Foundation (2015), *available at* flightsafety.org/asw-article/normalization-of-deviance/

29. Virginia State Crime Commission, *Texting While Driving*, Annual Report (2013)

30. David Robson, *The Intelligence Trap*, W.W. Norton & Company (2019)

31. *See* n.30, *supra.*

32. Allan McDonald, *Truth, Lies, and O-rings*, University Press of Florida (2012)

33. Society of American Archivists, *institutional memory*, at dictionary.archivists.org

34. Ben Cohen, *It's the Most Thankless Job in Banking. Silicon Valley Bank Didn't Fill It for Months*, WSJ (2023), *available at* wsj.com

35. *See* n.34, *supra.*

36. *Id.*

37. *Id.*

38. Greg Brown, *The Silicon Valley Bank Collapse: Takeaways and Lessons Learned*, Kenan Institute of Private Enterprise (2023), *at* kenaninstitute.unc.edu

39. Eliot Brown, *Silicon Valley Bank Dropped a Hedge Against Rising Rates in 2022*, WSJ, (2023), *available at* wsj.com

40. *See* n.34, *supra.*

41. *Id.*

42. Leslie Groves, *Now It Can Be Told: The Story of the Manhattan Project*, Da Capo Press (1983); Richard Rhodes, *The Making of the Atomic Bomb*, Simon & Schuster (2012); Kai Bird, *American Prometheus*, Vintage (2006); PBS, *The Bomb* (2015); Richard Rhodes, *Dark Sun: The Making of the Hydrogen Bomb*, Simon & Schuster (1996); Cynthia Kelly, *The Manhattan Project*, Black Dog & Leventhal (2020)

43. *Id.*

44. *Id.*

45. *Id.*

46. *Id.*

47. *Id.*

48. *Id.*

49. *Id.*

50. *Id.*

51. *Id.*

52. *Id.*

53. *Id.*

54. *Id.*

55. Baron Capital, *Ron Baron interview with Elon Musk*, NYC (2022); Walter Isaacson, *Elon Musk* Simon & Schuster, (2023)

56. *See* n.42, *supra.*

57. Burton Folson, *John D. Rockefeller and the Oil Industry*, FEE (1988), *available at* fee.org/articles/ john-d-rockefeller-and-the-oil-industry; Ron Chernow, *Titan: The Life of John D. Rockefeller, Sr.*, Vintage Books (1998)

58. *See* n.57, *supra,* also see Steven Loborec, *Patterning Your Department After Great Leaders: John D. Rockefeller*, NIH (2015), *available at* ncbi.nlm.nih.gov

59. *See* n.57, *supra*

60. *See* Chapter 3, n.11 and 13-15, *supra.*

61. Michael Junge, *Leadership and Decision*, Naval War College (2020)

62. Statement of Admiral H.G. Rickover, *Before the Subcommittee on Energy Research and Production of the Committee on Science and Tech*, *U.S. House of Representatives* (1979)

63. Hyman Rickover, *The Never-Ending Challenge of Engineering: Admiral H.G. Rickover in His Own Words*, American Nuclear Society (2015)

64. *See* n.63, *supra.*

65. *See* n.1, *supra.*

66. Jack Browning, *Top court filing statistics from around the country*, One Legal (2023) *available at* onelegal.com/blog/top-court-filing-statistics-united-states

67. *See* n.66, *supra.*

68. Federal Reserve System, *2024 Federal Reserve Stress Test Results* (2024); Ryan Tracy, *What Are the Fed's Stress Tests?*, WSJ (2017), *available at* wsj.com

69. *See* n.1, *supra.*

70. *Id.*

71. Robert Sternberg, *Thinking and Problem Solving*, Academic Press (1994)

72. *See* n.71, *supra.*

## Chapter Five: Supply Chain Actions

1. James Gleick, *Isaac Newton*, Vintage, (2004); Robert Lloyd, *Writing the Literature Review*, FHSU Pressbooks, (2023) *available at* fhsu.pressbooks.pub

2. *Id.*

3. *Id.*

4. *Id.*

5. FBI, *How does the FBI differ from the CIA?*, FAQs, *available at* fbi.gov; CIA, The *CIA's Updated Executive Order 12333 Attorney General Guidelines*, *at* cia.gov

6. Alex Gibney, *Zero Days – documentary film*, Magnolia Pictures (2016); Kim Zetter, *Countdown to Zero Day: Stuxnet and the Launch of the World's First Digital Weapon*, Crown (2015); Fred Kaplan, *Dark Territory*, Simon & Schuster (2016)

7. *Id.*

8. *Id.*

9. *Id.*

10. *Id.*

11. *Id.*

12. Jack Repcheck, *Copernicus' Secret: How the Scientific Revolution Began*, Simon & Schuster (2008); Douglas Brunt, *The Mysterious Case of Rudolf Diesel: Genius, Power, and Deception on the Eve of World War*, Atria (2023); Arthur Firstenberg, *The Invisible Rainbow: A History of Electricity and Life*, Chelsea Green Publishing (2020); James Mahaffey, *Atomic Awakening: A New Look at the History and Future of Nuclear Power*, Pegasus (2010); Office of Fossil Energy and Carbon Management, *DOE's Early Investment in Shale Gas Technology Producing Results Today*, DOE (2011) *available at* energy.gov/fecm/articles/does-early-investment-shale-gas-technology-produci ng-results-today; Eric Lax, *The Mold in Dr. Florey's Coat: The Story of the Penicillin Miracle*, Hot Paperbacks (2005)

13. Lise Meitner and O.R. Frisch, *Disintegration of Uranium by Neutrons: a New Type of Nuclear Reaction*, Nature (1939)

14. Malcom Gladwell, *Outliers: The Story of Success*, Back Bay Books (2011)

15. Biblehub, *New Living Translation*, Tyndale House Publishers, Inc. (2007), *available at* biblehub.com/nlt/job/38.htm

16. Google, *Talks at Google, available at* blog.google/inside-google/talks-google

17. *See* n.16, supra.

18. Goldman Sachs, *Goldman Sachs Talks, available at* goldmansachs.com/intelligence /series/goldman-sachs-talks; Standard Industries, *Standard Speaker Series, available at* standardindustries.com/in-the-news/speaker-series; Adobe, *Distinguished Lecture Series, available at* research.adobe.com/distinguished-lecture-series; Columbia Business School, *Distinguished Speaker Series, available at* business.columbia.edu/realestate/events

19. Rob McCarney, *The Hawthorne Effect: a randomized, controlled trial*, NIH (2007), *available at* ncbi.nlm.nih.gov/pmc/articles/PMC1936999; MBA Knowledge Base, *Elton Mayo's Hawthorne Experiment and It's Contributions to Management, available at* mbaknol.com/management-principles/elton-mayos-hawthorne-experiment-and-its-contributions-to-management; Western Electric, *Statistical Quality Control Handbook by Western Electric*, Mack (1958); Ayesh Perera, *Hawthorne Effect: Definition, How It Works, and How to Avoid It*, Simply Psychology, *available*

*at* simplypsychology.org/hawthorne-effect.html

20. *Id.*

21. *Id.*

22. California Restaurant Association, *Employee Theft: Why do employees steal?*, *available at* calrest.org/labor-employment/employee-theft-why-do-employees-steal

23. Chris Kolmar, *22 Stunning Employee Theft Statistics [2023]: Facts Every Employer Should Know*, *Zippa* (2023), available at zippia.com/advice/employee-theft-statistics; Zurich, *Three employee-related litigation trends companies need to manage*, (2024) *available at* zurichna.com/knowledge/articles/2024/01/three-employee-related-litigation-trends-companies-need-to-manage; U.S. DOJ, *Looking Out of Number One – Employee Theft*, Security World (1979), *available at* ojp.gov/ncjrs/virtual-library/abstracts/looking-out-number-one-employee-theft; Inge Black, *Internal controls and investigations*, Investigations and the Art of the Interview, Science Direct (2021), *available at* sciencedirect.com/topics/psychology /better-business-bureau; Ivy Walker, *Your Employees Are Probably Stealing From You. Here Are Five Ways To Put An End To It*, Forbes (2018), *available* forbes.com/sites/ivywalker/2018/12/28/your-employees-are-probably-stealing-from-you-here-are-five-ways-to-put-an-end-to-it

24. Association of Certified Fraud Examiners, *Report the the Nations: 2020 Global Study on Occupational Fraud and Abuse*, ACFE (2020)

25. Mary Goodman, *Employee Theft: Are You Blind to It?*, CBS News (2011), *available at* www.cbsnews.com/news/employee-theft-are-you-blind-to-it; Tessa Kaye, *Crime/Employee Theft*, Introduction to Risk Management Group Project, Temple University (2016); Willis North America, *Executive Risks Alert*, Willis (2008)

26. *See* Chapter 1, n.10, supra.

27. *See* n. 23, supra.

28. Etienne Romsom, *Countering global oil theft: responses and solutions*, Working Paper, (2022), *available at* wider.unu.edu

29. AP News, *Crude crime: 4 men accused of stealing $2.4 million of oil*, AP (2022), *available at* apnews.com/article/crime-north-dakota-theft

30. Western District of Oklahoma, *Federal Jury Convicts in Multi-Million-Dollar Oil and Gas Fraud*, DOJ (2019), *available at* justice.gov/usao-wdok/pr/federal-jury-convicts-multi-million-dollar-oil-and-gas-fraud; Western District of Oklahoma, *Oil and Gas Repairman Sentenced to 33 Months for $450,000 False Invoice Scheme*, DOJ (2018), *available at* justice.gov/usao-wdok/pr/oil-and-gas-repairman-sentenced-33-months-450000-false-invoice-scheme; Southern District of Texas, *Former Energy Company Executives Arrested for Embezzling more than $1 Million from Employer*, DOJ (2015), *available at* justice.gov/usao-sdtx/pr/former-energy-company-

executives-arrested-embezzling-more-1-million-employer; SEC, *SEC Charges UAE-Based Brooge Energy and Former Executives with Fraud,* SEC (2023), *available at* sec.gov/newsroom/press-releases/2023-256; U.S. Secret Service, *Oil & Gas Office Manager Pleads Guilty to Embezzling Over $1.2 Million,* U.S. *Attorney's Office,* (2019), *available at* secretservice.gov/press/releases/2019/10/oil-gas-office-manager-pleads-guilty-embezzling-over-12-million; Western District of Oklahoma, *Indictment Unsealed Charging Houston Man with Defrauding Oklahoma City Company with False Delivery Invoices,* DOJ (2015), *available at* justice.gov/usao-wdok/pr/indictment-unsealed-charging-houston-man-defrauding-oklahoma-city-company-false; Northern District of Oklahoma, *Drumright Man Pleads Guilty to Stealing More Than $400,000 from an Illinois Oil and Gas Company,* DOJ (2021) *available at* justice.gov/usao-ndok/pr/drumright-man-pleads-guilty-stealing-more-400000-illinois-oil-and-gas-company; Southern District of Texas, *Two Former Executives of Houston-Based Oil Supply Company Plead Guilty in Illegal Kickback Scheme,* DOJ (2016), *available at* justice.gov/usao-sdtx/pr/two-former-executives-houston-based-oil-supply-company-plead-guilty-illegal-kickback; Antitrust Division, *Investigation into Procurement Irregularities at the Department of Energy Strategic Petroleum Reserve Results in Guilty Plea and Indictment,* DOJ (2023), *available at* justice.gov; FBI New Orleans Division, *Former Shell Oil Executive Sentenced to 18 Months' Imprisonment for Income Tax Evasion and Mail Fraud,* FBI, (2009), *available at* archives.fbi.gov/archives/neworleans/press-releases/2009

31. CNBC, *Full Transcript: Billionaire Investor Warren Buffet Speaks with CNBC's Becky Quick on "Squawk Box" Today,* CNBC TV (2018), *available at* cnbc.com/2018/05/07/full-transcript-billionaire-investor-warren-buffett-speaks-with-cnbcs-becky-quick-on-squawk-box-today

32. *See* n. 31, supra.

33. Berkshire Hathaway Inc., Code of Business Conduct and Ethics, available at berkshirehathaway.com/govern/ethics.pdf; Berkshire Hathaway Inc., *Corporate Governance, available at* berkshirehathaway.com/govern/govern.html

34. NYT, *The Vanishing Salad Oil: A $100 Million Mystery,* NYT (1964), *available at* nytimes.com/1964/01/06/archives/the-vanishing-salad-oil-a-100-million-mystery.html; Matthew Partridge, *Great frauds in history: Tino De Angelis' salad-oil scam,* Money Week, *available at* moneyweek.com/investments/investment-strategy/601677/great-frauds-in-history-tino-de-angelis-salad-oil-scandal, Madmedic, *How Salad Oil Almost Crashed the U.S. Economy,* Medium (2020), *available at* medium.com/@madmedic11671/how-salad-oil-almost-crashed-the-u-s-economy-c3ed3c2cb797; Myles Udland, *Bill Ackman wants us to remember how a young Warren Buffett took advantage of the 'Great Salad Oil Scandal of 1963',* Business

Insider, *available at* finance.yahoo.com/news/bill-ackman-wants-us-remember-154454835.html

35. *Id.*
36. *Id.*
37. *Id.*
38. *See* n. 31, supra.
39. *See* Chapter 2 n. 2, supra.
40. Julie Steinberg, *Saudi Aramco Looks to Supply-Chain Finance to Free Up Cash*, WSJ (2021), *available at* www.wsj.com/articles/aramco-looks-to-supply-chain-finance-to-free-up-cash-11618577732
41. *See* n. 40, supra.

## Chapter Six: Pre-Spud Actions

1. BLM, *Surface Operating Standards and Guidelines for Oil and Gas Exploration and Development*, 4th Edition, U.S. DOI (2007)
2. Caterpillar, *Dual-fuel Engine Specs* (2024); Halliburton, *Dual Fuel Systems* (2024)

## Chapter Seven: Drilling Actions

1. Halliburton, *Halliburton Cementing Tables*, Halliburton Red Book (1999)
2. Vallourec, *Material Selection Guide*, Vallourec S.A. (2024)
3. API, *API Specification 5CT*, 9th Edition, American Petroleum Institute (2012)
4. Anne Willkomm, *The Impact of Too Many Meetings*, Drexel University (2023) *available at* drexel.edu/graduatecollege/professional- development/blog/2023
5. *See* Chapter 3, n.11 and 13-15, *supra.*
6. *Id.*
7. *Id.*
8. Matthew Hatami, *Oilfield Survival Guide*, Oilfield Books (2017)
9. Eliyahu Goldratt, *The Goal: A Process of Ongoing Improvement*, North River (2014)
10. *Id.*
11. John Maxwell, *The 360 Degree Leader*, Harper Collins (2011)
12. Truckstop, *Deadhead Miles*, *available at* truckstop.com/blog/deadhead-miles
13. Mushambi Mutuma, *Never Ask a Barber if You Need a Haircut* (2015), *available at* medium.com/@mushambimutuma
14. Danny Boyd, *Top Permian Operators Leveraging Technology To Drive Bottom-Line Success*, The American Oil & Gas Reporter (2022), *available at* aogr.com/magazine/cover-story/top-permian-operators-leveraging-technology-to-drive

-bottom-line-success

15. Weatherford, *Mobile Bucking Units*, Tubular Running Services (2014)

16. Schlumberger, *Managed pressure drilling services*, SLB (2024), *available at* slb.com

17. Schlumberger, *@balance Xpress Mobile MPD service*, SLB (2024), *available at* slb.com/products-and-services/innovating-in-oil-and-gas/well-construction/rigs-and-equipment/managed-pressure-drilling-services/atbalance-xpress-mobile-mpd-services

18. Purify Fuel, *nanO2 Combustion Catalysts, Case Studies, available at* purifyfuel.com

19. TechnipFMC, *Offline Cementing*, Tech Bulletin (2020), *available at* technipfmc.com

20. Nabors, *Integra Tubular Running Services* (2024), *available at* nabors.com

21. Tomax, *Anti Stick-Slip Tool (AST)* (2024), *available at* tomax.no/products

22. Nabors, *SmartSlide Automated Slide Drilling* (2024), *available at* nabors.com

23. Impulse, *ActiPulse Ball Drop Vibration Control, available at* impulsedownhole.com

24. Aggreko, *Microgrid Nat Gas Generators* (2024), *available at* aggreko.com

25. NOV, *Downhole Adjustable Motors* (2024), *available at* nov.com/products

26. *See* n.25, *supra.*

27. Halliburton, *Sperry Drilling Vibration Monitoring and Mitigation Guidelines*, Sperry Drilling, Halliburton (2024), *available at* halliburton.com

28. CAT, *Smart Power Management Revolutionizes Rig Performance* (2024), available at cat.com/en_US/articles/customer-stories/oil-gas

29. See n.27, supra.

30. Newpark, *Water-based fluid systems* (2024), *available at* newpark.com

31. Baker Hughes, *Drill bits* (2024), *available at* bakerhughes.com

32. TTS, *Casing XRV* (2024), *available at* ttsdrilling.com

33. See n.32, supra.

34. NCS Multistage, *Well Construction* (2024), *available at* ncsmultistage.com

35. H&P, *Drilling Automation* (2024), *available at* helmerichpayne.com

36. Peter Boul, *Nanotechnology Research and Development in Upstream Oil and Gas*, Wiley, (2019) *available at* onlinelibrary.wiley.com/doi/full/10.1002/ente.201901216

37. Halliburton, *Cerebro In-Bit Sensing* (2023), *available at* halliburton.com

38. NOV, *PosiTrack TVM Tool* (2024), *available at* nov.com/products

39. Drilling Tools International, *Drill-N-Ream* (2024), *available at* drillingtools.com

40. Baker Hughes, *Telemetry services* (2024), *available at* bakerhughes.com

41. Gyrodata, *Technologies & Services, available at* gyrodata.com

42. Enventure, *Expandable Liner Technology, available at* enventuregt.com

43. Trent Jacobs, *Why Shell Drilled a "Horseshoe" Well in the Permian Basin*, Journal of Petroleum Technology (JPT) (2020), *available at* jpt.spe.org/why-shell-drilled-horseshoe-well-permian-basin

44. *See* n.35, supra.

45. IBM, *What is computer vision?* (2024), *available at* ibm.com/topics/computer-vision

46. NOV, *IntelliServ Wired Drill Pipe* (2024), *available at* nov.com/products

## Chapter Eight: Completion Actions

1. John Charnes, *Financial Modeling with Crystal Ball and Excel,* Wiley (2012); Christian Robert, *Monte Carlo Statistical Methods,* Springer (2010); Samprit Chatterjee, *Regression Analysis by Example,* Wiley (2006)

2. *Id.*

3. *Id.*

4. *Id.*

5. *Id.*

6. Dark Vision, *Downhole Imaging* (2024), *available at* darkvisiontech.com; EV, *Technology* (2024), *available at* evcam.com; EXPRO, *Downhole Video* (2024), *available at* expro.com; DHVI, *Downhole Video* (2024), *available at* DHVI.net

7. *See* n.6, supra.

8. Halliburton, *ExpressFiber disposable fiber service* (2024), *available at* halliburton.com

9. Paramount, *The Godfather Trilogy: Francis Ford Coppola DVD commentary* (2023)

10. Seismos, *Products* (2024), available at seismos.com

11. GeoDynamics, *Technology & Services* (2024), *available at* perf.com

12. Enhanced Energetics, *Resources* (2024), *available at* enhancedenergetics.com

13. ThruTubing Solutions, *Slicfrac Zonal Isolation,* *available at* thrutubing.com

14. Colter Cookson, *Advanced Diverters Speed Operations,* The American Oil & Gas Reporter (2020), *available at* aogr.com/magazine/sneak-peek-preview/advanced-diverters-speed-operations

15. GR Energy Services, *Solutions* (2024), *available at* grenergyservices.com; Tejas Completion Services, *Services* (2024), *available at* tejascompletionservices.com

16. Kevin Creighton, *Get Back In The Action: The New York Reload,* Shooting Illustrated (2023), *available at* shootingillustrated.com/content/get-back-in-the-action-the-new-york-reload;

17. Merriam-Webster, *Hobson's choice* (2024), *available at* merriam-webster.com

18. SandPile, *Sand Pile Services* (2024), *available at* thesandpile.com

19. *See* n.17, supra.

20. Downing, *Completions (2024), available at* downingusa.com

21. *Id.*

22. Emerald Surf Sciences, *Technology & Services* (2024), *available at* emeraldsurf.net; Emerald Surf Sciences, *Scale Demonstrations, Laminar Flow vs Turbulent Flow* (2019)

23. RedZone, *Coil Tubing Research and Analysis, Presentation,* Oklahoma (2019)

24. *See* n.19-20, supra. *also see* Charles Pope, *Stop, Drop and Circulate, An Engineered Approach to Coiled Tubing Intervention in Horizontal Wells,* SPE Lecture (2020) *available at* streaming.spe.org/course-stop-drop-and-circulate-an-engineered-approach-to-coiled-tubing-intervention-in-horizontal-wells

25. NETL, *Unconventional Resources,* U.S. DOE (2019) *available at* osti.gov

26. Downing, *Completions* (2024) *available at* downingusa.com

27. Colter Cookson, *Modern Frac Fleets Deliver Wells Faster And Consume Less Fuel,* The American Oil and Gas Reporter (2023) *available at* aogr.com/magazine/sneak-peek-preview/modern-frac-fleets-deliver-wells-faster-and-consume-less-fuel

28. Hunting, *Preloaded Perforating Gun Service* (2024), *available at* huntingplc.com

29. Cactus, *Products* (2024) *available at* cactuswhd.com

30. Trent Jacobs, Chevron, *Oxy Share Details on Better Mouse Traps for Permian Completions,* JPT (2021) *available at* jpt.spe.org/chevron-oxy-share-details-on-better-mouse-traps-ffor-permian-completions

31. Liberty, *Greaseless Wireline Cables* (2024) *available at* libertyenergy.com

32. FET, *Float Equipment* (2024), *available at* f-e-t.com

33. Atlas Energy Solutions, *Dune Express* (2024), *available at* atlas.energy

34. *See* n.11, supra.

35. Cold Bore, *Our Products, available at* coldboretechnology.com; Corva, *Energy Solutions* (2024) *available at* corva.ai

36. DCS, *DRIFLOWX* (2024) *available at* downholechem.com

37. Halliburton, *Completions* (2024) *available at* halliburton.com

38. Core Laboratories, *Completions* (2024) *available at* corelab.com

39. G&H, *Yellow Jacket Products* (2024) *available at* ghdiv.com

40. Blake DeNoyer, Jack Owen, *An analysis of Fine-mesh Proppant,* Hart Energy (2019), *at* hartenergy.com/exclusives/analysis-fine-mesh-proppant-182211

41. *See* n.23, supra.

42. Amr Radwan, *Engineered Microparticles Improve Formation Breakdown,* Optimize Well Productivity, The American Oil & Gas Reporter (2023) *available at* aogr.com/magazine/frac-facts/engineered-microparticles-improve-formation-breakdown-optimize-well-productivity

43. Halliburton., *Operator Saves $125,000 with new High-viscosity Friction Reducer* (2023) *available at* halliburton.com/en/resources/operator-saves-125000-usd-with-new-high-viscosity-friction-reducer

44. Halliburton, *Auto Pumpdown Service* (2023) *available at* halliburton.com/en/products/auto-pumpdown-service

45. Stephen Whitfield, *Centralized start-stop system helps reduce fuel consumption, emissions*

*from idling engines in hydraulic fracturing operations*, Drilling Contractor (2023), *available at* drillingcontractor.org/centralized-start-stop-system-helps-reduce-fuel-consumption-emissions-from-idling-engines-in-hydraulic-fracturing-operations-64851

46. NexTier, *Services* (2024), *available at* nextierofs.com
47. Enercorp, *eFlowback* (2024), *available at* enercorp.net
48. ThruTubing Solutions, *Products* (2024) *available at* thrutubing.com

## Chapter Nine: Production Actions

1. Emily DiNuzzo, *This Is Why People Believe Pennies Bring Good Luck*, Reader's Digest, (2023) *available at* rd.com/article/why-are-pennies-lucky/#:~:text=All%20day%20long%20you'll,to%20protect%20people%20from%20evil.

2. Roger Lowenstein, *When Genius Failed: The Rise and Fall of Long-Term Capital Management*, Random House (2001)

3. *Id.*

4. *Id.*

5. *Id.*

6. Andreas Jager, *The Rise and Fall of Long-Term Capital Management*, Medium (2023), *available at* medium.com/@andreas.jaeger.rpllc/the-rise-and-fall-of-long-term-capital-management-34e16be931cd

7. Pumps & Pipes, *Overview* (2024) *available at* pumpsandpipes.org/about

8. API, *Standards and Manuals* (2024) *available at* api.org; API, *Manual of Petroleum Measurement Standards (MPMS), available at* api.org; *also see* requirements relative to your area of operations and regulatory body.

9. Cimarron, *Tankless Facilities* (2020) *available at* cimarron.com

10. Pioneer Energy, *Emissions Control Treater* (2024), *available at* pioneerenergy.com

11. Nidhi Subbaraman, Tornadoes, *Hurricanes and Wildfires Racked Up $165 Billion in Disaster Damage in 2022* (2023), WSJ, *available at* wsj.com/articles/tornadoes-hurricanes-and-wildfires-racked-up-165-billion-in-disaster-damage-in-2022-11673366441; Adam Smith, *2023: A historic year of U.S. billion-dollar weather and climate disasters*, NOAA (2024) *available at* climate.gov/news-features/blogs/beyond-data/2023-historic-year-us-billion-dollar-weather-and-climate-disasters#:~:text=In%202023%2C%20the%20United%20States,Consumer%20Price%20Index%2C%202023).

12. George Washington, *Letter From George Washington to John Trumbull, 25 June 1799*, National Archives (1799) *available at* founders.archives.gov/documents/Washington/06-04-02-0120

13. Chris Hoofnagle, Aniket Kesari, Aaron Perzanowski, *The Tethered Economy*, The George Washington Law Review (2019) *available at* gwlr.org/wp-content/uploads/2019/10/87-Geo.-Wash.-L.-Rev.-783.pdf

14. *Id.*

15. EOG Resources, *Closed Loop Gas Capture, IOGCC Annual Conference 2021*, OCC, *available at* oklahoma.gov/content/dam/ok/en/iogcc/documents/ac/2021/special_feature_-_iogcc_closedloopgascapture_final.pdf

16. Schlumberger, *VX Spectra Flowmeter Enables Real-Time Sand Quantifications During Drillout in the Eagle Ford Shale* (2017), *available at* slb.com

17. Estis Compression, *Compression Technologies, High Pressure Gas Lift* (2024) *available at* estiscompression.com; JW Power, *Discussions on HPGL* (2023)

18. Ambyint, *Rod Lift Optimization* (2024), *available at* ambyint.com

19. TenEx, *Products* (2024) *available at* tenextechnologies.com

20. NOV, *Hercules HP Universal Wellhead* (2024) *available at* nov.com

21. XStream Lift, *XStream Lift Scout* (2024) *available at* xstreamlift.com

22. Sage Rider, *Paraffin Melting Technology* (2024) *available at* sageriderinc.com

23. ChampionX, *Sucker rod failure analysis* (2024) *available at* championx.com/contents/NOR_Sucker%20Rod%20Failure%20Analysis_BR_0322.pdf

24. Weatherford, *Continuous Rod* (2024) *available at* weatherford.com

25. LSPI, *Capabilities and Applications* (2024) *available at* liquidpower.com

26. *Id.*

27. Philip Hart, *Techniques to Reduce Operating Costs for Increased Reserves and Profitability*, SPE Lecture, SPE (2020) *available at* streaming.spe.org/techniques-to-reduce-operating-costs-for-increased-reserves-and-profitability-philip-hart

28. Process Vision, *Products* (2024) *available at* processvision.com

29. American Tank Gauge, *Gauges* (2024) *available at* americantankgauge.com

30. *Id.*

31. Endurance Lift Solutions, *Blaze* (2024) *available at* endurancelift.com; *B4C Technologies, Properties, Advantages* (2024) *available at* b4ctechnologies.com

32. *Id.*

33. NOVOMET, *Products and Services* (2024) *available at* novometgroup.com; Baker Hughes, *Products and Services* (2024) *available at* bakerhughes.com

34. Heartland Water Technology, *Oil and Gas Produced Water Management Service* (2024) *available at* heartlandtech.com; Colter Cookson, *Disposal Constraints Drive Creative Research Into Water Treatment, Reuse* (2024) The American Oil & Gas Reporter, *available at* aogr.com/magazine/sneak-peek-preview/disposal-constraints-drive-creative-research-into-water-treatment-reuse

35. Spartan Environmental Technologies, *Ozone Water Treatment Systems* (2024),

*available at* spartanwatertreatment.com

36. *See* Chapter 8, n.23, supra.

37. TechTAC, *Products* (2024) *available at* techtac.com

38. ClearWELL, *Oil and Gas Applications* (2024) *available at* clearwellenergy.com

39. Renegade Services, *Local Expander* (2024) *available at* renegadewls.com

40. Aculon, *Applications – Oil and Gas* (2024) *available at* www.aculon.com/oil-gas

41. Timur Bikmukhametov, Johannes Jäschke, *First Principles and Machine Learning Virtual Flow Metering: A Literature Review*, Journal of Petroleum Science and Engineering, Volume 184, 2020, *available at* sciencedirect.com/science/article/pii/S0920410519309088

42. Nissan Chemical America Corporation, *Products* (2024) *available at* nissanchem-usa.com; Polymer Technologies, *Services and Case Studies* (2024) *available at* polymerior.com; Biolin Scientific, *Applications – Oil and Gas* (2024) *available at* biolinscientific.com/industries/oil-and-gas

43. Baker Hughes, *Onshore Composite Pipe* (2024) *available at* bakerhughes.com

44. Raise Production, *Product Information* (2024) *available at* raiseproduction.com; Endurance Lift, *Specialty Pumps* (2024) *available at* endurancelift.com

45. Rockwell Automation, *Products* (2024) *available at* rockwellautomation.com; Flowco, *Automation* (2024) *available at* flowcosolutions.com; OleumTech, *Products and Solutions* (2024) *available at* oleumtech.com

46. Baker Hughes, *ESP Systems - Motors* (2024) *available at* bakerhughes.com

47. Ingar Tyssen, *6 Operational Benefits of Multiphase Flow Meters in Unconventional Oil Production*, Emerson (2020), *available at* emersonautomationexperts.com/2020/measurement-instrumentation/flow/6-operational-benefits-of-multiphase-flow-meters-in-unconventional-oil-production

48. SilverJack Lift Systems, *Application, Optimization* (2024) *available at* silverjack.ca; Douglas Kinnairdi, *Pump Adjustments with Just a Click, Hart Energy* (2017) *available at* hartenergy.com/exclusives/pump-adjustments-just-click-176734

49. Weatherford, *Long-Stroke Pumping Unit* (2024) *available at* weatherford.com; Liberty Lift, *XL Long Stroke Units* (2024) *available at* libertylift.com

50. Sage Rider, *Electric Gas Lift System* (2024) *available at* sageriderinc.com

51. Superod, *Products* (2024) *available at* superod.com

## Chapter Ten: Tracking Progress

1. Jerry Muller, *The Tyranny of Metrics*, Princeton University Press (2018)

## Chapter Eleven: Moonshot Actions

1.  Alan Shepard, *Moon Shot: The Inside Story of America's Apollo Moon Landings*, Turner Publishing (1994); Cambridge Dictionary, *Moonshot* (2024), *available at* dictionary .cambridge.org/us/dictionary/english/moonshot; Merriam-Webster, *A New Meaning of 'Moonshot'* (2020), *available at* merriam-webster.com/wordplay /moonshot-words-were-watching; The Planetary Society, *How much did the Apollo program cost?* (2024) *available at* planetary.org/space-policy/cost-of-apollo

2.  Matthew Simmons, *Twilight in the Desert: The Coming Saudi Oil Shock and the World Economy*, John Wiley & Sons (2005); Kenneth Deffeyes, *Hubbert's Peak: The Impending World Oil Shortage*, Princeton University Press (2008); Reuters, *Peak oil closer than IEA forecasts show: report* (2009) *available at* reuters.com/article /idUSTRE5A85JT; Terry Macalister, *Key oil figures were distorted by US pressure, says whistleblower*, The Guardian (2009) available at theguardian.com/environment /2009/nov/09/peak-oil-international-energy-agency; Brian Schwartz, *Q&A: What is peak oil*, Johns Hopkins Bloomberg School of Public Health (2008), *available at* publichealth.jhu.edu/2008/schwartz-peak-oil-background

3.  Robert Draper, *Iraq, 20 Years Later: A Changed Washington and a Terrible Toll on America*, NYT (2023) *available at* nytimes.com/2023/03/20/us/politics/iraq-20-years.html; Gordon Lubold, *U.S. Spent $5.6 Trillion on Wars in Middle East and Asia: Study*, WSJ (2017) *available at* wsj.com/articles/study-estimates-war-costs-at-5-6-trillion-1510106400; Daniel Trotta, *Iraq war costs U.S. more than $2 trillion – study*, Reuters (2013), *at* reuters.com/article/iraq-war-anniversary-cost/iraq-war-costs-u-s-more-than-2-trillion-study-idINDEE92D0BR20130314

4.  Neta Crawford, *20 Years of War, A Costs of War Research Series*, Brown University, (2019) *available at* watson.brown.edu/costsofwar; U.S. Department of Defense, *Casualty Status* (2024) *available at* defense.gov/casualty.pdf

5.  Neta Crawford, *Costs of War – Iraqi Civilians*, Brown University (2023) *available at* watson.brown.edu/costsofwar; BBC News, *Iraqi study estimates war-related deaths at 461,000* (2013) *available at* bbc.com/news/world-middle-east-24547256

6.  Alan Greenspan, The Age of Turbulence: Adventure in a New World, Penguin Boks (2008); Antonia Juhasz, *Why the war in Iraq was fought for Big Oil*, CNN (2013), *available at* cnn.com/2013/03/19/opinion/iraq-war-oil-juhasz/index; Michael Moore, *Six Years Ago, Chuck Hagel Told the Truth About Iraq*, Huff Post (2013) *available at* huffpost.com/entry/chuck-hagel-iraq-oil_b_2414862; John Duffield, *Oil and the Decision to Invade Iraq*, Georgia State University, (2012) *available at* scholarworks.gsu.edu/cgi/viewcontent.cgi?article=1089&context= political_science_facpub

7. Christophe McGlade, *Can CO2-EOR really provide carbon-negative oil?*, IEA (2019) *available at* iea.org/commentaries/can-co2-eor-really-provide-carbon-negative-oil; Amrith Ramkumar, *A Long-Shot Climate Bet Suddenly Turns Hot*, WSJ, (2023) *available at* wsj.com/us-news/climate-environment/carbon-removal-credits-climate-startups-investments-4aa3ca70

8. Trent Jacobs, *Fighting Water With Water: How Engineers Are Turning the Tides on Frac Hits*, SPE, (2018) *available at* jpt.spe.org/twa/fighting-water-water-how-engineers-are-turning-tides-frac-hits; Science Direct, *Gas Injection* (2024) *available at* sciencedirect.com/topics/engineering/gas-injection

9. GA Drilling, *Plasmabit* (2024) *available at* gadrilling.com; Kocis, *Novel Deep Drilling Technology based on Electric Plasma Developed in Slovakia*, URSI, (2017) *available at* ursi.org/proceedings/procGA17/papers/Paper_H41P-6(1402).pdf

10. Kairos Energy Services, *Capabilities* (2024) *available at* kairosenergyservices.com

11. Southwest Research Institute, *Evaluating Natural Gas as a Hydraulic Fracturing Fluid*, SWRI (2018), *available at* www.swri.org/technology-today/evaluating-natural-gas-foam-hydraulic-fracturing-fluid

12. Natalia Cano, *Power from Geothermal Resources as a Co-product of the Oil and Gas Industry: A Review*, American Chemical Society (2022) *available at* ncbi.nlm.nih.gov/pmc/articles/PMC9670100

13. Pacific Northwest National Laboratory, *Generating Hydropower from Injection Wells*, PNNL (2024) *available at* pnnl.gov/sites/default/files/media/file/EED_2613_BROCH_AltOpsPocketGuide_final_web.pdf; American Water Works Association, *California utility turning water pressure into electricity*, AWWA, (2024) *available at* awwa.org/AWWA-Articles/california-utility-turning-water-pressure-into-electricity

14. Chris Carpenter, *New Closed Blender Reduces Footprint, Gasification of CO2 in Waterless Fracturing*, JPT (2020) *available at* jpt.spe.org/new-closed-blender-reduces-footprint-gasification-co2-waterless-fracturing

15. Particle Drilling Technologies, *Applications and Technology* (2024) *available at* particledrilling.com

16. Sven Olson Sr., *Multiphase Technologies Increase Production Rates, Improve Reserves Recovery*, The American Oil & Gas Reporter (2020) *available at* aogr.com/magazine/editors-choice/multiphase-technologies-increase-production-rates-improve-reserves-recovery

17. Nabors, *PACE-R800 rig* (2024) *available at* nabors.com

18. Encore Green Environmental, *Conservation By-Design* (2024) *available at* encoregreenenvironmental.com; Darren Smith, *Re-engineered Technology Cleans Produced Water for Green Use*, Oilman Magazine (2021) *available at* oilmanmagazine

.com/article/re-engineered-technology-cleans-produced-water-for-green-use

19. Novas Energy, *Plasma Pulse Technology* (2024) *available at* novasenergy.ca; Robert Rapier, *Plasma Pulse: A Clean Fracking Alternative That Requires No Water Or Chemicals*, Forbes (2022) *available at* forbes.com/sites/rrapier/2021/ 12/16/a-clean-alternative-to-fracking-that-requires-no-water-or-chemicals

20. Enteq Technologies, *Products* (2024) *available at* enteq.com

21. Parker Hannifin, *Dissolvable Materials Technology* (2024) *available at* parker.com; Damorphe, *Innovation* (2024) available at damorphe.com; Luxfer Mel Technologies, *Dissolving Alloys* (2024) *available at* luxfermeltechnologies.com

22. U.S. DOE, *Produced Water from Oil and Gas Development and Critical Minerals*, (2024) *available at* energy.gov/sites/default/files/2024-06/Produced%20Water%20fr om%20Oil%20and%20Gas%20Development%20and%20Critical%20Mineral s%20Fact%20Sheet_6.18.24.pdf; Nancy Luedke, *Mining The Treasures Locked Away In Produced Water*, Texas A&M Today, (2024) *available at* today.tamu.edu /2024/01/23/mining-the-treasures-locked-away-in-produced-water

23. Halliburton, *ExpressFiber disposable fiber service* (2024) *available at* halliburton.com

24. U.S. DOE, *Produced Water: Form a Waste to a Resource, Office of Fossil Energy and Carbon Management* (2020) *available at* energy.gov/fecm/articles/produced-water-waste-resource#:~:text=Produced%20water%20that%20has%20been, even%20non%2Dedible%20crop%20irrigation.

25. GasFrac Energy Services, *BlackBrush Extends GASFRAC Contract for Waterless Frac Process*, RigZone (2012) *available at* rigzone.com/news/oil_gas/a/116 000/blackbrush_extends_gasfrac_contract_for_waterless_frac_process

26. Bill Gates, *How to Avoid a Climate Disaster: The Solutions We Have and the Breakthroughs We Need, Vintage* (2022); WWF, Mangroves for Community and Climate (2024) *available at* worldwildlife.org

27. Nancy Stauffer, *Turning carbon dioxide into valuable products*, MIT News (2022), *available at* news.mit.edu/2022/turning-carbon-dioxide-valuable-products-0907; Jade Boyd, *New catalyst can turn smelly hydrogen sulfide into a cash cow*, Rice News, (2022) *available at* news.rice.edu/news/2022/new-catalyst-can-turn-smelly-hydrogen-sulfide-cash-cow

28. Fortify, *4 Reasons Why Composites Are Replacing Traditional Materials* (2018) *available at* 3dfortify.com/composites-replace-traditional-materials

29. Markforged, *3D Printers* (2024) *available at* markforged.com

30. Quaise, *Millimeter Wave Drilling – Gyrotron* (2024) *available at* quaise.energy

31. Trent Jacobs, *With Over 200 Shale Well Installs, This Slug-Smoothing Lift Technology Is Boosted By Schlumberger JV*, JPT (2017) *available at* jpt.spe.org/twa/slug-smoothing-technology-sees-over-200-shale-installs-gets-boost-schlumberger-jv

32. Hydrozonix, *Technologies* (2024) *available at* hydrozonix.com

33. Leistritz, *Multiphase Pumping Systems* (2024) *available at* leistritzcorp.com

34. Trent Jacobs, *Aramco Explores Expired Medications To Treat a Billion-Dollar Oilfield Problem*, JPT (2023), *available at* jpt.spe.org/aramco-explores-expired-medications-to-treat-a-billion-dollar-oilfield-problem

35. Atria, *Transparent metals* (2024) *available at* atriainnovation.com

36. Diversified Well Logging, *RoboLogger* (2024) *available at* dwl-usa.com

37. Hashed Ahmed, *Producer-To-Injector Conversion to Enhance Oil Productivity and Profitability*, International Journal of Innovative Technology and Exploring Engineering (2019) *available at* ijitee.org/wp-content/uploads/papers/v8i4s/DS2847028419.pdf

38. Al-Hameedi, *Insights into the applications of waste materials in the oil and gas industry: state of the art review, availability, cost analysis, and classification*, Journal of Petroleum Exploration and Production Technology (2020) *available at* doi.org/10.1007/s13202-020-00865-w

39. Serdar Bender, *Flue gas injection for EOR and sequestration: Case study*, Journal of Petroleum Science and Engineering (2017) *available at* doi.org/10.1016/j.petrol.2017.07.044

40. Upwing Energy, *Subsurface Compressor System* (2024) *at* upwingenergy.com

41. Brendan Lynch, *New technology could 'transform' hydraulic fracturing, make unconventional reservoirs development more efficient*, Smart MicroChip Proppants, KU News (2019) *available at* news.ku.edu/news/article/2019/08/27/smart-microchip-proppants-technology-could-transform-hydraulic-fracturing-making

42. Iofina Resources, *Operations* (2024) *available at* iofina.com; Lei Hu, *Simultaneous recovery of ammonium, potassium and magnesium from produced water by struvite precipitation*, Chemical Engineering Journal (2020) *available at* sciencedirect.com/science/article/abs/pii/S1385894719324118

43. Fishbones, *Technology* (2024) *available at* fishbones.as

44. Quidnet Energy, *Our Solution* (2024) *available at* quidnetenergy.com

45. Trent Jacobs, *Bringing the Heat: Aramco Field Tests High-Temperature Chemistry To Slash Tight-Gas Completion Costs*, JPT (2020) *at* jpt.spe.org/bringing-heat-aramco-field-tests-high-temperature-chemistry-slash-tight-gas-completion-costs

46. Matt Schwartz, *Wireline-Deployed Acid Improves Frac Efficiency, Reduces Costs In Niobrara*, The American Oil and Gas Reporters (2021) *available at* https://www.aogr.com/magazine/editors-choice/wireline-deployed-acid-improves-frac-efficiency-reduces-costs-in-niobrara; StimStixx, *Technology*, (2024) *available at* stimstixx.com

47. Dushyant Singh Shekhawat, *Magnetic Recovery-Injecting Newly Designed Magnetic*

*Fracturing Fluid with Applied Magnetic Field for EOR*, SPE (2016), *available at* doi.org/10.2118/181853-MS

48. Patricia Craig, *Industrial waste can be engineered into proppants for shale gas and oil recovery*, Penn State (2014) *available at* psu.edu/news/research/story/industrial-waste-can-be-engineered-proppants-shale-gas-and-oil-recovery

49. Juan Du, *Advances in nanocomposite organic coatings for hydraulic fracturing proppants*, Gas Science and Engineering (2023) *available at* sciencedirect.com/science/article/abs/pii/S2949908923002315

50. Veronique Greenwood, *This Creature Eats Stone. Sand Comes Out the Other End*, NYT, (2019) *at* nytimes.com/2019/06/18/science/shipworm-rocks-sand

51. Marwell, *Products* (2024) *available at* marwell-tech.no

52. DOE, *How Maglev Works* (2024) *available at* energy.gov/articles/how-maglev-works

53. Seismos, *Products* (2024) *available at* seismos.com

54. Weatherford, *Continuous Foamer Injection Via Capillary Line* (2023) *available at* weatherford.com

55. Changhee Kim, *Hydraulic transport of sand-water mixtures in pipelines, Part I. Experiment*, Journal of Mechanical Science and Technology (2008) *available at* link.springer.com/article/10.1007/s12206-008-0811-0

56. Modern version of an "Oklahoma Jack" style multi-well rod lift setup.

57. Sun Specialty Products, *Production Enhancement* (2024) *available at* sunspecialtyproducts.com; NRG, *Products and Services* (2024) *available at* nrgproppants.com

## Chapter Twelve: Perseverance

1. Burton Folson, *John D. Rockefeller and the Oil Industry*, FEE (1988) *available at* fee.org/articles/ john-d-rockefeller-and-the-oil-industry

2. Grant Segall, *John D. Rockefeller: Anointed with Oil*, Oxford University Press (2012)

3. *See* n.1, *supra*; *see also* David Henderson, *Great Moments in Cost Cutting: Rockefeller Edition*, Econlog Post (2013) *available at* econlib.org/archives

4. Ron Chernow, *Titan: The Life of John D. Rockefeller, Sr.*, Vintage Books (1998); Daniel Yergin, *The Prize: The Epic Quest for Oil, Money & Power*, Simon & Schuster (2012); Burton Folson, *John D. Rockefeller and the Oil Industry*, FEE (1988)

5. Ron Chernow, *Titan: The Life of John D. Rockefeller, Sr.*, Vintage Books (1998)

6. Orison Marden, *How They Succeeded*, Lothrop Publishing (1901)

7. Lynn Cook, *An Old Fracking Hot Spot Makes a Comeback*, WSJ (2017)

8. Robert Rapier, *The Permian Basin is Now the World's Top Oil Producer*, Forbes (2019)

ENGINEERING

IS

COST REDUCTION

# ABOUT THE AUTHOR

**Matthew J. Hatami** is an engineer and entrepreneur, with a degree in Petroleum and Natural Gas Engineering from West Virginia University and an MBA from Columbia University. He is a licensed Professional Petroleum Engineer in the State of Oklahoma. Matthew started his career 24 years ago as a field engineer in the Permian Basin. He has worked in the oil and gas industry across multiple basins in the United States and overseas, primarily focusing on shale development. He has worked on the oilfield services side of the industry and the operating side, including positions working with geology, land, drilling, completions, production, reservoir, regulatory, legal, accounting, finance, and corporate strategy. Before venturing into the entrepreneurial arena, Matthew was the Vice President of Resource Development for American Energy Partners and American Energy Global Partners. He also worked as a Senior Asset Manager at Chesapeake Energy Corporation, as a Financial Analyst in Corporate Strategy and Planning at Hess Corporation, and as an Engineer at Halliburton Company.